D1801720

LONDON MATHEMATICAL SOCIETY LECTURE NOTE SERIES

Managing Editor: Professor I.M. James,
Mathematical Institute, 24-29 St Giles, Oxford

1. General cohomology theory and K theory, P.HILTON
4. Algebraic topology, J.F.ADAMS
5. Commutative algebra, J.T.KNIGHT
8. Integration and harmonic analysis on compact groups, R.E.EDWARDS
9. Elliptic functions and elliptic curves, P.DU VAL
10. Numerical ranges II, F.F.BONSALL & J.DUNCAN
11. New developments in topology, G.SEGAL (ed.)
12. Symposium on complex analysis, Canterbury, 1973, J.CLUNIE & W.K.HAYMAN (eds.)
13. Combinatorics: Proceedings of the British Combinatorial Conference 1973, T.P.McDONOUGH & V.C.MAVRON (eds.)
15. An introduction to topological groups, P.J.HIGGINS
16. Topics in finite groups, T.M.GAGEN
17. Differential germs and catastrophes, Th.BROCKER & L.LANDER
18. A geometric approach to homology theory, S.BUONCRISTIANO, C.P. BOURKE & B.J.SANDERSON
20. Sheaf theory, B.R.TENNISON
21. Automatic continuity of linear operators, A.M.SINCLAIR
23. Parallelisms of complete designs, P.J.CAMERON
24. The topology of Stiefel manifolds, I.M.JAMES
25. Lie groups and compact groups, J.F.PRICE
26. Transformation groups: Proceedings of the conference in the University of Newcastle-upon-Tyne, August 1976, C.KOSNIOWSKI
27. Skew field constructions, P.M.COHN
28. Brownian motion, Hardy spaces and bounded mean oscillations, K.E.PETERSEN
29. Pontryagin duality and the structure of locally compact Abelian groups, S.A.MORRIS
30. Interaction models, N.L.BIGGS
31. Continuous crossed products and type III von Neumann algebras, A.VAN DAELE
32. Uniform algebras and Jensen measures, T.W.GAMELIN
33. Permutation groups and combinatorial structures, N.L.BIGGS & A.T.WHITE
34. Representation theory of Lie groups, M.F. ATIYAH *et al.*
35. Trace ideals and their applications, B.SIMON
36. Homological group theory, C.T.C.WALL (ed.)
37. Partially ordered rings and semi-algebraic geometry, G.W.BRUMFIEL
38. Surveys in combinatorics, B.BOLLOBAS (ed.)
39. Affine sets and affine groups, D.G.NORTHCOTT
40. Introduction to Hp spaces, P.J.KOOSIS
41. Theory and applications of Hopf bifurcation, B.D.HASSARD, N.D.KAZARINOFF & Y-H.WAN
42. Topics in the theory of group presentations, D.L.JOHNSON
43. Graphs, codes and designs, P.J.CAMERON & J.H.VAN LINT
44. Z/2-homotopy theory, M.C.CRABB
45. Recursion theory: its generalisations and applications, F.R.DRAKE & S.S.WAINER (eds.)
46. p-adic analysis: a short course on recent work, N.KOBLITZ
47. Coding the Universe, A.BELLER, R.JENSEN & P.WELCH
48. Low-dimensional topology, R.BROWN & T.L.THICKSTUN (eds.)

London Mathematical Society Lecture Note Series. 70

Stochastic Differential Equations on Manifolds

K.D. ELWORTHY
Professor of Mathematics
University of Warwick

CAMBRIDGE UNIVERSITY PRESS

Cambridge

London New York New Rochelle

Melbourne Sydney

Published by the Press Syndicate of the University of Cambridge
The Pitt Building, Trumpington Street, Cambridge CB2 1RP
32 East 57th Street, New York, NY 10022, USA
296 Beaconsfield Parade, Middle Park, Melbourne 3206, Australia

© Cambridge University Press 1982

First published in 1982

Printed in Great Britain at the University Press, Cambridge

Library of Congress catalogue card number 82-4426

British Library cataloguing in publication data

Elworthy, K.D.
Stochastic differential equations on manifolds.
-(London Mathematical Society lecture note
series, ISSN 0076-0552; 70)
1. Stochastic differential equations
I. Title II. Series
519.2 QA274.23

ISBN 0 521 28767 7

CONTENTS

PREFACE

You have not played as yet? Do not do so: above all avoid a martingale if you do. *W.M. Thackeray*

ABOUT THESE NOTES. ACKNOWLEDGEMENTS

These are a much expanded and revised version of notes of seminars given at the University of Warwick during the year 1973/74. These were given and written up jointly with J. Eells. The audience consisted of non-probabilists with a reasonably good background in manifold theory. The aim was to go through, from the beginning, the basic properties of stochastic differential equations, extend the theory to manifolds and in particular use this in the proof that by polygonal approximation the Cartan development or 'rolling' has a 'stochastic extension' which gives a geometric construction of Brownian motion on Riemannian manifolds (see §§11 A,B Chapter VII). Another aim was to describe, from this point of view, the stochastic parallel transport discussed in Itô's Stockholm article (1963).

The previous year we had been on leave separately: myself at the University of California at Santa Cruz, and Aarhus University, beginning to learn some Stratonovich calculus after earlier suggestions from R. Curtain; and Eells at I.A.S. Princeton, and I.H.E.S. Bures-sur-Yvette, where he collaborated with P. Malliavin (1972/3) in an examination of diffusions on vector bundles, horizontal lifts, etc. using a different approach. We met up at I.H.E.S., and all these institutions deserve thanks for their hospitality.

In the seminars we used McShane's belated (Riemann) integrals because of their straightforward definition and the clear exposition in McShane's book (1974) of the main technical results we needed. The same integration theory is used in these notes on the grounds that it allows a more rapid advance into the subject than the more sophisticated and general theories, especially for those with little or no probabilistic background and/or those who are more interested in the statements of theorems than all the details of the proofs: but see the "Suggestions on

How to Use Them" below.

The notes are still designed for non-probabilists. There are appendices giving outlines of the necessary background from manifold theory and differential geometry: however complete ignorance of these two topics will make the text heavy going. The book (Choquet-Bruhat et al. 1977) may be a help here. In general the proofs are given in complete detail: especially the 'globalizations' needed to work on manifolds. This is not just from pedantry: it is a subject in which it is very easy to make wrong statements, and also, not having a probabilistic background, I have often found probabilistic proofs difficult to follow. The main parts of Chapters I - VII are very close to our original notes. The principal chunks of additional material are the rather messy treatment of integrals over random time intervals, the covariant equations on vector bundles, the flow theorems of Chapter VIII and the discussion of the B.R. bridge in Chapter IX which is taken from, as yet unpublished, joint work of myself and A. Truman.

The main technical difference to more abstract treatments of manifold valued processes, e.g. (Meyer 1981, 1981a,b) is that we restrict ourselves to manifold valued processes which are either solutions of stochastic differential equations or smooth functions of such solutions. The state spaces are also allowed to be infinite dimensional. This causes little extra difficulty since we restrict ourselves to separable Hilbert spaces as models; it is also very useful in proving some of the flow theorems in Chapter VIII, and is presently being applied in Euclidean gauge field theory. We only consider finite dimensional Brownian motions explicitly, to save space, but the infinite dimensional versions fit in perfectly well to our framework.

References to the literature are indicated by author and a date. For a published article the date usually refers to the publication date, otherwise it is a guess at how recently it was possible to get hold of it, or when it may be published. Historical significance is therefore quite absent. In any case although I have been free in including references which may be found useful lots have been missed; (Ikeda & Watanabe 1981) and Lecture Notes in Maths 851 are a good source.

The exposition owes a lot to my contact with C. DeWitt-Morette, in particular to writing up notes with her on a course I gave in the Physics Department at the University of Texas at Austin in the Autumn of

1978. Her influence was important in the choice (even the existence?) of the examples. Her encouragement has certainly been important for the publication of these notes at all.

Apart from the debts owed to those already mentioned many others have helped with suggestions and major or minor corrections to this and previous versions. In particular I wish to thank C.J. Atkin, P.J. Baxendale, I.M. Caldwell, R.W. Dowell, S.E.A. Mohammed, M. Pollicott, S.J. Rogerson, and K.Y. Woo. L.C.G. Rogers furnished useful references, and several times his expertise saved me from despair. Conversations with G.F. Vincent-Smith led to some last minute improvements.

Finally, the typing was by my secretary, Terri Moss: it is difficult to imagine how it could ever have been finished had it been otherwise.

SUGGESTIONS ON HOW TO USE THEM

If you know about stochastic differential equations go straight to Chapter VII, and from then on try to avoid getting irritated with the lack of probabilistic sophistication. If you want to allow infinite dimensional state spaces, for example to follow the proofs of the flow theorems in Chapter VIII, you will first need to convince yourself that all the relevant facts you need about stochastic differential equations are true equally well in Hilbert spaces: for a sophisticated treatment of this see (Metivier & Pellaumail 1980).

If you don't know about stochastic differential equations the main steps are the basic facts about Brownian motion in Chapter II, the definition and existence of stochastic integrals, and substitution theorem, in Chapter III, and the rules coming from the special cases in Chapter IV. Even if you are not particularly interested in mathematical details, to appreciate the vital role played by the filtrations it is necessary to go through at least one proof: a good idea would be to prove the existence of first order integrals, rather than second order ones, based on the proof of Proposition 5B of Chapter IV. If you can dip into the proofs of Chapter IV and prove for yourself the rules $dB_t\ dB_t = dt$, $dB_t\ dt = 0$, $dB_t\ dB_t\ dB_t = 0$, and $dB_t\ dB'_t = 0$, for B_t and B'_t independent Brownian motions, you can feel pretty confident. If you want to stay in finite dimensions and don't like $\otimes$ you can first consider one-dimensional processes, interpreting $\otimes$ as ordinary multiplication, and then work in coordinates for the multidimensional theory. We have not taken the shortest route through this material, but rather have given results whose statements at least are clear (in general) and which should give a good idea of the behaviour of the integrals.

In any case there is no point in worrying about the precise conditions on the coefficients, or 'noises' dz, that are needed to make the results go through. *This is a very important point.* The approach here is a crude first approach, and if you want a more refined machine you should consult the standard texts: e.g. (Arnold 1974, Friedman 1975) when the noise is Brownian motion or (Meyer 1976, Metivier & Pellaumail 1980) to get full power. However the crude appraoch is adequate for what we want.

The Itô formulae in Chapter V are absolutely fundamental. The existence theorem for integral equations in Chapter VI is standard. In that Chapter the differentiability results start a thread which continues in generalizations later as does the piecewise linear approximation theorem. The latter in particular can be omitted entirely although it is used as a tool in the proof used for the Markov property.

In Chapter VII the basic results are the existence and uniqueness theorem, Theorem 2E, the theorem about submanifolds in §3 which is an important tool, and the 'main theorem for compact manifolds'. The non-explosion criterion in §6 is used frequently later, and convergence to the coffin state, as in Remark 6, is also important. Again the Itô formulae in §9 are of fundamental importance but the formulae for derivative processes are a side issue: this is also true for the complicated formulae for the derivative process of Brownian motion. It is in the construction of Brownian motion via the stochastic development that the differential geometry and stochastic analysis begin to work together. The section on vector bundles can be omitted: only some very special cases are used later.

Chapter VIII is also mostly not essential. The section on flows is probably of most use to those interested in stochastic differential equations as perturbed dynamical systems: if the global analysis here is too heavy refer to (Carverhill & Elworthy 1982) for a simplified version which also includes many recent results. The regularity results of §8 are useful, but if you are interested in non-degenerate diffusions on finite dimensional manifolds much stronger theorems come from partial differential equation theory.

We can go straight to Chapter IX once the Itô formulae on manifolds have been mastered. It contains various applications to some partial differential equations of the type occurring in differential geometry and in some branches of mathematical physics. For many of these it is not necessary to have gone through §5 on Markov processes.

For an overview of Chapters I-VIII see (Elworthy 1978) and for a survey of topics close to those of Chapter IX see (Elworthy 1981).

INTRODUCTION

It looked insanely complicated, and this was one of the reasons why the snug plastic cover it fitted into had the words DON'T PANIC printed on it in large friendly letters.

Douglas Adams: The Hitch-Hikers Guide to the Galaxy.

(A) There are at least two fairly distinct ways of looking at stochastic differential equations. One is that they provide a tool in the theory of certain second order partial differential operators. The other is that they give a model for ordinary dynamical systems perturbed by 'noise'.

(B) The first takes off from the Feynman-Kac formula: for suitable $V:\mathbb{R}^n \to \mathbb{R}$ the solution $g_t:\mathbb{R}^n \to \mathbb{R}$, $t \geq 0$, to

$$\frac{\partial g_t}{\partial t} = \tfrac{1}{2} \Delta g_t + V g_t \tag{1}$$

for g_0 given, is

$$g_t(x) = \int_{C_0(\mathbb{R}^n)} e^{\int_0^t V(x+\sigma(s))ds} g_0(x+\sigma(t)) \, d\gamma(\sigma)$$

where d refers to a certain measure on the space $C_0(\mathbb{R}^n)$ of continuous paths $\sigma:[0,\infty) \to \mathbb{R}^n$ with $\sigma(0) = 0$: the Wiener measure, which is associated to Brownian motion on $\mathbb{R}^n$. The book (Simon 1979) shows how this formula, together with properties of Brownian motion, can be used to get insight into the operator $\tfrac{1}{2}\Delta + V$. The use of stochastic differential equations gives constructions of diffusions more general than Brownian motion, and these are associated to more general operators. Usually when there is some geometry involved in the operator this can be reflected in the construction of the diffusion. This gives a direct link between the geometry and the analysis, and at the very least provides valuable intuition. Examples here are where the Euclidean Laplacian is replaced by the Laplace-Beltrami operator of a Riemannian manifold, or by a gauge invariant Laplacian. The methods also apply to certain differential operators on infinite spaces: i.e. infinite systems; and have been applied to the non-linear problems of harmonic maps between Riemannian manifolds, see (Kendall 1982). Our final chapter is particularly concerned with introducing these ideas.

(C) The taking off point for the second way of looking at stochastic differential equations is the Langevin equation for the Ornstein-Uhlenbeck velocity process. This process was constructed so as to give a better model of physical Brownian motion than that given by the standard, Einstein-Smulochowski, model. It is discussed in detail in (Nelson 1967). The equation is

$$dv_t(\omega) = \gamma dB_t(\omega) - \beta v_t(\omega)dt \qquad t \geq 0. \tag{2}$$

Here $t \mapsto v_t(\omega)$ and $t \mapsto B_t(\omega)$ are paths in $\mathbb{R}^n$, parametrized by $\omega \in \Omega$, where Ω is a space with a probability measure μ on it. Both γ and β are constants. The idea is that having solved the equation for v the measure of the set of Ω such that $v_\cdot(\omega)$ lies in subsets A_i of $\mathbb{R}^n$ at given times t_i, $\mu\{\omega \in \Omega : v_{t_i}(\omega) \in A_i,\ i = 1 \text{ to } m\}$, should be the probability that our physical Brownian motion will have velocity which behaves in the same way when it starts with the given initial velocity v_0. Similarly for the position $x_t(\omega)$ given by

$$x_t(\omega) = x_0(\omega) + \int_0^t v_s(\omega)ds.$$

The probabilistic behaviour of B is assumed known: it is taken so that $\mu\{\omega \in \Omega : B_{t_i}(\omega) \in A_i,\ i = 1 \text{ to } m\}$ are those forecast by the Einstein-Smulochowski idealized model of Brownian motion, and we simply call $\{B_t : t \geq 0\}$ a 'Brownian motion'. The equation is supposed to represent Newton's equations for the velocity of a particle moving with friction β under a random battering. For our purposes it is convenient to think of γ as a measure of its susceptibility to the battering. The representation of the battering by 'dB_t' is because disjoint movements of (idealized) Brownian motion are independent of each other, so 'dB_t' is 'purely random' in some sense, also the increments are Gaussian and stationary in time (i.e. their probabilistic behaviour is invariant in time): in engineers language 'dB_t' represents 'Gaussian white noise'.

In mathematicians language it is not clear that it represents anything at all. The first thing anyone who had done any mathematical analysis would do to (2) would be to divide through by 'dt' and interpret

$\frac{dB_t(\omega)}{dt}$ as a derivative. This is no help however, for almost all ω the derivatives do not exist: a typical Brownian particle in the Einstein-Smulochowski model performs a path which is not differentiable. (This was a good reason for attempting to find a more realistic model of physical Brownian motion). In fact this is not a severe problem for (2) since we can also interpret it as the integral equation

$$v_t(\omega) = v_0(\omega) + \gamma B_t(\omega) - \beta \int_0^t v_s(\omega)ds. \qquad (2)'$$

This has solution

$$v_t(\omega) = e^{-\beta t} v_0(\omega) + \gamma B_t(\omega) - \gamma\beta e^{-\beta t} \int_0^t e^{\beta s} B_s(\omega)ds,$$

(the paths $B_.(\omega)$ can be taken to be continuous and with $B_0(\omega) = 0$). However suppose that the susceptibility of the particle to the battering depended on its velocity and perhaps also on position, time, and direction: think of cycling through a hail-storm. The constant γ then has to be replaced by an $n \times n$-matrix $\gamma(x_t, v_t, t)$. The integral equation would then be replaced by the coupled equations

$$\begin{aligned} v_t(\omega) &= v_0(\omega) + \int_0^t \gamma(x_s(\omega), v_s(\omega), s)dB_s(\omega) - \beta \int_0^t v_s(\omega)ds \\ x_t(\omega) &= x_0(\omega) + \int_0^t v_s(\omega)ds. \end{aligned} \qquad (2)''$$

This time we still have problems: a typical Brownian path $B_s(\omega)$ does not even have bounded variation, so $\int_0^t \gamma(x_s(\omega), v_s(\omega), s)dB_s(\omega)$ cannot even be interpreted as a Stieltjes integral. To make sense of it we have to develop a new calculus, a calculus which does not work with the individual paths $B_.(\omega)$ or $v_.(\omega)$ but with the whole parametrized family, and in which second order increments of B have to be kept in. The basic rules turn out to be fairly straightforward, although the mathematics is fairly complicated. However careless use can very easily lead to mistakes.

It turns out that the probabilistic behaviour of solutions of equations like (2)" are governed by partial differential equations like (1). In the first approach this is used to get information about the analogues of (1), but this can be applied the other way round and the partial differential equation can be used to obtain information about our noisy dynamical system. In these notes we do not go into any details about this: but the result on exit times in Proposition 8B of Chapter IX is a good example. There is a very helpful discussion of the fundamental problems in modelling noisy dynamical systems in the introduction of (McShane 1974).

(D) Of course these are not the only ways of looking at stochastic differential equations. One unifying theme of many of the articles in (DeWitt-Morette & Elworthy 1981) is a relationship between diffusion processes, and their generalizations, and quantum physics which is more direct than going via partial differential equations: the heuristics of Feynmann path integration are an example of an analogous situation. Some of the most sophisticated stochastic analysis appears in filtering theory. From the point of view of pure probability theory stochastic differential equations play an important role in the general theory of stochastic processes, and even here it is natural to consider them a manifold with related differential geometry: see (Meyer 1981a,b). On the other hand the rich interaction with differential geometry is a growing and fruitful field of study in its own right and is likely to enrich both disciplines.

CHAPTER I : PRELIMINARIES AND NOTATION

DON'T PANIC

The Hitch-Hikers Guide
to the Galaxy

§1 MEASURE THEORETIC BACKGROUND

(A) A family $\mathcal{A}$ of subsets of a set X is a *σ-algebra* if it is closed under countable unions and complementation, and if $X \in \mathcal{A}$. We call $(X,\mathcal{A})$ a *measurable space*. A *measure* μ on $\mathcal{A}$ is a non-negative real valued (we permit $+\infty$ also) σ-additive function on $\mathcal{A}$; a *probability* is a measure with $\mu(X) = 1$.

If a non-negative function is σ-additive on an algebra $\mathcal{A}_0$ of subsets of X, then it has a canonical extension to a measure on the σ-algebra generated by $\mathcal{A}_0$.

Let μ be a finitely additive non-negative finite valued function on $\mathcal{A}_0$, and suppose μ satisfies the following semi-continuity property.

If $\{A_n\}_{n\geq 1} \subset \mathcal{A}_0$ is a decreasing sequence

$(A_1 \supset A_2 \supset \ldots)$ with $\cap \{A_n : n \geq 1\} = \emptyset$, then $\mu(A_n) \to 0$.

Then μ is σ-additive on $\mathcal{A}_0$.

If $(X,\mathcal{A})$ and $(Y,\mathcal{B})$ are two measurable spaces, and $f:X \to Y$ a map, say that f is *measurable* if $f^{-1}(B) \in \mathcal{A}$ for all $B \in \mathcal{B}$. Let $\mathcal{L}^0(X,\mathcal{A};Y,\mathcal{B})$ denote the totality of such maps.

If μ is a measure on $\mathcal{A}$ and $f \in \mathcal{L}^0(X,\mathcal{A};Y,\mathcal{B})$ then $f\mu$ is the measure on $\mathcal{B}$ defined by $(f\mu)(B) = \mu(f^{-1}(B))$. It follows that if $h:Y \to R$ is measurable then

$$\int_X h\circ f \, d\mu = \int_Y h d(f\mu)$$

when either side exists.

(B) Let Y be a metrisable space, and $\mathcal{B}$ = Borel Y the σ-algebra generated by the open (equivalently the closed) subsets of Y; elements of $\mathcal{B}$

are the *Borel sets* of Y. A measure on $\mathcal{B}$ is a *Borel measure*.

Every Borel probability μ is *regular:* For every $A \in \mathcal{B}$.

$$\mu(A) = \sup \{\mu(C): C \subset A \text{ and } C \text{ is closed}\}$$

$$= \inf \{\mu(U): A \subset U \text{ and } U \text{ is open}\}.$$

If Y is complete and separable and if μ is a Borel probability, then μ is *tight* i.e. for every $\varepsilon > 0$, there is a compact $Y_\varepsilon \subset Y$ such that

$$\mu(Y - Y_\varepsilon) < \varepsilon$$

The *support of* μ is

$$\text{spt } \mu = \{x \in X: \mu(U) > 0 \text{ for every open neighbourhood } U \text{ of } x\}.$$

Thus spt μ is closed.

Let d be a metric on Y and $\mathcal{B}$ = Borel Y. For a measure μ on $(X,\mathcal{A})$, a sequence $\{f_i\}_{i \geq 1} \subset \mathcal{L}^0(X,\mathcal{A};Y,\mathcal{B})$ *converges in measure* to f if for every $\varepsilon > 0$

$$\lim_{i\to\infty} \mu\{x \in X: d(f_i(x),f(x)) \geq \varepsilon\} = 0.$$

This comes from a pseudo-metric on $\mathcal{L}^0(X,\mathcal{A};Y,\mathcal{B})$ which is complete if (Y,d) is complete. Convergence in measure implies convergence almost everywhere of some subsequence. On the other hand if $\mu(X) < \infty$ then convergence almost everywhere implies convergence in measure.

(C) Consider a finite measure space $(\Omega,\mathcal{F},\mu)$, i.e. a measurable space $(\Omega,\mathcal{F})$ with a measure μ on $\mathcal{F}$ such that $\mu(\Omega) < \infty$. Let $(E,|\ |)$ be a real separable Banach space, considered with its Borel σ-algebra. A *simple function* $f:\Omega \to E$ is one expressible as a finite sum

$$f = \sum_{i=1}^{n} x_i\, \chi_{A_i}$$

where $\{x_i\}_{i=1}^n \subset E$ and $\{A_i\}_{i=1}^n$ is a disjoint collection of elements in $\mathcal{F}$. We use χ_A to denote the characteristic function of a set A.

In the separable case under consideration the following conditions on $f:\Omega \to E$ are equivalent:

(i) f is measurable

(ii) f <u>is strongly measurable</u> i.e. f is a limit almost everywhere of simple functions

(iii) f <u>is weakly measurable</u> i.e. for each $h \in E^*$ (the dual space of E), $h \circ f: \Omega \to R$ is measurable.

In particular (for example by (iii) and the classical result) <u>a pointwise limit of measurable maps is measurable</u>.

We shall use Bochner integration for vector valued maps: $f:\Omega \to E$ is *Bochner integrable* if there exists a sequence $\{f_n\}_{n=1}^{\infty}$ of simple functions with

$$f = \lim f_n \quad \text{a.e.}$$

and

$$\lim_{n\to\infty} \int_\Omega |f_n(\omega) - f(\omega)| d\mu = 0$$

The (Bochner) *integral* is then defined by

$$\int_\Omega f\, d\mu = \lim \int_\Omega f_n\, d\mu$$

where the integrals of simple functions have their obvious meanings.

In our separable situation $f:\Omega \to E$ is *integrable* iff f is measurable and $\int_\Omega |f| d\mu < \infty$. We have the spaces $\mathcal{L}^p(\Omega,\mu;E)$ of measurable maps $f:\Omega \to E$ with

$$\|f\|_{\mathcal{L}^p} = \left(\int_\Omega |f|^p\, d\mu \right)^{1/p} < \infty$$

and the corresponding Banach spaces of equivalence classes $L^p(\Omega,\mu;E)$, $p \geq 1$. In general we shall use $|\ |$ to denote norms of the initial Banach spaces we work with and $\|\ \|$ to denote $\|\ \|_{\mathcal{L}^2}$ for $\mathcal{L}^2$-spaces.

(D) Let ν and μ be finite measures on (Ω,F). The measure ν is *absolutely continuous* with respect to μ, symbolically $\nu << \mu$, if whenever $B \in F$ satisfies $\mu(B) = 0$ then $\nu(B) = 0$. The measures are *equivalent*, $\nu \approx \mu$, if $\nu << \mu$ and $\mu << \nu$. The *Radon-Nikodym theorem* says that if $\nu << \mu$ there is an element

$\frac{d\nu}{d\mu} \in L^1(\Omega,F,\mu;R)$, the *Radon-Nikodym derivative of* ν *with respect to* μ, such that for all $B \in F$

$$\nu(B) = \int_B \frac{d\nu}{d\mu}\, d\mu$$

(E) Apart from standard references for the basic measure theory e.g. Kingman & Taylor, the following references may be useful for the case of infinite dimensional E: K.R. Parthasarathy (1967), L. Schwartz (1973), Dunford & Schwartz (1957) and Hille & Phillips (1957).

§2 MULTILINEAR MAPS AND TENSOR PRODUCTS

(A) For Banach spaces $G_1,\ldots,G_q$, E let $\mathbb{L}(G_1,\ldots,G_q;E)$ denote the Banach space of continuous q-linear maps

$$B: G_1 \times \ldots \times G_q \to E$$

with $\quad |B| = \sup\{|B(x_1,\ldots,x_q)|: |x_i| \leq 1,\ i = 1,\ldots,q\}.$

The algebraic tensor product $\bigotimes_{1 \leq i \leq q} G_i$ of the vector spaces G_i can be completed using the norm

$$|z| = \inf\{\sum_j |x_1^j|\,|x_2^j|\ldots|x_q^j| : z = \sum_j x_1^j \otimes \ldots \otimes x_q^j\}.$$

The resulting Banach space will be written

$$\hat{\bigotimes}_{1 \leq i \leq q} G_i.$$

The assignment $B \to \tilde{B}$ where $\tilde{B}(x_1 \otimes \ldots \otimes x_n) = B(x_1,\ldots,x_n)$ defines an isomorphism $\mathbb{L}(G_1,\ldots,G_q;E) \to \mathbb{L}(\hat{\bigotimes}_{1 \leq i \leq q} G_i;E)$.

(B) For separable Banach spaces $G_1,\ldots,G_q$, E a map $B:\Omega \to \mathbb{L}(G_1,\ldots,G_q;E)$ will be called *F-random* if for each $\underline{x} = (x_1,\ldots,x_q) \in G_1 \times \ldots \times G_q$ the map $B_{\underline{x}}:\Omega \to E$, $\omega \to B(\omega)(x_1,\ldots,x_q)$ is measurable with respect to the σ-algebra F. For the corresponding $\tilde{B}:\Omega \to L(\hat{\otimes} G_i;E)$ and $z \in \hat{\otimes} G_i$ since the map $\omega \to \tilde{B}(\omega)(z)$ is a pointwise limit of sums of maps of the form $B_{\underline{x}}$ we see that B is *F-random* iff $\tilde{B}$ *is F-random*. Also by the separability of $G_1,\ldots,G_q$ if B *is F-random then* $|B|$ *is measurable*.

If B, as above, is F-random and $f_i:\Omega \to G_i$, $i = 1,\ldots,q$, are simple, clearly the map $B(f_1,\ldots,f_n):\Omega \to E$ given by $\omega \to B(\omega)(f_1(\omega),\ldots,f_n(\omega))$ is measurable. Approximating measurable functions by simple ones we have: <u>if</u> $B:\Omega \to \mathbb{L}(G_1,\ldots,G_q:E)$ <u>is</u> F-<u>random and</u> $f_i \in \mathcal{L}^0(\Omega,F;G_i)$, $i = 1,\ldots q$,

<u>then</u>

$$B(f_1,\ldots,f_q) \in \mathcal{L}^0(\Omega,F;E).$$

§3 CONDITIONAL EXPECTATIONS

(A) The proof of the existence of conditional expectations for vector valued random variables given below (and taken from Chatterji (1968) is necessarily more complicated than the proof for real functions because of the lack of a Radon-Nikodym theorem for vector valued set functions (the standard counter example is the map Borel $([0,1]) \to L^1([0,1];R)$ given by $A \to \chi_A$, which is certainly 'absolutely continuous' with respect to Lebesgue measure). For simplicity later we consider only separable Banach spaces G, and take (Ω,F,μ) to be a probability space.

<u>Proposition</u>. <u>To each</u> σ-subalgebra F_s <u>of</u> F <u>there exists a unique continuous linear map</u>

$$\mathbb{E}_s : L^1(\Omega,F,\mu;G) \to L^1(\Omega,F_s,\mu|F_s;G)$$

<u>such that</u>

$$\int_A f\, d\mu = \int_A \mathbb{E}_s(f)d\mu \tag{1}$$

<u>for all</u>

$$f \in L^1(\Omega,F,\mu;G) \text{ and } A \in F_s.$$

<u>Moreover</u>

$$|\mathbb{E}_s(f)(\omega)|_G \leq \mathbb{E}_s(|f|_G)(\omega) \qquad \text{a.e.} \tag{2}$$

<u>and</u>

$$|\mathbb{E}_s| = 1. \tag{3}$$

<u>Proof</u> (a) Suppose $G = R$ and $f \in \mathcal{L}^1(\Omega,F,\mu;G)$ is non-negative. Define $\mu_f : F_s \to R$ by

$$\mu_f(A) = \int_A f\, d\mu.$$

Then μ_f is a finite measure on F_s and $\mu_f \ll \mu|F_s$.

Set $\mathbb{E}_s(f) = \dfrac{d\mu_f}{d(\mu|F_s)}$. Then $\mathbb{E}_s(f)$ certainly satisfies (1).

Moreover $E_s(f) \geq 0$, a.e. .

(b) For general G, let $S \subset L^1(\Omega,F,\mu;G)$ be the dense linear subspace consisting of all simple functions, and given its induced norm. If $f \in S$ and is represented as such by

$$f = \sum_{i=1}^{n} x_i \, \chi_{A_i}$$

with $\{A_i\}_{i=1}$ a disjoint family in F, we set

$$\mathbb{E}_s(f) = \sum_{i=1}^{n} x_i \, \mathbb{E}_s(\chi_{A_i})$$

where $\mathbb{E}_s(\chi_{A_i})$ is defined as in (a).

This gives a well defined linear map

$$\mathbb{E}_s : S \to L^1(\Omega,F_s,\mu|F_s;G), \text{ satisfying (1)}$$

Moreover

$$|\mathbb{E}_s(f)(\omega)| \leq \sum_{1}^{n} |x_i \, \mathbb{E}_s(\chi_{A_i})(\omega)|$$

$$= \sum_{1}^{n} \mathbb{E}_s(\chi_{A_i})(\omega) \, |x_i| \qquad \text{a.e.}$$

$$= \mathbb{E}_s(|f|)(\omega)$$

since $|f|(\omega) = \sum_{1}^{n} |x_i| \chi_{A_i}(\omega)$ because the $\{A_i\}$ are disjoint.

Thus

$$\|\mathbb{E}_s(f)\|_{L^1} \leq \int |\mathbb{E}_s(f)| d\mu \leq \int \mathbb{E}_s(|f|) d\mu = \|f\|_{L^1}$$

by (1), and we see $\mathbb{E}_s$ extends to a continuous linear map:

$$\mathbb{E}_s : L^1(\Omega,F,\mu;G) \to L^1(\Omega,F_s,\mu|F_s;G) \text{ satisfying (3).}$$

By continuity it also satisfies (1). Using almost everywhere convergence we see (2) is satisfied also. //

Remarks (i) The application of $\mathbb{E}_s$ is often called *conditioning with respect to* F_s, and $\mathbb{E}_s(f)$ is the *conditional expectation of* f *with respect to* F_s, often written $\mathbb{E}(f|F_s)$.

(ii) The map $\mathbb{E}_s$ can be considered as a projection of norm one, of

$L^1(\Omega,F,\mu;G)$ onto its linear subspace $L^1(\Omega,F_s,\mu|F_s;G)$. As such it is characterised by saying that it commutes with integration over each $A \in F_s$:

$$\begin{array}{ccc} L^1(\Omega,F,\mu;G) & \xrightarrow{\int_A} & R \\ \mathbb{E}_s \downarrow & \nearrow \int_A & \\ L^1(\Omega,F_s,\mu|F_s;G) & & \end{array}$$

(iii) When $F_s = \{\emptyset,\Omega\}$ then $\mathbb{E}_s(f)$ is just the *expectation* i.e. the integral, of f; and then is often written $\mathbb{E}(f)$ so

$$\mathbb{E}(f) = \int f \, d\mu.$$

(iv) In general conditional expectations behave very much like integrals. For example using Proposition 3B below and the usual method for integrals we have *Hölder's inequality:*

<u>If</u> $f_1,\ldots,f_n \in L^1(\Omega,F,\mu;R)$ <u>and if</u> $p_1,\ldots,p_n$ <u>are positive reals with</u> $\Sigma^1/p_i = 1$, <u>then</u>

$$\mathbb{E}_s\Big(\prod_{k=1}^{n} |f_k|\Big) \le \prod_{k=1}^{n} \big(\mathbb{E}_s(|f_k|^{p_k})\big)^{\frac{1}{p_k}}$$

(v) Let P be the orthogonal projection of the Hilbert space $L^2(\Omega,F,\mu;\mathbb{R})$ onto its closed linear subspace $L^2(\Omega,F_s,\mu|F_s;\mathbb{R})$. Then $P(f) = \mathbb{E}_s(f)$ whenever f is a real valued $\mathcal{L}^2$ function. In fact:

(a) $P(f)$ is F_s-measurable

and (b) for $A \in F_s$, since P is self-adjoint

$$\int_A P(f)d\mu = \int_\Omega \chi_A \, P(f)d\mu = \int_\Omega P(\chi_A)f \, d\mu$$
$$= \int_\Omega \chi_A \, f \, d\mu = \int_A f \, d\mu.$$

Thus, if the 'events' F_s are assumed 'known', $\mathbb{E}_s(f)$ is the best mean square estimate to f in terms of what is known.

This interpretation can be seen more clearly when conditioning with respect to a *σ-algebra* $\sigma\{h\}$ *generated by some measurable function* $h:\Omega \to X$ i.e. the smallest σ-algebra on Ω for which h is measurable. For f real valued, in the notation of the proof of the proposition we obtain measures $h\mu$ and $h\mu_f$ on X with $h\mu_f << h\mu$. Set

$$F = \frac{d(h\mu_f)}{d(h\mu)} \in L^1(X,A,h\mu;\mathbb{R})$$

where A is the given σ-algebra on X. Then

$$\mathbb{E}(f|\sigma\{h\}) = F\circ h \qquad \text{a.s.}$$

as is easily seen using §1A. Thus the conditional expectation really is a function of the given function h which may be considered as 'known' or 'observed'. In particular h could correspond to a collection of functions, by taking X to be a product space, or the history up to time t of some $x:[0,\infty) \times \Omega \to M$, by taking X equal to a suitable space of maps of $[0,t]$ into M and $h(\omega) = x(-,\omega)|[0,t]$.

The notation $\mathbb{E}(f|h = x)$ is often used for the value $F(x)$ defined for $h\mu$ -almost all x in X. The corresponding results and notation hold for Banach space valued f, as in the proof of the proposition.

(B) Proposition 3B Suppose $T:\Omega \to \mathbb{L}(G;H)$ is F_s-random, where G and H are separable Banach spaces and F_s is a σ-subalgebra of F. Then for each $f \in L^1(\Omega,F,\mu;G)$ with $T(f) \in L^1(\Omega,F,\mu;H)$ we have

$$\mathbb{E}_s(T(f)) = T(\mathbb{E}_s(f)).$$

Proof. Note first that if $A \in F$ and $B \in F_s$ then

$$\mathbb{E}_s(\chi_B\chi_A) = \chi_B\, \mathbb{E}_s(\chi_A). \tag{1}$$

For n = 0,1,..., set $\Omega_n = \{\omega \in \Omega: n \leq |T(\omega)| < n + 1\}$, and for each $\varepsilon > 0$ choose a countably valued $f_\varepsilon \in L^1(\Omega,F,\mu;G)$ such that for each n

$$\int_{\Omega_n} |f_\varepsilon - f|\, d\mu \leq \frac{\varepsilon}{(n+1)2^n}.$$

Then

$$\int |T(f_\varepsilon - f)|\,d\mu \leq \sum_n \int_{\Omega_n} |T|\; |f_\varepsilon - f|\,d\mu$$

$$< \varepsilon.$$

In particular $T(f_\varepsilon) \in L^1(\Omega,F,\mu;H)$.

Suppose $f_\varepsilon = \sum_{i=1}^{\infty} x_i\, \chi_{A_i}$, $x_i \in G$, $\{A_i\}_i$ disjoint elements of F. Approximating

$T(x_i)$ in $\mathcal{L}^1(\Omega,F_s,\mu|F_s;H)$ by simple functions and using (1) we see

$$\mathbb{E}_s(T(x_i\chi_{A_i})) = \mathbb{E}_s(T(x_i)\chi_{A_i})$$
$$= T(x_i)\,\mathbb{E}_s(\chi_{A_i}).$$

Thus, by continuity of $\mathbb{E}_s$,

$$\mathbb{E}_s(T(f_\varepsilon)) = T(\mathbb{E}_s(f_\varepsilon)).$$

Letting $\varepsilon \to 0$ and using the continuity of $\mathbb{E}_s$ again completes the proof. //

Corollary 3B Suppose $h \in L^0(\Omega,F_s,\mu;H)$ and $f \in L^1(\Omega,F,\mu;G)$ with $h \otimes f \in L^1(\Omega,F,\mu; H \hat{\otimes} G)$. Then $\mathbb{E}_s(h \otimes f) = h \otimes \mathbb{E}_s(f)$.

Proof. Define $T:\Omega \to \mathbb{L}(G;H \hat{\otimes} G)$ by

$$T(\omega)(x) = h(\omega) \otimes x.$$

(C) Two σ-subalgebras A,B of F are said to be *conditionally independent as conditioned by the* σ*-subalgebra* F_s *of* F if for all $A \in A$, $B \in B$

$$\mathbb{E}_s(\chi_A\chi_B) = \mathbb{E}_s(\chi_A)\,\mathbb{E}_s(\chi_B)$$

When $F_s = \{\emptyset,\Omega\}$ this reduces to

$$\mu(A \cap B) = \mu(A)\,\mu(B)$$

and then A,B are simply called *independent*.

Two measurable functions f, g on Ω are said to be *conditionally independent as conditioned by* F_s if their σ-algebras $\sigma\{f\}$, $\sigma\{g\}$ are. By approximating by simple functions we see that if f,g are conditionally independent separable Banach space valued functions then

$$\mathbb{E}_s(f \otimes g) = \mathbb{E}_s(f) \otimes \mathbb{E}_s(g)$$

when defined.

Composition on the left by measurable functions clearly preserves conditional independence, so for example with f,g as above the maps $|f|$, $|g|$ will also be conditionally independent.

Exercise 3C(i) If $f \in L^1(\Omega,F,\mu;G)$ is independent of a sub-σ-algebra A of F show that

$$\mathbb{E}(f|A) = \mathbb{E}(f)$$

(ii) If $f_i : \Omega \to M_i$ for $i = 1,2$ are measurable and $\underline{f}:\Omega \to M_1 \times M_2$ is given by

$$\underline{f}(\omega) = (f_1(\omega),f_2(\omega))$$

show that f_1 and f_2 are independent if and only if

$$\underline{f}(\mu) = f_1(\mu) \otimes f_2(\mu).$$

The definitions and results above can be extended in the obvious ways to the notion of (conditional) independence of finite collections of sub-algebras or measurable functions.

(D) The following lemma will enable us to enlarge our σ-algebras in a technically very useful way, but without changing anything of substance:

Lemma 3D For a σ-subalgebra G of F let $\tilde{G}$ consist of those subsets F in F for which there exist G in G with $\mu(F \Delta G) = 0$ i.e. with $\mu(F-G) = 0$ and $\mu(G-F) = 0$. Then $\tilde{G}$ is a σ-subalgebra of F and if $f \in \mathcal{L}^1(\Omega,F,\mu;G)$

$$\mathbb{E}(f|\tilde{G}) = \mathbb{E}(f|G) \qquad \tilde{G}\text{-almost surely.}$$

Proof It is easy to see that $\tilde{G}$ is a σ-algebra. Also $\mathbb{E}(f|G)$ is $\tilde{G}$-measurable and for any $F \in \tilde{G}$, taking $G \in G$ with $\mu(G \Delta F) = 0$ we have

$$\int_F \mathbb{E}(f|G)d\mu = \int_G \mathbb{E}(f|G)d\mu = \int_G f\,d\mu = \int_F d\mu.$$

Thus $\mathbb{E}(f|G)$ is a version of $\mathbb{E}(f|\tilde{G})$. //

§4. ISONOMY

(A) Let $\{X,A\}$ be a measurable space and $\{\Omega_i,F_i,\mu_i\}$, $i = 1,2$, both probability spaces. For a set T consider maps

$$z^i : T \to \mathcal{L}^0(\Omega_i,F_i;X) \qquad i = 1,2$$

(often called *processes*). Let $FP(T)$ denote the collection of finite subsets of T. If $\underline{t} = \{t_1,\ldots,t_m\}$ is in $FP(T)$ set

$$X^{\underline{t}} = X \times \ldots \times X \qquad \text{(m-times)}$$

and define

$$z^i_{\underline{t}} : \Omega \to X^{\underline{t}}$$

by

$$z^i_{\underline{t}}(\omega) = (z^i(t_1)(\omega),\ldots,z^i(t_m)(\omega)).$$

The *finite dimensional distributions* of z^i are the measures $z^i_{\underline{t}}(\mu)$ on $X^{\underline{t}}$ for $\underline{t} \in FP(T)$ and z^1 and z^2 are said to be *isonomous* or *equally distributed* if $z^1_{\underline{t}}(\mu) = z^2_{\underline{t}}(\mu)$ for all $\underline{t} \in FP(T)$.

If so we shall write

$$z^1 \stackrel{\cdot}{\underset{\cdot}{\sim}} z^2.$$

For sets $T_1,\ldots,T_p$ and measurable spaces $X_1,\ldots,X_p$ with maps

$$\alpha^r_i : T_r \to \mathcal{L}^0(\Omega,F;X_r) \qquad r = 1,\ldots,p, \quad i = 1,2$$

we shall say that $\alpha^1_1, \alpha^2_1,\ldots,\alpha^p_1$ *are jointly isonomous to* $\alpha^1_2,\alpha^2_2,\ldots,\alpha^p_2$, and write

$$\{\alpha^1_1,\ldots,\alpha^p_1\} \stackrel{\cdot}{\underset{\cdot}{\sim}} \{\alpha^1_2,\ldots,\alpha^p_2\}$$

if the maps

$$\alpha_i : T_1 \times \ldots \times T_p \to \mathcal{L}^0(\Omega,F;X_1 \times \ldots \times X_p) \qquad i = 1,2$$

$$\alpha_i(t^1,\ldots,t^p)(\omega) = (\alpha^1_i(t^1)(\omega),\ldots,\alpha^p_i(t^p)(\omega))$$

are isonomous.

For example, the proof of the following lemma is immediate from Exercise 3C(ii):

<u>Lemma 4A</u> <u>Suppose</u> $\{\alpha^1_1, \alpha^2_1\} \stackrel{\cdot}{\underset{\cdot}{\sim}} \{\alpha^1_2, \alpha^2_2\}$. <u>Then</u> α^1_1 <u>and</u> α^2_1 <u>are independent if and only if</u> α^1_2 <u>and</u> α^2_2 <u>are independent</u>. //

(B) Since experientially it is usually the isonomy class of the process which is considered rather than the process itself, it is extremely important to the physical relevance of a theory that the results be isonomy invariant. There will be no real loss of generality in assuming that there is only one basic probability space: if $\alpha_i:T \to \mathcal{L}^0(\Omega_i,F_i;M)$, $i = 1,2$, for probability spaces $\{\Omega_i,F_i,\mu_i\}$, we can replace them by

$$\tilde{\alpha}_i:T \to \mathcal{L}^0(\Omega_1 \times \Omega_2,\ F_1 * F_2;M)$$

$$\tilde{\alpha}_i(t)(\omega_1,\omega_2) = \alpha_i(t)(\omega_i)$$

with $\{\Omega_1 \times \Omega_2,\ F_1 * F_2\}$ furnished with the product measure $\mu_1 \otimes \mu_2$. Then α_1 and α_2 are defined to be isonomous if and only if $\tilde{\alpha}_1$ and $\tilde{\alpha}_2$ are.

Lemma 4B For a set T and measurable spaces M,N suppose we have

$$\Phi:T \times M \to N$$

measurable in M, and isonomous maps

$$\alpha_i\ :\ T \to \mathcal{L}^0(\Omega,F;M) \qquad i = 1,2.$$

Then the maps

$$\phi_i:T \to \mathcal{L}^0(\Omega,F;N)$$

$$\phi_i(t)(\omega) = \Phi(t,\alpha_i(t)(\omega)) \qquad i = 1,2,$$

are isonomous.

Proof. Let $\underline{t} = \{t_1,\ldots,t_m\} \subset T$ and suppose $B \subset N^{\underline{t}}$ is measurable. Then

$$\{\omega \in \Omega:\phi_{i,\underline{t}}(\omega) \in B\} \quad = \{\omega \in \Omega:\alpha_{i,\underline{t}}(\omega) \in \tilde{B}\}$$

where

$$\tilde{B} = \{\underline{x} \in M^{\underline{t}}:(\Phi(t_1,x_1),\ldots,\Phi(t_m,x_m)) \in B\}$$

and the result follows. //

From Lemma 4B we can deduce, for example, that if α_j^i have values in a vector space, $i,j = 1,2$, and

$$\{\alpha_1^1,\ \alpha_1^2\} \ \dot{\sim}\ \{\alpha_2^1,\ \alpha_2^2\}$$

then

$$\{\alpha_1^1,\ \alpha_1^2,\ \alpha_1^1 + \alpha_1^2\} \ \dot{\sim}\ \{\alpha_2^1,\ \alpha_2^2,\ \alpha_2^1 + \alpha_2^2\},$$

with corresponding results for products when defined.

(C) When M is a metric space and T has a topology, a map $z:T \to \mathcal{L}^0(\Omega,F;M)$ is said to be *continuous in measure*, or *continuous in probability* if it is continuous into $\mathcal{L}^0(\Omega,F;M)$ given the topology of convergence in measure determined by the measure μ on $\{\Omega,F\}$.

Exercise 4C. Show that continuity in measure depends only on the isonomy class of z.

§5 GRONWALL'S LEMMA

The following extension of the 'integrating factor' method of solving ordinary differential equations will be used extensively:

Gronwall's Lemma Let ϕ,α be real valued Lebesgue integrable functions on the interval $[0,T]$ such that for some $L > 0$

$$\phi(t) \leq \alpha(t) + L\int_0^t \phi(s)ds \qquad t \in [0,T] \qquad (1)$$

Then

$$\phi(t) \leq \alpha(t) + L\int_0^t e^{L(t-s)}\alpha(s)ds, \text{ almost all } t \in [0,T].$$

Proof. Set $\Psi(t) = \int_0^t \phi(s)ds$. Then Ψ is differentiable with $\Psi'(s) = \phi(s)$ almost everywhere on $[0,T]$, so (1) becomes

$$\Psi'(t) \leq \alpha(t) + L\,\Psi(t)$$

giving

$$\frac{d}{dt}(e^{-Lt}\Psi(t)) \leq e^{-Lt}\alpha(t) \qquad \text{a.e.}$$

whence

$$\Psi(t) \leq e^{Lt}\int_0^t e^{-Ls}\alpha(s)\,ds.$$

Substituting for $\Psi(t)$ in (1) gives the result. //

CHAPTER II : KOLMOGOROV'S THEOREM, TOTOKI'S THEOREM, AND BROWNIAN MOTION

§1 KOLMOGOROV'S CONSISTENCY CONDITIONS

(A) For a measurable space M and a set T let $\rho(t):M^T \to M$ be the evaluation map at $t \in T$ on the space M^T of all maps $\rho:T \to M$ i.e.

$$\rho(t)(\sigma) = \sigma(t).$$

Let $\bar{C}$ be the σ-algebra generated by $\{\rho(t):t \in T\}$. Then we have

$$\rho:T \to \mathcal{L}^0(M^T,\bar{C};M).$$

In fact $\bar{C}$ is the smallest σ-algebra containing all the ρ-*cylinder sets* C of M^T where C consists of those subsets of M^T of the form

$$\{\rho_{\underline{t}}^{-1}(B):B \subset M^{\underline{t}} \text{ measurable, } \underline{t} \in FP(T)\},$$

(for example the subset $\{\sigma:T \to M:\sigma(t_i) \in B_i,\ i = 1,\dots,m\}$ where $B_1,\dots,B_m$ are measurable subsets of M).

A probability measure μ on $\{M^T,\bar{C}\}$ has 'finite-dimensional distributions' $\mu_{\underline{t}} = \rho_{\underline{t}}(\mu)$ on $M^{\underline{t}}$ for each $\underline{t} \in FP(T)$. These are the finite dimensional distributions of the process ρ. If $\underline{s} \in FP(T)$ and $\underline{s} \subset \underline{t}$ there is the projection (or restriction) map

$$\pi_{\underline{s},\underline{t}}:M^{\underline{t}} \to M^{\underline{s}}$$

such that

$$\pi_{\underline{s},\underline{t}} \circ \rho_{\underline{t}} = \rho_{\underline{s}}.$$

The finite dimensional distributions must therefore satisfy the *consistency conditions:* $\pi_{\underline{s},\underline{t}}(\mu_{\underline{t}}) = \mu_{\underline{s}}$ $\underline{s} \subset \underline{t}$.

Conversely there is the Daniell-Kolmogorov theorem:

<u>Theorem 1A. Given a family</u> $\{\mu_{\underline{t}} : \underline{t} \in FP(T)\}$ <u>of tight probability measures on a topological space</u> M <u>satisfying the consistency conditions there is a unique probability measure</u> μ on $\{M^T, \bar{C}\}$ <u>with them as finite dimensional distributions.</u>

The proof is a matter of extending the set function $\tilde{\mu}$ defined on C by $\tilde{\mu}(\rho_{\underline{t}}^{-1}(A)) = \mu_{\underline{t}}(A)$ to a measure on $\bar{C}$. We are mainly interested in the case $M = \mathbb{R}^n$, and a proof can be found in (Bauer), (Parthasarathy 1967), or gleaned from (Totoki). //

(B) In Theorem 1A we can take $M = \mathbb{R}^n$ and $T = (0,R]$ or $(0,\infty)$ with the family $\{\gamma_{\underline{t}} : \underline{t} \in FP(T)\}$ of probability measures defined by

$$\gamma_{\underline{t}}(B) = \int_B h(0,0;t_1,x_1)h(t_1,x_1;t_2,x_2)\ldots h(t_{m-1},x_{m-1};t_m,x_m)\, dx_1,\ldots,dx_m$$

where $h(s,x;t,y) = (2\pi)^{-\frac{n}{2}}(t-s)^{-\frac{n}{2}}\exp(-\frac{1}{2}\frac{|y-x|^2}{t-s}) \qquad s < t$

and $0 < t_1 < \ldots < t_m$.

When $B = B_1 \times \ldots \times B_m$ for $B_j \subset \mathbb{R}^n$ then $\gamma_{\underline{t}}(B)$ is the 'probability that a Brownian particle starting from the origin in $\mathbb{R}^n$ is in B_j at time t_j for $j = 1,\ldots,m$' according to the Einstein-Smulochowski model. See (Nelson, 1967).

After checking the consistency of $\{\gamma_{\underline{t}} : \underline{t} \in FP(T)\}$, Theorem 1A ensures that there is at least one process with these as its finite dimensional distributions: namely we could take the measure γ on M^T determined by the $\{\gamma_{\underline{t}}\}$ and work with $\{M^T, \bar{C}, \gamma\}$ as our basic probability space. However, working with the space M^T of all paths will turn out to be much too general for our purposes (for example, the space of continuous paths is not measurable in $\{M^T, \bar{C}\}$, see e.g. Kingman & Taylor).

§2 SAMPLE CONTINUITY

(A) Let T be a topological space and M a metric space. For a probability space (Ω, F, μ) a process (i.e. a map)

$$z : T \to \mathcal{L}^0(\Omega, F; M)$$

is said to be *sample continuous*, or to have *continuous sample paths* if for almost all $\omega \in \Omega$ the *sample function*

$$T \to M \qquad t \mapsto z(t,\omega)$$

is continuous. If so z will be continuous in measure. However, the converse is false: take $\Omega = [0,1]$ with Lebesgue measure and take $T = [0,1]$ with

$$z(t,\omega) = 0 \quad \text{if } 0 < t < \omega$$
$$= 1 \quad \text{otherwise.}$$

(B) Two maps $z_i : T \to \mathcal{L}^0(\Omega,F;M)$, $i = 1,2$ are said to be *stochastically equivalent* if for each $t \in T$

$$z_1(t,\omega) = z_2(t,\omega) \quad \text{for almost all } \omega \in \Omega.$$

For example if $\Omega = T = [0,1]$ with Lebesgue measure take

$$z_i : T \times \Omega \to \mathbb{R} \qquad i = 1,2$$

with $z_1(t,\omega) = 1$ all (t,ω)
and $z_2(t,\omega) = 1$ if $t \neq \omega$
$= 0$ if $t = \omega$.

When z_1 and z_2 are sample continuous and stochastically equivalent they satisfy the stronger condition:

$$z_1(t,\omega) = z_2(t,\omega) \qquad \text{for all } t \in T, \text{ almost all } \omega \in \Omega,$$

provided T is separable. This is because stochastic equivalence implies equality on any countable subset of T, for almost all $\omega \in \Omega$. Our example does not satisfy this stronger condition. Stochastically equivalent processes are also called *versions* of each other. Clearly stochastic equivalence implies isonomy. Usually one works with the version of a map which has the best sample properties; for example a sample continuous version if one exists.

(C) The following is a standard result, due to Kolmogorov, when p = 1, which is the case we are mainly interested in. A proof in that case is in (Parthasarathy 1967) or can be found in most standard texts, e.g. (Bauer). The general case is proved in (Totoki). See Appendix C.

For a compact space T and a metric space (M,d) we shall let C(T;M) denote the metrizable space of continuous maps $f:T \to M$ with a metric $\underline{d}$ given by

$$\underline{d}(f_1,f_2) = \sup\{d(f_1(t),f_2(t)): t \in T\}.$$

If $0 \in T \subset \mathbb{R}$ and $x_0 \in M$ we shall let $C_{x_0}(T;M)$ denote the subspace of those f with $f(0) = x_0$. We shall let $\rho(t)$ denote the evaluation map at t on either of these spaces:

$$\rho(t)(f) = f(t).$$

Theorem 2C. (Totoki-Kolmogorov) Let (M,d) be a metric space and T a compact subset of $\mathbb{R}^p$. Let $\{\mu_{\underline{t}}:\underline{t} \in FP(T)\}$ be a family of tight probability measures on $\{M^{\underline{t}}:\underline{t} \in FP(T)\}$, satisfying the consistency consistency conditions. Suppose they also satisfy *Condition 2C:*

There exists $\alpha,\beta,\gamma > 0$ such that whenever $\delta > 0$ and $\underline{t} = \{t_1,t_2\}$ we have

$$\mu_{\underline{t}}\{(x_1,x_2) \in M \times M: d(x_1,x_2) > \delta\} < \beta\delta^{-\alpha}|t_1-t_2|^{p+\gamma}.$$

Then there is a unique Borel measure μ on $C(T;M)$ with $\rho_{\underline{t}}(\mu) = \mu_{\underline{t}}$ for each $\underline{t} \in FP(T)$. //

(D) It is easy to check that the Brownian motion probabilities $\{\gamma_{\underline{t}}:\underline{t} \in FP(\mathbb{R}(>0))\}$ of §1B satisfy Condition 2C. In fact Theorem 2C can be used to show that for any $R > 0$ there is a measure γ^R on $C_0([0,R];\mathbb{R}^n)$ with $\rho_{\underline{t}}(\gamma^R) = \gamma_{\underline{t}}$ when $\underline{t} \subset (0,R]$. This shows that the following definition is not vacuous:

Let $T = [0,R]$ or $[0,\infty)$. A map

$$z:T \to \mathcal{L}^0(\Omega,F;\mathbb{R}^n)$$

for some probability space $\{\Omega,F,\mu\}$ will be called an *n-dimensional Brownian motion* if

(i) it has continuous sample paths, and

(ii) its finite dimensional distributions $z_{\underline{t}}(\mu)$ are equal to $\gamma_{\underline{t}}$ whenever $\underline{t} \subset \mathbb{R}(>0)$.

It follows from (i) and (ii) that $z(0,\omega) = 0$ for almost all $\omega \in \Omega$.

We will set $\Omega^R = C_0([0,R];\mathbb{R}^n)$ with F^R = Borel Ω^R. Then if

$$z = \rho:T \to \mathcal{L}^0(\Omega^R,F^R;\mathbb{R}^n)$$

i.e. $\quad z(t)(\sigma) = \sigma(t) \qquad \sigma \in \Omega^R$

with the probability space $\{\Omega^R, F^R, \gamma^R\}$, we will call z a *standard model* of Brownian motion.

§3 SOME BASIC PROPERTIES OF BROWNIAN MOTION

(A) We continue with the notation of §2.

Proposition 3A (i) Suppose $0 < R < S$. Then the restriction map $\Omega^S \to \Omega^R$ maps γ^S to γ^R.

(ii) For positive R and S the map

$$\Phi : \Omega^R \times \Omega^S \to \Omega^{R+S}$$

$$\Phi(\sigma_1,\sigma_2)(t) = \sigma_1(t) \qquad \text{if } 0 \leq t \leq R$$

$$= \sigma_1(R) + \sigma_2(t-R) \text{ otherwise}$$

is a homeomorphism with

$$\Phi(\gamma^R \otimes \gamma^S) = \gamma^{R+S}.$$

Proof. Part (i) is immediate from the fact that the finite dimensional distributions determine the measure uniquely. This is also used to prove part (ii): after observing that Φ is a homeomorphism it is enough to verify that $\Phi(\gamma^R \otimes \gamma^S)$ has the correct finite dimensional distributions. For this suppose $\underline{t} = \{t_1,\ldots,t_m\}$ with

$$0 < t_1 < \ldots < t_p = R < t_{p+1} < \ldots < t_m \leq R+S$$

and suppose $B = B_1 \times \ldots \times B_m \subset (\mathbb{R}^n)^{\underline{t}}$, (in order to include $t_p = R$ we can, if necessary, take $B_p = \mathbb{R}^n$). Set $C = \{\sigma_1 \in \Omega^R : \sigma_1(t_i) \in B_i \text{ for } 1 \leq i \leq p\}$ and set $t_0 = 0$, $x_0 = 0$. Then

$$\rho_{\underline{t}}\Phi(\gamma^R \otimes \gamma^S)(B) = \int_{\Omega^R} \chi_C(\sigma_1)\gamma^S\{\sigma_2 \in \Omega^S : \sigma_2(t_j - R) + \sigma_1(R) \in B_j, j=p+1,\ldots,m\}d\gamma^R(\sigma_1)$$

$$= \int_{\mathbb{R}^n \times \ldots \times \mathbb{R}^n} \{\prod_{i=1}^{p} \chi_{B_i}(x_i)h(t_{i-1},x_{i-1};t_i,x_i)\}I(x_p,\ldots,x_m)dx_1 \ldots dx_m$$

where

$$I(x_p,\ldots,x_m) = \chi_{B_{p+1}}(x_{p+1} + x_p)\prod_{j=p+2}^{m} \chi_{B_j}(x_j + x_p)h(t_{j-1}-R,x_{j-1};t_j-R,x_j).$$

After substituting the values of h and making the change of variable

$x_j \mapsto x_j - x_p$ for j = p+1 to m we get the required result:

$$\rho_{\underline{t}}\, \Phi(\gamma^R \otimes \gamma^S)(B) = \gamma_{\underline{t}}(B). \quad //$$

Corollary 3A. Let $z:[0,\infty) \to \mathcal{L}^0(\Omega,F;\mathbb{R}^n)$ be an n-dimensional Brownian motion. Then for $R \geq 0$ the process $t \mapsto z(t+R) - z(R)$ is isonomous to z.

Proof. It suffices to consider $0 \leq t \leq S$ for arbitrary $S > 0$, and to take a standard model. In fact define

$$f_2:\Omega^{R+S} \to \Omega^S$$

by

$$f_2(\sigma)(t) = \sigma(t+R) - \sigma(R).$$

We must show $f_2(\gamma^{R+S}) = \gamma^S$. However f_2 is the composition

$$\Omega^{R+S} \xrightarrow{\Phi^{-1}} \Omega^R \times \Omega^S \xrightarrow{\text{proj.}} \Omega^S$$

and the result follows from the proposition. //

(B) Let $z:[0,\infty) \to \mathcal{L}^0(\Omega,F;\mathbb{R}^n)$ be an n-dimensional Brownian motion. For $t > 0$ let F_t be the σ-sub-algebra of F generated by $\{z(s):0 \leq s \leq t\}$ and let G^t be the σ-sub-algebra generated by $\{z(s)-z(t):s \geq t\}$.

Proposition 3B. F_t and G^t are independent.

Proof. Set G^t_n equal to the σ-algebra generated by $\{z(s)-z(t):t+n \geq s \geq t\}$ for each n = 1,2,... . Since G^t is generated by G^t_{n} it suffices to prove that G^t_n and F_t are independent (the measure ν_B on G^t given by $\nu_B(A) = \mu(A\cap B)$ for a given $B \in F_t$ is determined by its values on G^t_n). By isonomy invariance, Lemma 4A of Chapter I, we can take $\Omega = \Omega^{n+t}$ and $\mu = \gamma^{n+t}$, with $z = \rho$. The result reduces to showing that the maps

$$f_1:\Omega^{n+t} \to \Omega^t \qquad f_1(\sigma) = \sigma|[0,t]$$

and

$$f_2:\Omega^{n+t} \to \Omega^n \qquad f_2(\sigma)(s) = \sigma(t+s) - \sigma(t)$$

are independent. But if $\underline{f}:\Omega^{n+t} \to \Omega^t \times \Omega^n$ is given by $\underline{f}(\sigma) = (f_1(\sigma),f_2(\sigma))$

then $\underline{f}$ is the inverse of the homeomorphism Φ of Proposition 3A. It follows that $\underline{f}(\gamma^{n+t}) = \gamma^t \otimes \gamma^n$. Since $\gamma^t = f_1(\gamma^{n+t})$ and $\gamma^n = f_2(\gamma^{n+t})$ by Proposition 3A(i) and Corollary 3A, the proof is completed by Exercise 3C(ii) of Chapter I. //

(C) For $s > 0$ let $\mathbb{E}_s$ denote the conditional expectation operator with respect to the σ-algebra F_s defined above.

Proposition 3C. *If z is an n-dimensional Brownian motion then for* $t > s$

$$\mathbb{E}_s(z(t) - z(s)) = 0 \qquad \text{a.s.}$$

and

$$\mathbb{E}_s(|z(t)-z(s)|^{2p}) = [(2p-2+n)(2p-4+n)\ldots n]|t-s|^p \qquad \text{a.s.}$$

for $p = 1,2,\ldots$.

Proof. By Exercise 3C(i) of Chapter I and the previous proposition the conditional expectations are the same as the expectations, i.e. the integrals over Ω. Moreover by Corollary 3A we have

$$\int_\Omega (z(t) - z(s))d\mu = \int_\Omega z(t-s)d\mu$$

and

$$\int_\Omega |z(t)-z(s)|^{2p}d = \int_\Omega |z(t-s)|^{2p} d\mu.$$

However

$$\int_\Omega z(t-s)d\mu = \int_{\mathbb{R}^n} (2\pi(t-s))^{-n/2} x \exp\left(\frac{-|x|^2}{2(t-s)}\right) dx = 0$$

while

$$\int_\Omega |z(t-s)|^{2p} d\mu = \int_{\mathbb{R}^n} |x|^{2p} (2\pi(t-s))^{-n/2} \exp\left(\frac{-|x|^2}{2(t-s)}\right) dx$$

$$= (2\pi)^{-n/2}(t-s)^p \int_{\mathbb{R}^n} |y|^{2p} \exp\left(\frac{-|y|^2}{2}\right) dy.$$

Now for small α

$$\int_{\mathbb{R}^n} \exp(\alpha|y|^2) \exp\left(\frac{-|y|^2}{2}\right) dy = (2\pi)^{n/2} (1 - 2\alpha)^{-n/2}.$$

Expanding $(1-2\alpha)^{-n/2}$ by the binomial theorem and equating coefficients of α yields

$$\int_{\mathbb{R}^n} |y|^{2p} \exp\left(\frac{-|y|^2}{2}\right) dy = (2\pi)^{n/2}2^p \frac{n}{2} \left(\frac{n}{2} + 1\right) \ldots \left(\frac{n}{2} + p-1\right),$$

and the result follows. //

Remark. The property $\mathbb{E}_s\, z(t) = z(s)$ a.s. for $s > t$ is called the *martingale property*.

Exercise 3C. For z an n-dimensional Brownian motion prove

$$\mathbb{E}\ \langle z(t), z(s)\rangle = n \min\{s,t\}.$$

(D) Let z be an n-dimensional Brownian motion. Then we can write $z(t)(\omega) = (z^1(t)(\omega),\dots,z^n(t)(\omega))$ where $z^i : T \to \mathcal{L}^0(\Omega,\mathcal{F};\mathbb{R})$.

Proposition 3D. (i) The z^i are independent Brownian motions.

(ii) For each t and s with $t > s > 0$, the measurable functions $z^1(t) - z^1(s),\dots,z^n(t) - z^n(s)$ are conditionally independent as conditioned by $\mathcal{F}_s$.

Proof. It is easy to see that the natural isomorphism of $C_0([0,R];\mathbb{R}^n)$ with the direct sum of n copies of $C_0([0,R];\mathbb{R})$ maps the Wiener measure γ^R onto the product of the Wiener measures on each $C_0([0,R];\mathbb{R})$. This with the generalisations of Exercise 3C, Chapter I and Lemma 4A, Chapter I proves part (i).

For (ii) let A^i be in the σ-algebra generated by $z^i(t) - z^i(s)$. By Exercise 3C, Chapter I, and Proposition 3B:

$$\begin{aligned}\mathbb{E}_s(\chi_{A^1}\cdots\chi_{A^n}) &= \mathbb{E}(\chi_{A^1}\cdots\chi_{A^n})\\ &= \mathbb{E}(\chi_{A^1})\dots\mathbb{E}(\chi_{A^n}) \qquad \text{by part (i)}\\ &= \mathbb{E}_s(\chi_{A^1})\dots\mathbb{E}_s(\chi_{A^n}). \qquad //\end{aligned}$$

CHAPTER III : THE INTEGRAL: ESTIMATES AND EXISTENCE

§1 McSHANE'S INTEGRAL

We shall generally be concerned with the following objects:

A subset $T \subset \mathbb{R}$ with an interval $[a,b] \subset T$;

A probability space (Ω, F, μ) with a family, or *filtration*, of σ-algebras $\{F_\tau : \tau \in T\}$ such that $F_\sigma \subset F_\tau \subset F$ whenever $\sigma < \tau$;

Banach spaces $G_1, \ldots, G_q$ and a Hilbert space H, all separable;

Maps $z^\rho : [a,b] \to \mathcal{L}^0(\Omega, F; G_\rho)$ $\quad \rho = 1, \ldots, q$

and a map $B : T \times \Omega \to \mathbb{L}(G_1, \ldots, G_q; H)$.

To save space, and brackets, we will sometimes use the notation z^ρ_t for $z^\rho(t)$ and B_τ, or $B(\tau)$, for $B(\tau,-)$, with the corresponding convention for other processes.

We shall usually require our processes to be *adapted* to the filtration: for z^ρ and similar processes this means that each $z^\rho(t)$ is F_t-measurable, and for B it means that $B(\tau)$ is F_τ-measurable for each τ in T. The σ-algebra F_τ can be thought of as consisting of those events which are known about at time τ, or the 'past' at time τ. The term *non-anticipating* is sometimes used instead of 'adapted', but it also has a more technical definition; a process adapted to $\{F_\tau : \tau \in T\}$ is also called an *$F_\star$-process*.

From time to time our maps will be subjected to some of the following conditions (for positive integers r and p):

Condition A(r)

Each z^ρ is adapted and there exist constants $K > 0$, $\delta > 0$ such that if $a \leq s < t \leq b$ and $t-s < \delta$ then, almost everywhere

$$|\mathbb{E}_s(z^\rho(t) - z^\rho(s))| \leq K\,(t-s)$$

and

$$\mathbb{E}_s(|z^\rho(t) - z^\rho(s)|^{2p}) \leq K(t-s) \qquad p = 1,\ldots,r$$

Condition B(p)

B is adapted and is *continuous in* $\mathcal{L}^{2p}$ *norm at each point of* [a,b] i.e. if $t \in [a,b]$ then $\|B(t)\|_{\mathcal{L}^{2p}} < \infty$ and as $\tau \to t$ in T so $\|B(\tau) - B(t)\|_{\mathcal{L}^{2p}} \to 0$.

Condition B_o

(i) B is adapted.

(ii) B is *norm continuous in measure at each point of* [a,b] i.e. for each $t \in [a,b]$

$$|B(\tau) - B(t)| \to 0 \text{ in measure as } \tau \to t \text{ in } T.$$

(iii) B has *bounded sample paths almost surely* i.e. there is an $\mathcal{F}$-measurable $b:\Omega \to R$ with

$$|B(\tau,\omega)| < b(\omega) \text{ all } \tau \in T, \qquad \text{a.e.}$$

By a *partition of* [a,b] *in* T we mean an ordered set Π,

$$\Pi = (t_1,\ldots, t_{m+1}; \tau_1,\ldots,\tau_m)$$

with $a = t_1 < t_2 < \ldots < t_{m+1} = b$ and each $\tau_j \in T$. It is *belated* if $\tau_j \leq t_j$ for each j and *Cauchy* if $\tau_j = t_j$ for each j. We write

$$\text{mesh } \Pi = \max \{t_{j+1} - \min (t_j,\tau_j):1 \leq j \leq m\},$$

$$\Delta_j t = t_{j+1} - t_j, \quad \Delta_j z^\rho = z^\rho(t_{j+1}) - z^\rho(t_j) \qquad \text{and}$$

$$\mu(\Pi) = \max \{\Delta_j t : 1 \leq j \leq m\}$$

Given Π and the data described above there is the *Riemann sum:*

$$S(\Pi) \equiv S(\Pi;B;z^1,\ldots,z^q) : \Omega \to H$$

$$S(\Pi) = \sum_{j=1}^{m} B(\tau_j)(\Delta_j z^1,\ldots,\Delta_j z^q).$$

For simplicity we shall use the "Riemann" version of the stochastic integral using McShane's approach (McShane 1974). As with the classical Riemann integral this puts more restrictions on the integrands B than those needed for a Lebesgue type definition. Also McShane's conditions A(q) are slightly more restrictive than necessary on the 'noises' z^p. For the general theory with Brownian motion as noise see for example (Gikhman & Skorohod 1972), (Arnold 1974), or (Friedman 1975), and for more general noises (Meyer 1976), (Metivier & Pellaumail 1980) where infinite dimensional spaces are allowed, or (Bichteler 1981). For a survey of the general theory with historical comments and a fine bibliography see (Dellacherie 1980) or for a brief discussion (Rogers 1981). McShane's approach and his conditions are put in the context of the general theory in (Protter 1979).

Definition B *has a (belated) integral with respect to* $(z^1,\dots,z^q)$ *over* $[a,b]$ if there exists $J \in \mathcal{L}^0(\Omega,F;H)$ such that $S(\Pi; B;z^1,\dots,z^q) \to J$ in measure as mesh $\Pi \to 0$ through belated partitions Π. Each such J will be called (a version of) the integral and written

$$\int_a^b B(t)(dz^1(t),\dots,dz^q(t))$$

or just

$$\int_a^b B(dz^1,\dots,dz^q).$$

Thus the integral is only defined up to equivalence in $\mathcal{L}^0$. After some estimates in the next section, we will show that conditions A(q) together with either B(1) or B_0 suffice to ensure that the integrals exist; and that then, if the z^ρ have continuous sample paths, integrals with $q \geq 3$ vanish.

The fact that we use belated rather than Cauchy partitions will be of little importance for our main purposes. However, as will become apparent in Chapter V, §2, the limits of Riemann sums using partitions with $t_j < \tau_j < t_{j+1}$ will not in general exist, even for integrals like $\int_a^b z(t)dz(t)$ when z is a Brownian motion on $\mathbb{R}$.

§2 THE FUNDAMENTAL ESTIMATE

Almost all the later results depend on the following lemma of McShane. Here, and hereafter, if f has values in a normed vector space we use $\|f\|$ to denote the real number

$$\|f\| = \sqrt{\int |f|^2 \, d\mu}$$

when defined i.e. if $|f|$ is measurable even if f itself is not measurable. The corresponding meanings are also attached to $\|f\|_{\mathcal{L}^p}$.

Lemma 2 Given: (a) a probability space (Ω,F,μ) with σ-algebras $F_1 \subset F_2 \subset \ldots \subset F_m \subset F_{m+1} = F$

(b) A Banach space G, and a Hilbert space H, both separable

(c) F_j-random maps

$$u_j : \Omega \to \mathbb{L}(G;H) \qquad j = 1,\ldots,m$$

(d) $\Delta_j : \Omega \to G$ each F_{j+1}-measurable, $j=1,\ldots m$, with constants C_j, D_j, satisfying almost everywhere

$$|\mathbb{E}_j(\Delta_j)| \leq C_j \qquad \mathbb{E}_j(|\Delta_j|^2) \leq D_j \qquad j = 1,\ldots,m$$

where $\mathbb{E}_j$ denotes conditioning with respect to F_j.

Then

$$\left\|\sum_{j=1}^m u_j(\Delta_j)\right\| \leq 2 \sum_{j=1}^m C_j\|u_j\| + \left(\sum_{j=1}^m D_j\|u_j\|^2\right)^{\frac{1}{2}}.$$

When G is a Hilbert space the factor 2 in the right hand side is not needed.

Proof. We may assume the right hand side is finite i.e. that each $\|u_j\| < \infty$. Note that (d) implies that each $\|\Delta_j\|^2 \leq D_j < \infty$.

Further, for $R > 0$ set

$$\Omega_R = \{\omega \in \Omega : |u_j(\omega)| \leq R\} \in F_j$$

Then

$$\begin{aligned}\int_{\Omega_R} |u_j(\Delta_j)|^2 \, d\mu &\leq \int_{\Omega_R} |u_j|^2 \, |\Delta_j|^2 \, d\mu \\ &= \int_{\Omega_R} \mathbb{E}_j(|u_j|^2 \, |\Delta_j|^2) d\mu \\ &= \int_{\Omega_R} |u_j|^2 \, \mathbb{E}_j(|\Delta_j|^2) d\mu \\ &\leq D_j \|u_j\|^2 .\end{aligned}$$

Since $\Omega_R \nearrow \Omega$ as $R \to \infty$ we see, that for $j = 1,\ldots,m$

$$\|u_j(\Delta_j)\|^2 \leq D_j \|u_j\|^2 < \infty \qquad (1)$$

Now

$$\|\sum_{j=1}^{m} u_j(\Delta_j)\|^2 = \int_\Omega |\sum_j u_j(\Delta_j)|^2 \, d\mu$$

$$= \sum_j \int_\Omega |u_j(\Delta_j)|^2 \, d\mu + 2 \sum_{i<j} \int_\Omega \langle u_i(\Delta_i), u_j(\Delta_j)\rangle \, d\mu$$

$$\leq \sum_j D_j \|u_j\|^2 + 2 \sum_{i<j} \int_\Omega \langle u_i(\Delta_i), u_j(\Delta_j)\rangle \, d\mu.$$

Assume temporarily that $C_j = 0$ for all j.

Then, if $i < j$, $\int_\Omega \langle u_i(\Delta_i), u_j(\Delta_j)\rangle \, d\mu = \int_\Omega \mathbb{E}_j \langle u_i(\Delta_i), u_j(\Delta_j)\rangle d\mu$

$$= \int_\Omega \langle u_i(\Delta_i), u_j(\mathbb{E}_j \Delta_j)\rangle d\mu$$

$$= 0,$$

proving the result in this special case.

When the C_j do not all vanish set:

$$\tilde{\Delta}_j = \Delta_j - \mathbb{E}_j(\Delta_j)$$

Then

$$\mathbb{E}_j(\tilde{\Delta}_j) = 0$$

and when G is a Hilbert space

$$\mathbb{E}_j(|\tilde{\Delta}_j|^2) = \mathbb{E}_j(|\Delta_j|^2) - |\mathbb{E}_j(\Delta_j)|^2$$

$$\leq D_j.$$

Consequently by the first part,

$$\|\Sigma\, u_j(\Delta_j)\| = \|\Sigma u_j(\tilde{\Delta}_j) + \Sigma\, u_j(\mathbb{E}_j \Delta_j)\|$$

$$\leq \|\Sigma u_j(\tilde{\Delta}_j)\| + \Sigma \|(|u_j| \,.\, |\mathbb{E}_j \Delta_j|)\|$$

$$\leq \{\Sigma\, D_j \|u_j\|^2\}^{\frac{1}{2}} + \Sigma\, C_j \|u_j\|$$

as required.

When G is not Hilbert more care is required if all the C_j

do not vanish: for example see the original method in Lemma 1.1 of (McShane 1974). However we shall not need to use this particular case, and so leave it to the reader. //

The estimate in Lemma 2 is the basis of all the work that follows. It is for this reason that we work with Hilbert spaces H rather than general Banach spaces.

Corollary 2 Suppose $z^1,\ldots,z^q$ are as in §1 and satisfy condition A(q) with constants, K,δ. Let $\Pi = (t_1,\ldots,t_{m+1};\tau_1,\ldots,\tau_m)$ be a partition of [a,b] with $\mu(\Pi) < \delta$. For j = 1,...,m let

$$B_j:\Omega \to \mathbb{L}(G_1,\ldots,G_q;H)$$

be F_{t_j}-random. Then

$$\|\sum_{j=1}^{m} B_j(\Delta_j z^1,\ldots,\Delta_j z^q)\|$$

$$\leqslant \beta(\sum_{j=1}^{m} \|B_j\|^2 \Delta_j t)^{\frac{1}{2}}$$

where $\beta = 2K(b-a)^{\frac{1}{2}} + K^{\frac{1}{2}}$.

Proof. First suppose $q \geqslant 2$ and set $G = \hat{\bigotimes}_{1\leqslant\rho\leqslant q} G_\rho$. As in Chapter I, §2, let $\tilde{B}_j:\Omega \to \mathbb{L}(G;H)$ correspond to B_j, j = 1,...,m. Define

$$\Delta_j:\Omega \to G \text{ by } \Delta_j(\omega) = \bigotimes_{1\leqslant\rho\leqslant q} \Delta_j z^\rho.$$

By the generalised Hölder inequality, §3A of Chapter I, for almost all ω,

$$\mathbb{E}_{t_j}(|\Delta_j|^2) \leqslant \mathbb{E}_{t_j}(\prod_{1\leqslant\rho\leqslant q} |\Delta_j z^\rho|^2)$$

$$\leqslant \prod_{1\leqslant\rho\leqslant q} [\mathbb{E}_{t_j}(|\Delta_j z^\rho|^{2q})]^{1/q}$$

$$\leqslant K\Delta_j t \qquad \text{by A(q).}$$

Similarly $|\mathbb{E}_{t_j}(\Delta_j)| \leqslant \mathbb{E}_{t_j}(|\Delta_j|) \leqslant K\Delta_j t$, (when $q \geqslant 2$). By the lemma, with $\tilde{B}_j:\Omega \to \mathbb{L}(G;H)$ replacing u_j, and with $C_j = D_j = K\Delta_j t$:

$$\|\sum_{j=1}^{m} \tilde{B}_j(\Delta_j)\| < 2K \sum_{j=1}^{m} \|B_j\| \ (\Delta_j t)^{\frac{1}{2}}(\Delta_j t)^{\frac{1}{2}}$$

$$+ \left(\sum_{1}^{m} \|B_j\|^2 K\Delta_j t\right)^{\frac{1}{2}}$$

$$< 2K \{(\sum_{1}^{m} \|B_j\|^2 \Delta_j t \ (b-a)\}^{\frac{1}{2}}$$

$$+ \left(\sum_{1}^{m} \|B_j\|^2 \ K\Delta_j t\right)^{\frac{1}{2}}$$

by the Cauchy-Schwarz inequality. This is just the inequality we wanted.

For $q = 1$ we have only to apply the Cauchy-Schwarz inequaltiy after a direct application of the Lemma. //

Remark. For $q > 2$ note that we have a stronger inequality than necessary for the lemma: $\mathbb{E}_{t_j}(|\Delta_j|) < K(t_{j+1}-t_j)$. An obvious modification of (1) in the proof of the lemma shows that in these circumstances it is not essential to insist that H is Hilbert rather than just Banach.

3 THE MAIN EXISTENCE THEOREM

The proof of our main existence theorem is extremely simple in the case $q = 1$. The reader may find it more pleasant, and equally instructive, to work it through in that case rather than in the case $q = 2$ which we concentrate on. (See the proof of Proposition 5B of Chapter IV.)

Theorem 3 Let $z^1,\ldots,z^q$ satisfy A(q) and let B satisfy B(1). Then the integral $\int_c^e B(dz^1,\ldots,dz^q)$ exists for every $[c,e] \subset [a,b]$ and has an F_e-measurable version. In fact, given $\varepsilon > 0$, there exists $\delta^1 > 0$ such that for all $[c,e] \subset [a,b]$ and for all belated partitions Π of $[c,e]$ with mesh $\Pi < \delta^1$ we have

$$\|S(\Pi) - \int_c^e \dot{B}(dz^1,\ldots,dz^q)\| < \varepsilon.$$

Proof. We shall work only with partitions of [c,e] which are belated and have mesh less than the constant δ of condition A(q).

(i) Suppose $\Pi = (t_1,\ldots,t_{m+1}; \tau_1,\ldots,\tau_m)$

and $\Pi' = (t_1,\ldots,t_{m+1};\tau'_1,\ldots,\tau'_m)$.

Then

$$\|S(\Pi)-S(\Pi')\| = \|\sum_{j=1}^{m}(B(\tau_j)-B(\tau_j'))(\Delta_j z^1,\ldots,\Delta_j z^q)\|,$$

and by the corollary:

$$\|S(\Pi)-S(\Pi')\| \prec \beta(\sum_1^m \|B(\tau_j)-B(\tau_j')\|^2 \Delta_j t)^{\frac{1}{2}}$$

(1) $$\prec \beta\ (a-b)^{\frac{1}{2}} \max_j \|B(\tau_j)-B(\tau_j')\| \qquad (1)$$

which can be made arbitrarily small by choosing mesh Π sufficiently small.

(ii) For $q = 2$ and Π as above consider a partition $\Pi" = (t_1",\ldots,t_{n+1}";$ $\tau_1",\ldots,\tau_n")$. Say $\Pi"$ *is a division of* Π if $\{t_1,\ldots,t_{m+1}\} \subset \{t_1",\ldots,t_{n+1}"\}$ and if $[t_j", t_{j+1}"] \subset [t_i, t_{i+1}]$ implies $\tau_i = \tau_j"$. Say $\Pi"$ is a *simple division* of Π if also each (t_i, t_{i+1}) either contains one division point of $\Pi"$, which we will call s_i, or none, in which case we will set $s_i = t_i$. Assuming $\Pi"$ is a simple division of Π, set

$$J = \{j \text{ s.t. } s_j \neq t_j\}$$

and

$$\mu(\Pi,\Pi") = \max\ \{(t_{j+1}-t_j) : j \in J\}.$$

Then $S(\Pi) - S(\Pi")$

$$= \sum_{j\in J} \tilde{B}(\tau_j)\ \{[z^1(s_j)-z^1(t_j) \otimes z^2(t_{j+1})-z^2(s_j)] + [z^1(t_{j+1})-z^1(s_j)\otimes z^2(s_j)-z^2(t_j)]\}.$$

In order to apply our estimates to this:

$$|\mathbb{E}_{t_j}\{(z^1(s_j)-z^1(t_j)) \otimes (z^2(t_{j+1})-z^2(s_j))\}|$$

$$= |\mathbb{E}_{t_j}\mathbb{E}_{s_j}\{(z^1(s_j)-z^1(t_j)) \otimes (z^2(t_{j+1}) - z^2(s_j))\}|$$

$$= |\mathbb{E}_{t_j}\{(z^1(s_j)-z^1(t_j)) \otimes \mathbb{E}_{s_j}(z^2(t_{j+1})-z^2(s_j))\}|$$

by Corollary 3B of Chapter I, since $z^1(s_j)-z^1(t_j)$ is F_{s_j}-measurable.

This last quantity is majorised by

$$\mathbb{E}_{t_j}\Big(|z^1(s_j)-z^1(t_j)|K(t_{j+1}-s_j)\Big)$$

$$< K(t_{j+1}-s_j)\Big\{\mathbb{E}_{t_j}(|z^1(s_j)-z^1(t_j)|^2\Big\}^{\frac{1}{2}}$$

$$< \sqrt{K}^3(t_{j+1}-s_j)(s_j-t_j)^{\frac{1}{2}} > \sqrt{K}^3\mu(\Pi,\Pi")^{\frac{1}{2}}\Delta_j t.$$

Similarly

$$\mathbb{E}_{t_j}\Big(|(z^1(s_j)-z^1(t_j)) \otimes (z^2(t_{j+1})-z^2(s_j))|^2\Big)$$

$$< K^2\mu(\Pi,\Pi")\Delta_j t$$

and we have the corresponding estimates with z^1 and z^2 interchanged. Applying the lemma:

$$\|S(\Pi) - S(\Pi")\| < 4 \sum_{j=1}^{m} \sqrt{K}^3\mu(\Pi,\Pi")^{\frac{1}{2}}\Delta_j t \,\|B(\tau_j)\|$$

$$+ 2\Big\{\sum_1^m K^2\mu(\Pi,\Pi")\Delta_j t \,\|B(\tau_j)\|^2\Big\}^{\frac{1}{2}}$$

$$< \mu(\Pi,\Pi")^{\frac{1}{2}}C_0 \tag{2}$$

where $C_0 = \sup_j \|B(\tau_j)\| \Big(4\sqrt{K}^3(e-c) + 2\,K\sqrt{(e-c)}\Big) < C$ when

$$C = (\sup_{a<t<b} \|B(t)\| + 1)\Big(4\sqrt{K}^3(e-c) + 2K\sqrt{(e-c)}\Big) \text{ if mesh } \Pi \text{ is small}$$

enough.

For the case $q = 1$ we would have $S(\Pi) = S(\Pi")$, and for $q > 3$ we could obtain a similar estimate to (2) by analogous, but more complicated, computations.

(iii) To complete the proof suppose we have two partitions Π_0, Π'. Take a Cauchy partition Π_0' with subdivision points the union of those of Π_0 and Π'. Let Π_1 be the simple division of Π_0 obtained by adding to Π_0 the mid-points of all those intervals of Π_0 which contain more than one division point of Π_0' in their interior; and let Π_1' be the simple division of Π_0' obtained by adding these division points to Π_0'. Repeat this procedure until we obtain a partition Π_p with each interval containing at most one division point of Π_0', (and hence of Π_p'), in its interior.

Next take the partition $\tilde{\Pi}_p$ which has the same subdivision points as Π_p and has Π_p' as a simple subdivision.

$$\begin{array}{ccccccc}
\Pi_o & \xrightarrow{\text{s.d}} & \Pi_1 & \xrightarrow{\text{s.d}} & \cdots & \xrightarrow{\text{s.d}} & \Pi_p \\
 & & & & & & \downarrow \tau\text{-change} \\
 & & & & & & \tilde{\Pi}_p \\
 & & & & & & \downarrow \text{s.d} \\
\Pi_0' & \xrightarrow{\text{s.d}} & \Pi_1' & \xrightarrow{\text{s.d}} & \cdots & \xrightarrow{\text{s.d}} & \Pi_p'
\end{array}$$

Note that $\mu(\Pi_i,\Pi_{i+1}) \leq \frac{1}{2}\mu(\Pi_{i-1},\Pi_i) \leq \ldots \leq 2^{-i}\mu(\Pi_0)$ while

$\mu(\Pi_i',\Pi_{i+1}') \leq \mu(\Pi_i,\Pi_{i+1})$ and

$\mu(\Pi_p,\Pi_p') \leq \mu(\Pi_0)$.

Thus, by (2),

$$\|S(\Pi_0) - S(\Pi_p)\| \leq C \sum_{i=1}^{p} 2^{-\frac{1}{2}i} \mu(\Pi_0)^{\frac{1}{2}}$$

$$\|S(\Pi_p') - S(\tilde{\Pi}_p)\| \leq C\, \mu(\Pi_0)^{\frac{1}{2}}$$

$$\|S(\Pi_p') - S(\Pi_0') \leq C \sum_{i=1}^{p} 2^{-\frac{1}{2}i} \mu(\Pi_0)^{\frac{1}{2}}$$

while by (1), $\|S(\Pi_p) - S(\tilde{\Pi}_p)\|$ may be made arbitrarily small by taking mesh (Π_0) sufficiently small. Thus we can estimate $\|S(\Pi_0) - S(\Pi_0')\|$ and similarly $\|S(\Pi_0') - S(\Pi')\|$ to complete the proof. //

Remarks (i) McShane observes that the result still holds if we only insist that B is bounded in $\mathcal{L}^2$-norm on T and continuous in $\mathcal{L}^2$-norm at almost all points of [a,b]. This is used in Theorem 7A below.

(ii) Under the hypothesis of the theorem, for the belated partition Π of [c,e] the corollary gives

$$\|S(\Pi;B;z^1,\ldots,z^q)\| \leq \beta\left(\sum_1^m \|B(\tau_j)\|^2 \Delta_j t\right)^{\frac{1}{2}}.$$

But $\int_c^e \|B(\tau)\|^2 \, d\tau$ exists as a Riemann integral, so taking Cauchy partitions Π and letting mesh $\Pi \to 0$ the theorem shows:

Corollary 3 Under the conditions of the theorem

$$\left\|\int_c^e B(dz^1,\ldots,dz^q)\right\| \leq \beta\left\{\int_c^e \|B(t)\|^2 \, dt\right\}^{\frac{1}{2}}$$

where $\beta = 2K(b-a)^{\frac{1}{2}} + K^{\frac{1}{2}}$.

In particular if $F(t) = \int_a^t B(dz^1,\ldots,dz^q)$ then $F:[a,b] \to \mathcal{L}^2(\Omega,\mu;H)$ is $\frac{1}{2}$-Hölder continuous:

$$\|F(s) - F(t)\| \leq \text{const. } |s-t|^{\frac{1}{2}} \text{ for } s,t \in [a,b]. \quad //$$

§4 OTHER EXISTENCE THEOREMS: SUBSTITUTION

(A) Theorem 4A With the usual notation suppose $[c,e] \subset [a,b]$, $z^1,\ldots,z^q$ satisfy A(q) and B satisfies condition B_0. Then $\int_c^e B(dz^1,\ldots,dz^q)$ exists and the belated Riemann sums $S(\Pi)$ converge in measure uniformly with respect to the subintervals $[c,e]$ of $[a,b]$ i.e. given $\varepsilon > 0$ and $\alpha > 0$ there is a $\delta_1 > 0$ such that for all subintervals $[c,e]$ of $[a,b]$

$$\mu\{\omega \in \Omega : |S(\Pi)(\omega) - \int_c^e B(dz^1,\ldots,dz^q)(\omega)| > \alpha\} < \varepsilon$$

whenever Π is a belated partition of $[c,e]$ with mesh $\Pi < \delta_1$.

Proof. The theorem is a direct consequence of the following two simple lemmas. The technique involved will be used frequently.

Lemma 4A(i) Suppose there exists a sequence $\{\Omega_i\}_{i=1}^{\infty}$ in F and a sequence $B_i : T \times \Omega \to \mathbb{L}(G^1,\ldots,G^k;F)$ of F-random functions, where $G^1,\ldots,G^k$ and F are separable Banach spaces, such that

(a) $\Omega_i \subset \{\omega \in \Omega : B_i(\tau,\omega) = B(\tau,\omega) \text{ all } \tau \in T\}$

(b) $\lim_{i\to\infty} \mu(\Omega_i) = 1$

and (c) $\int_c^e B_i(dz^1,\ldots,dz^q)$ exists for each i.

Then $\int_c^e B(dz^1,\ldots,dz^q)$ exists

<u>and</u> $$\int_c^e B(dz^1,\ldots,dz^q) = \int_c^e B_i(dz^1,\ldots,dz^q) \quad \text{a.s. on } \Omega_i.$$

<u>Proof</u>. Suppose $\varepsilon > 0$. Choose i so that $\mu(\Omega_i) > 1 - \varepsilon/2$. For $\alpha > 0$ choose δ_1 so that if Π and Π' are belated partitions of $[c,e]$ with mesh less than δ_1 then

$$\mu\{\omega \in \Omega: |S(\Pi,B_i) - S(\Pi',B_i)| > \alpha\} < \varepsilon/2.$$

But then $$\mu\{\omega \in \Omega: |S(\Pi,B) - S(\Pi',B)| > \alpha\}$$

$$\leq \mu\{\omega \in \Omega: |S(\Pi,B_i) - S(\Pi',B_i)| > \alpha\} + \mu(\Omega-\Omega_i)$$

$$< \varepsilon.$$

Thus $\{S(\Pi,B)\}$ is Cauchy in measure and our integral exists. Since $S(\Pi,B_i) = S(\Pi,B)$ on Ω_i the corresponding integrals certainly agree almost surely on Ω_i. //

<u>Lemma 4A(ii)</u> <u>If</u> $B: T \times \Omega \to \mathbb{L}(G^1,\ldots,G^k;F)$ <u>satisfies condition</u> B_0 <u>then for each</u> $N > 0$ <u>there exists</u> $B_N: T \times \Omega \to \mathbb{L}(G^1,\ldots,G^k;F)$ <u>and</u> $\Omega_N \in F$ <u>such that</u>

(a) B_N <u>satisfies condition</u> B(1)

(b) $|B_N(t,\omega)| \leq N$ <u>for all</u> $t \in T$, <u>almost all</u> $\omega \in \Omega$

(c) $\Omega_N \subset \{\omega \in \Omega: B_N(\tau,\omega) = B(\tau,\omega) \text{ all } \tau \in T\}$

<u>and</u> (d) $\mu(\Omega_N) \to 1$ <u>as</u> $N \to \infty$.

<u>Proof</u>. There is an F-measurable map $b:\Omega \to \mathbb{R}$ and an $\tilde{\Omega} \in F$ with $\mu(\tilde{\Omega}) = 1$ and $|B(\tau,\omega)| < b(\omega)$ for all $(\tau,\omega) \in T \times \tilde{\Omega}$. For each positive N let

$$\Omega_N = \{\omega \in \tilde{\Omega}: |b(\omega)| \leq N\}.$$

Then $\Omega_N \uparrow \tilde{\Omega}$ as $N \to \infty$, whence $\mu(\Omega_N) \to 1$.

Also define a retraction $r_N: \mathbb{L}(G^1,\ldots,G^k;F) \to \mathbb{L}(G^1,\ldots,G^k;F)$ by

$$r_N(S) = \theta(|S|)S$$

where $\theta: \mathbb{R}(> 0) \to \mathbb{R}$

is given by
$$\theta(x) = \begin{cases} 1 & |x| \leq N \\ \dfrac{N}{|x|} & |x| \geq N. \end{cases}$$

Set $\qquad B_N = r_N \circ B.$

Using the uniform continuity of r_N we see that B_N remains norm continuous in measure and so by the dominated convergence theorem it is norm continuous in $\mathcal{L}^2$ (at each point of $[a,b]$). Thus (a) is true. Assertions (b) and (c) are clear. //

(B) Our next main result does two things. It shows that the classes of noises z against which we can integrate is sufficiently wide, and it also shows that 'substitutions' are valid. The latter part bears the main brunt of the proof of the change of variable formula (Itô's theorem) in Chapter V.

Lemma 4B(i) Suppose $A:T \times \Omega \to \mathbb{L}(F;H)$ and $B:T \times \Omega \to \mathbb{L}(G;F)$ satisfy B_0 where F,G, and H can be separable Banach spaces. Then

$$A\cdot B:T \times \Omega \to \mathbb{L}(G;H)$$

given by $\qquad A\cdot B(\tau,\omega) = A(\tau,\omega) \circ B(\tau,\omega)$

also satisfies B_0.

Proof. That $A\cdot B$ is adapted follows from §2B of Chapter I. It is also clear that $A\cdot B$ has almost surely bounded sample paths. For its norm continuity we may as well use Lemma 4A(ii) to obtain A_N and B_N as described there, for $N > 0$. Then $A_N\cdot B_N$ is easily seen to be norm continuous in $\mathcal{L}^2$ at each point of $[a,b]$, and therefore is so in measure, but given $\alpha > 0$ and $s,t \in T$

$$\{\omega \in \Omega: |A\cdot B(t,\omega) - A\cdot B(s,\omega)| > \alpha\}$$

$$\subset \{\omega \in \Omega: |A_N\cdot B_N(t,\omega) - A_N\cdot B_N(s,\omega)| > \alpha\}$$

$$\cup \{\omega \in \Omega: A_N\cdot B_N(\tau,\omega) \neq A\cdot B(\tau,\omega) \text{ some } \tau \in T\}$$

and the result follows easily. //

Lemma 4B(ii) For $r = 1,\dots,k$ let G^r, F^r be separable Banach spaces and let $\tilde{B}_r:T \times \Omega \to \mathbb{L}(G^r;F^r)$ satisfy condition B(k). Define $B:T \times \Omega \to \mathbb{L}(G^1,\dots,G^k;F^1 \hat{\otimes} \dots \hat{\otimes} F^k)$ by $B(t,\omega)(x^1,\dots,x^k) = \tilde{B}_1(x_1)\otimes\dots\otimes\tilde{B}_k(x^k)$. Then B satisfies condition B(1). If instead, each B_r satisfied condition B_0 then so would B.

Proof. Straightforward and essentially the same as for the case when each $F^r = R$. //

Most of the proof of the following is taken up with notation. It is especially simple if, say, $k = 1$ and $q_1 = 1$.

Proposition 4B For $r = 1,\ldots,k$ let $G_1^r,\ldots,G_{q_r}^r$ and F^r be separable Banach spaces, and let H be a separable Hilbert space. Suppose

$$z_\alpha^r : [a,b] \to \mathcal{L}^2(\Omega,\mu;G^r) \quad \alpha = 1,\ldots,q_r$$

satisfy Condition A $\left(\sum_{r=1}^{k} q_r\right)$ and assume the maps

$$B_r : T \times \Omega \to \mathbb{L}(G_1^r,\ldots,G_{q_r}^r;F^r)$$

satisfy one of the Conditions B_o, $B(k)$, $r = 1,\ldots,k$.

Let $\quad A : T \times \Omega \to \mathbb{L}(F^1,\ldots,F^k;H)$

satisfy Condition B_o.

Then, for $[c,e] \subset [a,b]$:

(i) The integral

$$I_1 = \int_c^e A(B_1,\ldots,B_k)(dz_1^1,dz_2^1,\ldots,dz_{q_k}^k)$$

exists.

(ii) If the integrals

$$u^r(t) = \int_c^t B_r(dz_1^r,\ldots,dz_{q_r}^r)$$

exist for $t \in [c,e]$, $r = 1,\ldots,k$, (e.g. if the F^r are Hilbert spaces) then the integral

$$I_2 = \int_c^e A(du^1,\ldots,du^k)$$

exists, and

$$I_1 = I_2.$$

Proof. First suppose each B_r satisfies B_o.

Define $$G^r = G_1^r \hat{\otimes} \dots \hat{\otimes} G_{q_r}^r$$

$$G = G_1 \hat{\otimes} \dots \hat{\otimes} G^k$$

and $$F = F_1 \hat{\otimes} \dots \hat{\otimes} F^k.$$

Then as in the lemma above the maps B_r combine to give a map B and corresponding $\tilde{B} : T \times \Omega \to \mathbb{L}(G;F)$ satisfying Condition B_0.

Furthermore A determines

$$\tilde{A} : T \times \Omega \to \mathbb{L}(F;H)$$

and I_1 may be written

$$I_1 = \int_c^e \tilde{A} \cdot \tilde{B}(dz_1^1 \otimes \dots \otimes dz_{q_k}^k).$$

Since A, and hence $\tilde{A}$, satisfies condition B_0, so does $\tilde{A}\cdot\tilde{B}$ by the first lemma. Therefore (i) follows directly from Theorem 4A.

For part (ii) take modifications A_N and $B_{r,N}$ of A and B_r as in Lemma 4A(ii) and set

$$u_N^r(t) = \int_c^t B_{r,N}(dz_1^r,\dots,dz_{q_r}^r)$$

and $$I_{2,N} = \int_c^e A_N(du_N^1,\dots,du_N^k)$$

if they exist.

For N sufficiently large our assumption that u^r exists implies that u_N^r exists except possibly on a set of small measure (*independent of* t) and that the Riemann sums for $I_{2,N}$, defined outside of that set, agree with those if I_2 outside another set of small measure. The corresponding statement holds for I_1 and $I_{1,N}$ where $I_{1,N}$ is defined using A_N and $B_{r,N}$. Thus by the argument of Lemma 4A(i) we can assume that A and $\{B_r\}$ all have all sample paths bounded by some fixed positive N, (the possibility that the Riemann sums for u_N^r may not converge on some set of small measure will be seen not to matter).

With this assumption, let

$$\Pi = (t_1,\dots,t_{m+1}; \tau_1,\dots,\tau_m)$$

be a belated partition of [c,e], and for each j = 1,...,m let

$$\Pi_j = (s^j_1,\ldots,s^j_{m_{j+1}};s^j_1,\ldots,s^j_{m_j})$$

be a Cauchy partition of $[t_j,t_{j+1}]$.

Set $$\Delta_{j,i}\, z^r = \bigotimes_{1\le\alpha\le q_r} (z^r_\alpha\,(s^j_{i+1}) - z^r_\alpha\,(s^j_i)) : \Omega \to G^r$$

and $$\Delta_{j,i}\, z = \bigotimes_{1\le r\le k} \Delta_{j,i}\, z^r : \Omega \to G.$$

Consider the sum

$$S(\Pi;\Pi_1,\ldots,\Pi_m) : \Omega \to H$$

given by

$$S(\Pi;\Pi_1,\ldots,\Pi_m) = \sum_j \tilde{A}\,(\tau_j)\,(\sum_j \tilde{B}(s^j_i)\Delta_{j,i}z).$$

We have $$S(\Pi;\Pi_1,\ldots,\Pi_m) = J + R$$

where $$J = \sum_{i,j} \tilde{A}\,(s^j_i)\tilde{B}(s^j_i)\Delta_{j,i}z$$

and

$$R = \sum_{i,j} (\tilde{A}(\tau_j) - \tilde{A}(s^j_i))\tilde{B}(s^j_i)\Delta_{j,i}z.$$

Since J is a Riemann sum for $\int_c^e \tilde{A}\cdot B(dz^1_1,\ldots,dz^k_{qk})$, it converges to that integral in measure as mesh $\Pi \to 0$ by part (i). Also by the fundamental Corollary there exists $\beta < \infty$, and $\delta > 0$, such that if mesh $\Pi < \delta$ then

$$\|R\| \le \beta \surd \sum_{i,j} \{\|\tilde{A}(\tau_j)-\tilde{A}(s^j_i))\tilde{B}(s^j_i)\|^2\,(s^j_{i+1}-s^j_i)\}.$$

Since $$\|(\tilde{A}(\tau_j) - \tilde{A}(s^j_i))\tilde{B}(s^j_i)\| \le N\,\|\tilde{A}(\tau_j)-\tilde{A}(s^j_i)\|$$

using the fact that $\tilde{A}$ satisfies B(1) we see

$$\|R\| \to 0 \text{ as mesh } \Pi \to 0.$$

Thus

$$S(\Pi;\Pi_1,\ldots,\Pi_m) \to \int_c^e \tilde{A}\cdot B(dz^1_1,\ldots,dz^k_{q_k})$$

in measure, as mesh $\Pi \to 0$.

However, for $j = 1,\ldots,m$, and $1 > \varepsilon_1 > 0$, we can choose Π_j so that

$$\left|\bigotimes_{r=1}^{k}\int_{t_j}^{t_{j+1}} B_r(dz_1^r,\ldots,dz_{q_r}^r) - \bigotimes_{r=1}^{k} S(\Pi_j;B_r;z_1^r,\ldots,z_{q_r}^r)\right| < \varepsilon_1/Nm$$

except possibly on a set of measure less than ε/m, since the integrals are assumed to exist and the natural k-linear map $F^1 \times \ldots \times F^k \to F$ is continuous. (If u_N^r does not exist on all of Ω we lose an extra set of small measure here in our replacement of u^r by u_N^r).

Then $$|S(\Pi;\Pi_1,\ldots,\Pi_m) - S(\Pi;A;u^1,\ldots,u^k)| < \varepsilon_1$$

on a set of measure greater than $1 - \varepsilon$.

It follows that $S(\Pi;A;u^1,\ldots,u^k)$ converges in measure to $\int_c^e \tilde{A} \cdot B(dz_1^1,\ldots,dz_{q_k}^k)$, whence $S(\Pi;A;u^1,\ldots,u^k)$ converges in measure to I_1, as required.

When some or all of the B_r satisfy Condition B(k) rather than Condition B_0 the proofs are essentially the same using Lemma 4B(ii) and replacing A by A_N. In any case we will not need to use those conditions.//

(C) Examples:

(i) For z a 1-dimensional Brownian motion if,

$$u(t) = \int_0^t z(s)dz(s) \qquad t > 0$$

then $$\int_a^b du(t)du(t) = \int_a^b z(t)^2\, dz(t)dz(t) \qquad 0 \leqslant a \leqslant b.$$

(ii) For z an n-dimensional Brownian motion if

$$u(t) = \int_0^t \langle z(s),dz(s)\rangle \qquad t > 0$$

then $$\int_a^b t^2 du(t) = \int_a^b t^2 \langle z(t),dz(t)\rangle \qquad 0 \leqslant a \leqslant b.$$

Remark 4C. Let $B:G_1 \times G_2 \to G_1 \hat{\otimes} G_2$ be the 'universal' bilinear map $(g_1,g_2) \mapsto g_1 \otimes g_2$. Given processes z_1 and z_2 as in Proposition 4B set

$$u(t) = \int_a^t B(dz_1,dz_2) = \int_a^t dz_1 \otimes dz_2$$

if it exists. Note that our existence theorems only apply here when both G_1 and G_2 are Hilbertable and at least one of them finite dimensional: otherwise $G_1 \hat{\otimes} G_2$ will not be Hilbertable. However see (Metivier & Pellaumail 1980) §3.6 Chapter 2.

Given $C:T \times \Omega \to \mathbb{L}(G_1,G_2;H)$ satisfying Condition B_0 we have the corresponding map

$$\tilde{C}:T \times \Omega \to \mathbb{L}(G_1 \hat{\otimes} G_2;H)$$

as in §2 of Chapter I. Taking $A = \tilde{C}$ in Proposition 4B we see that if $u(t)$ exists then

$$\int_c^e C(dz_1,dz_2) = \int_c^e \tilde{C}\, du.$$

Thus the second order integral is reduced to a first order integral.

When z_1 and z_2 have continuous sample paths the process $u(t) = \int_a^t dz_1 \otimes dz_2$ is often written as $\langle z_1,z_2\rangle_t$ and called the *angle bracket* process or *quadratic variation* if $z_1 = z_2$. To see that this agrees with the standard definition, e.g. (Meyer 1976), consider the Itô formula for $z(t)^2$ in Example (ii), §1, of Chapter V.

Clearly we can do the same for higher order stochastic integrals, again at least in the finite dimensional cases. Many of the results of the next chapter could be rephrased in terms of the corresponding processes $u(t)$.

When z is 1-dimensional

$$\langle z,z\rangle_t = \int_a^t dzdz = \lim_{\Pi} \Sigma(\Delta_j z)^2.$$

It follows that for each s and t with $a \leq s < t \leq b$ we have

$$\langle z,z\rangle_s \leq \langle z,z\rangle_t \qquad \text{almost surely.}$$

When also z has continuous sample paths we can choose a sample continuous version of $\langle z,z\rangle$, see Theorem 5C of Chapter IV below. For such a version we see that almost all the sample paths will be increasing functions of t. When z_1 and z_2 are 1-dimensional, and sample continuous, since

$$\langle z_1,z_2\rangle_t = \frac{1}{4}\langle z_1+z_2,z_1+z_2\rangle_t - \frac{1}{4}\langle z_1-z_2,z_1-z_2\rangle_t$$

we see that $\langle z_1,z_2\rangle$ has a version whose sample paths are the difference of two increasing functions i.e. of bounded variation on [a,b]. Thus, by taking components with respect to any bases, for general sample continuous, finite dimensional z_1 and z_2 (satisfying A(2)) we see that $\langle z_1,z_2\rangle$ has a version whose sample paths are of bounded variation on [a,b]. In particular when $\tilde{C}$ is sample continuous the integration $\int_c^e \tilde{C}d\langle z_1,z_2\rangle$ may be performed for each sample path as a classical Riemann-Stieltjes integral.

§5 RANDOM TIMES

(A) We wish to consider stochastic integrals over intervals [a,b] which are themselves functions of ω. It will be convenient to introduce here and in §6 the standing assumption (see §5C of Chapter IV);

Each σ-algebra F_τ *for* $\tau \in T$ *contains all sets in* F *which have measure zero.*

Let R* denote the extended real line $R^* = R \cup \{-\infty,\infty\}$, and take $[a,b] \subset R^*$ with $-\infty < a \leq b \leq \infty$.

A map $\xi:\Omega \to [a,b]$ will be called a *stopping* (or *Markov*) *time* if $\{\omega : t < \xi(\omega)\} \in F_t$, for all $t \in T$. If so, and if $\xi(\omega) \geq a$ for all ω, set

$$[a,\xi) \times \Omega \;= \{(t,\omega) : a \leq t < \xi(\omega)\}$$

and

$$\Omega_t = \Omega_t(\xi) = \{\omega : t < \xi(\omega)\}.$$

Then $\Omega_t \in F_t$.

We shall consider processes $x : [a,\xi) \times \Omega \to M$ where M is a metrisable space and ξ is a stopping time (the *lifetime* of x). Such a process will be called *admissable* if

(i) it is an F_*-*process* i.e. $x(t)|\Omega_t : \Omega_t \to M$ is F_t-measurable for each $t \in [a,b)$,

(ii) it has *continuous sample paths* almost surely i.e. given $t \in [a,b)$, for almost all $\omega \in \Omega_t$

$$x(\omega)|[a,t] : [a,t] \to M$$

is continuous.

Note that $x(a,\omega)$ need not be defined when $\xi(\omega) = a$. However it will often be given anyway as some initial condition. Also two such processes

$x_i : [a,\xi_i) \times \Omega \to M$, $i = 1,2$, will be called *equivalent* if $\xi_1 = \xi_2$ a.s. and if $x_1(\omega)|[a,t] = x_2(\omega)|[a,t]$ for almost all ω in $\Omega_t(\xi_1) \cap \Omega_t(\xi_2)$. As in §2B of Chapter II if the processes are admissible, equivalence follows if for each $t \in [a,b)$, $x_1(t)|\Omega_t(\xi_1) = x_2(t)|\Omega_t(\xi_2)$, a.s. Equivalence of x_1 with x_2 will be denoted by

$$x_1 \sim x_2.$$

(B) Proposition 5B. Let $\xi_i : \Omega \to [a,b]$ be stopping times, $i = 1,2,\ldots$. Then

(i) $\min(\xi_1,\xi_2)$ is a stopping time,

(ii) $\sup_i \xi_i$ is a stopping time.

Proof. (i) $\{\omega : s < \min(\xi_1(\omega), \xi_2(\omega))\} = \{\omega : s < \xi_1(\omega)\} \cap \{\omega : s < \xi_2(\omega)\}$

(ii) $\{\omega : s < \sup_i \xi_i\} = \bigcup_{i=1}^{\infty} \{\omega : s < \xi_i(\omega)\} \in F_s$. //

Notation: We will often write $\xi \wedge \eta$ for $\min\{\xi,\eta\}$ when ξ and η are R^* valued functions.

(C) Let $\xi : \Omega \to [a,b]$ be a stopping time. For a fixed $x : [a,\xi) \times \Omega \to M$ and $K \subset M$ define $W(K) : \Omega \to [a,b]$

$$\text{by } W(K) = \begin{cases} b, & \text{if } x(s,\omega) \in K, \text{ all } a \leqslant s < \xi(\omega) \\ \inf\{s : a \leqslant s < \xi(\omega) \text{ and } x(s,\omega) \notin K\}, & \text{otherwise.} \end{cases}$$

Proposition 5C. Let $x : [a,\xi) \times \Omega \to M$ be admissible. If M is a separable metric space and $U \subset M$ is open there is a stopping time

$$\tau(U) : \Omega \to [a,b] \text{ with } \tau(U) = W(U) \quad \text{a.s.}$$

and such that $x(\omega)([a,t]) \subset U$ for almost all ω satisfying $t < \tau(U)(\omega)$. [The map $\tau(U)$ will be called the *(first) exist time* from U, or the *(first) hitting time* of $M-U$.]

Proof. We can assume that all the sample paths of x are continuous and set $\tau(U) = K(U)$.

Let $\{t_i\}_{i=1}^{\infty}$ be dense in $[a,\infty)$ and $\{a_j\}_{j=1}^{\infty}$ dense in $M-U$. Then, if $t \in [a,b)$

$$\{\omega:\tau(U) < t\} = \bigcap_{n=1}^{\infty} \bigcup_{t_i<t} \{\omega: d(x(t_i,\omega), M-U) < \frac{1}{n}\}$$

$$= \bigcap_{n=1}^{\infty} \bigcup_{t_i<t} \bigcup_{j=1}^{\infty} \{\omega: d(x(t_i,\omega), a_j) < \frac{1}{n}\}$$

$$\in F_t. \quad //$$

§6 AMALGAMATION OF PROCESSES

(A) The following lemmas will be used to amalgamate possibly uncountable many processes defined up to different lifetimes but mutually equivalent on their common domains of definitions. This will be needed both in the construction of some 'improper' integrals and in the construction of solutions to stochastic differential equations on manifolds. First we have a measure theoretic lemma on the existence of 'least upper bounds' taken from (Nelson 1958); Nelson ascribes it to I. Segal.

Lemma 6A. For every subset A of $\mathcal{L}^0(\Omega,F;R^*)$ there is an element ξ of $\mathcal{L}^0(\Omega,F;R^*)$, uniquely determined up to equality almost everywhere, satisfying

(i) If $\alpha \in A$ then $\xi \geq \alpha$ a.e.

(ii) If $\eta \in \mathcal{L}^0(\Omega,F;R^*)$ has $\eta \geq \alpha$ a.e. for each $\alpha \in A$ then $\eta \geq \xi$ a.e.

(In other words $\mathcal{L}^0(\Omega,F,\mu;R^*)$ is a complete lattice under its natural order. We shall write $\xi = \sup\{\alpha: \alpha \in A\}$.)

Moreover if the supremum of each finite subset of A lies in A then ξ can be chosen so that $\xi = \sup_i \alpha_i$ for some sequence $\{\alpha_i\}_{i=1}^{\infty}$ in A.

Proof. By composing with an order preserving homeomorphism $q: R^* \to [0,1]$ we can assume that each $\alpha \in A$ has $\alpha(\Omega) \subset [0,1]$. We can also replace A by the subset B of $\mathcal{L}^0$ where

$$B = \{\sup_{\alpha \in F} \alpha : F \subset A \text{ and } F \text{ finite}\}.$$

Let $c = \sup\{\int \beta \, d\mu : \beta \in B\}$, so $\beta \leq 1$. By construction of B there exists a sequence $\{\beta_i\}_{j=1}^{\infty}$ in B with $c = \lim \int \beta_j \, d\mu$, and $\beta_j(\omega) \leq \beta_{j+1}(\omega)$ for all $\omega \in \Omega$, $j = 1,2,\ldots$. Set $\xi = \sup_j \beta_j$.

To check (i): If $\beta \in \mathcal{L}^0$ has $\beta(\omega) > \xi(\omega)$ on some set of positive measure then, using the monotone convergence theorem,

$$\sup_j \int \max(\beta,\beta_j)d\mu = \int \max(\beta,\xi)d\mu > \int \xi\, d\mu = c$$

whence $\max(\beta,\beta_j) \notin B$ for some j. Consequently $\beta \notin B$.

For (ii): If $\eta \geq \beta$ a.e. for each $\beta \in B$ then, a.e.,

$$\eta(\omega) \geq \sup \beta_j(\omega) = \xi(\omega)$$

as required.

The uniqueness of ξ follows from (i) and (ii). //

(B) Lemma 6B. Let A be a family of stopping times with values in $[a,\infty]$ and suppose that for each $\alpha \in A$ there is an F_*-process

$$I(\alpha) : [a,\alpha) \times \Omega \to P$$

into some topological space P. Assume that if $\alpha_1,\alpha_2 \in A$ and $a \leq t < \infty$ then

$$I(\alpha_1)(t,\omega) = I(\alpha_2)(t,\omega) \text{ almost surely on } \{\omega : t < \eta_1(\omega) \wedge \eta_2(\omega)\}$$

Then there exists an F_*-process

$$\bar{I} : [a, \sup\{\alpha : \alpha \in A\}) \times \Omega \to P$$

such that for each $t \in [a,\infty)$ and each $\alpha \in A$

$$\bar{I}(t,\omega) = I(\alpha)(t,\omega) \quad \text{almost surely on } \{\omega : t < \alpha(\omega)\}$$

and this defines $\bar{I}$ uniquely up to sets of measure zero for each $t \in [a,\infty)$.

Moreover if each $I(\alpha)$ is admissible then $\bar{I}$ can be chosen to be admissible.

Proof.

Case (i) : $A = \{\alpha_1,\alpha_2\}$.

$$\text{Define } \bar{I}(t,\omega) = \begin{cases} I(\alpha_1)(t,\omega) & t < \alpha_1(\omega) \\ I(\alpha_2)(t,\omega) & \alpha_1(\omega) \leq t < \alpha_2(\omega) \end{cases}$$

Case (ii) : A finite.

Use case (i) to proceed by induction, replacing

$$A = \{\alpha_1,\ldots,\alpha_{n+1}\} \text{ by } A' = \{\sup\{\alpha_1,\ldots,\alpha_n\}, \alpha_{n+1}\}.$$

By case (ii) we can always assume that the supremum of every finite subset of A lies in A.

Case (iii) : A countable, $A = \{\alpha_1,\alpha_2,\ldots\}$ say.

Using case (ii) replace A by $A' = \{\alpha_1',\alpha_2',\ldots\}$ where $\alpha_n' = \sup\{\alpha_i : 1 \leqslant i \leqslant n\}$. Define $\bar{I}$ by

$$\bar{I}(t,\omega) = I(\alpha_n')(t,\omega) \text{ if } \alpha_{n-1}'(\omega) \leqslant t < \alpha_n'(\omega)$$

where $\alpha_0'(\omega) \equiv a$.

Case (iv) : Arbitrary A.

Set $\bar{\alpha} = \sup\{\alpha : \alpha \in A\}$. Choose an increasing sequence $\{\alpha_i\}_{i=1}^{\infty}$ in A with $\alpha_i \to \bar{\alpha}$ almost surely, and define $\bar{I}$ from $\{\alpha_i\}_{i=1}^{\infty}$ as for case (iii). To see this behaves as required, suppose $\alpha \in A$. Then $\alpha \leqslant \bar{\alpha}$ almost surely and for $\varepsilon > 0$ we may take i with

$$\mu(\{\omega : t < \alpha(\omega)\} - \{\omega : t < \alpha(\omega) \wedge \alpha_i(\omega)\}) < \varepsilon .$$

Since $t < \alpha(\omega) \wedge \alpha_i(\omega)$ implies almost surely that

$$\bar{I}(t,\omega) = I(\alpha_i)(t,\omega) = I(\alpha)(t,\omega)$$

this gives $\mu(\{\omega : t < \alpha(\omega) \text{ and } \bar{I}(t,\omega) \neq \bar{I}(\alpha)(t,\omega)\}) < \varepsilon$

which proves what we wanted. A similar argument proves the uniqueness of $\bar{I}$. //

§7 STOCHASTIC INTEGRALS OVER A RANDOM INTERVAL

It is here that the lack of sophistication of our machine becomes slightly irksome, and some fine tuning is necessary to get up the power we want. The results here are used in the local uniqueness theorem for stochastic differential equations (in Chapter VI), needed for the construction of solutions on manifolds, and for the Itô formula for stochastic differential equations on manifolds when there is a possibility of explosion.

Some readers may prefer to gloss over the details.

We shall continue with the previous notation taking a closed subinterval $[c,d]$ of $[a,b] = T$, with $d < \infty$ and maps

$$z^\rho : [a,b] \to \mathcal{L}^0(\Omega,F;G_\rho) \qquad \rho = 1,2,\dots,q.$$

(A) Let $\xi,\xi_1 : \Omega \to [c,d]$ be stopping times with $\xi \leqslant \xi_1$. For

$$B : [a,\xi_1) \times \Omega \to \mathbb{L}(G_1,\dots,G_\rho;H)$$

consider

$$\bar{B} : [a,b] \times \Omega \to \mathbb{L}(G_1,\dots,G_\rho;H)$$

given by

$$\bar{B}(t,\omega) = \begin{cases} B(t,\omega) & t < \xi(\omega) \\ 0 & \text{otherwise.} \end{cases}$$

We *define*

$$\int_c^\xi B(t)(dz^1(t),\dots,dz^\rho(t)) = \int_c^d \bar{B}(t)(dz^1(t),\dots,dz^q(t))$$

whenever the right hand side exists.

Set $\Omega_\tau^1 = \{\omega \in \Omega \mid \tau < \xi_1(\omega)\}$, $a \leqslant \tau \leqslant d$. Unfortunately we need the following theorem.

<u>Theorem 7A</u> <u>Suppose</u> $z^1,\dots,z^q$ <u>satisfy</u> $A(q)$ <u>and</u> B <u>is</u> F-<u>random</u> <u>on</u> Ω_τ^1 <u>for each</u> $a \leqslant \tau \leqslant d$. <u>Assume</u>

(i) <u>there exists a measurable function</u> $m : \Omega \to \mathbb{R}$ <u>such that almost surely</u>

$$|B(\tau,\omega)| < m(\omega) \qquad a \leqslant \tau < \xi_1(\omega)$$

and (ii) <u>the map</u> $s \mapsto B(s,-)|\Omega_\tau^1$ <u>is norm-continuous in measure on</u> $[a,\tau]$, <u>each</u> $\tau \in T$.

<u>Then</u> $\bar{B}$ <u>satisfies condition</u> B_0 (<u>except for at most a countable set of discontinuities) of §1, and</u>

$$\int_c^\xi B(t)(dz^1(t),\dots,dz^q(t))$$

<u>exists</u>.

If also m *is bounded on* Ω *the integral exists in the sense of* L^2 *convergence and*

$$\left\| \int_c^{\xi} B(t)(dz^1(t),\dots,dz^q(t)) \right\| \leq \beta\left\{\int_c^d \|\bar{B}(t)\|^2\, dt\right\}^{\frac{1}{2}}$$

where β *depends only on* $[a,b]$ *and* $z^1,\dots,z^q$.

Moreover if $\{\xi^i\}_{i=1}^{\infty}$ *is a sequence of stopping times converging to* ξ *in measure and with* $c \leq \xi^i \leq \xi_1$ *for each* i, *then*

$$\int_c^{\xi_i} B(t)(dz^1(t),\dots,dz^q(t)) \to \int_c^{\xi} B(t)(dz^1(t),\dots,dz^q(t))$$

in measure and even in L^2 *if* m *is bounded*.

Proof. Let A denote the set of atoms of the measure $\xi(\mu)$ on $\mathbb{R}$. Then A is a countable set and if $a \leq s < t < b$

$$\mu\{\omega : s < \xi(\omega) \leq t\} \to 0 \text{ as } t \to s,$$

and as $s \to t$ if $t \notin A$.

Consequently if $\delta > 0$

$$\begin{aligned}\mu\{\omega: |\bar{B}(t,\omega) - \bar{B}(s,\omega)| > \delta\} &\leq \mu\{\omega \in \Omega_t : |B(t,\omega) - B(s,\omega)| > \delta\} \\ &\quad + \mu\{\Omega_s - \Omega_t\} \\ &\quad \to 0\end{aligned}$$

as $t \to s$, and as $s \to t$ provided $t \notin A$, by (ii).

Thus $\bar{B}$ is norm continuous in measure on T - A. It is easy to see that $\bar{B}(\tau)$ is F_τ-random for each τ and $\bar{B}$ has almost surely bounded sample paths. If also m is bounded we can therefore apply the modification of our Main Existence Theorem described in Remark (i), §3 above and proved in (McShane 1974) to get L^2 existence of the integral and also the estimate. The existence in measure for general m follows as usual.

When we are given the sequence $\{\xi^i\}_{i=1}^{\infty}$ assume first that m is bounded by $\bar{m} \in R$, say. For $\tau \in T$ set

$$\bar{B}_1(\tau,\omega) = \begin{cases} B(\tau,\omega) & \omega \in \Omega^1_\tau \\ 0 & \text{otherwise} \end{cases}$$

and

$$\bar{B}^i(\tau,\omega) = \chi_{[0,\xi^i)}(\tau)\bar{B}_1(\tau,\omega)$$

and write $I_i = \int_c^{\xi^i} B(t)(dz^1(t),\ldots,dz^q(t))$.

Then $$I_i = \int_c^d \bar{B}^i(t)(dz^1(t),\ldots,dz^q(t))$$

and
$$\|I_i - \int_c^{\xi} B(t)(dz^1(t),\ldots,dz^q(t))\|^2$$
$$\leq \beta^2 \int_c^d \|\bar{B}^i(t) - \chi_{[0,\xi)}(t)\bar{B}_1(t)\|^2 \, dt$$
$$\leq \beta^2 \int_c^d \int_\Omega |\chi_{[0,\xi^i)}(t) - \chi_{[0,\xi)}(t)|\bar{m}^2 \, d\mu \, dt$$
$$= \bar{m}^2 \beta^2 \int_\Omega |\xi^i - \xi| \, d\mu \to 0 \text{ as } i \to \infty,$$

by the dominated convergence theorem.

Thus $\{I_i\}_{i=1}^{\infty}$ converges as required when m is bounded. Lemmas 4A(i) and (ii) give the result for general m. //

<u>Remarks 7A</u> Under the conditions of the theorem, for a process

$$x : [c,\xi) \times \Omega \to H$$

the statement

$$x(t) = \int_c^{t\wedge\xi} B(s)(dz^1,\ldots,dz^q(s)) \qquad \text{a.s. on } \Omega_t$$

is equivalent to the statement

$$x(t) = \int_c^t B(s)(dz^1,\ldots,dz^q) \qquad \text{a.s. on } \Omega_t$$

where the interpretation of the second statement is that the right hand side exists as a limit in measure, on Ω_t, of the Riemann sums considered as maps $S(\Pi) : \Omega_t \to H$, and that the right and left hand sides are equal for almost all ω in Ω_t.

To see this let $\alpha(-)(\omega)$ be the characteristic function of $[c,\xi(\omega))$. Multiply both sides of both equations by $\alpha(t)$ and observe that $\alpha(t) = \alpha(t)\,\alpha(s)$ in order to convert the right hand side of the second equation into that of the first equation. //

Suppose now that $\xi_1 \equiv d$ and write

$$I(t) = \int_c^t B(t)(dz^1(t),\dots,dz^q(t)) \qquad c < t < d.$$

The following version of (Friedman 1975) vol. I Lemma 4.4., shows that our definition agrees with that of Friedman, at least in fairly general circumstances, (see also Remark 7B below):

Corollary 7A *Suppose* B *satisfies the conditions of the theorem and that there is a version* $\{I(t)\}_{c<t<d}$ *with right continuous sample paths. Then*

$$\int_c^\xi B(t)(dz^1(t),\dots,dz^q(t)) = I(\xi) \qquad \text{almost surely.}$$

Proof. Following Friedman define a sequence $\{\xi^i\}_{i=1}^\infty$ of stopping times by

$$\xi^i(\omega) = c \text{ if } \xi(\omega) = c$$

and

$$\xi^i(\omega) = \frac{n+1}{i} \wedge d \text{ if } \frac{n}{i} < \xi(\omega) < \frac{n+1}{i} \qquad n = 1,2,\dots .$$

Then each ξ^i is simple while $\xi^i \to \xi$ almost surely. Each ξ^i is a stopping time since, for $t < d$

$$\{\omega:\xi^i(\omega) < t\} = \{\omega:\xi(\omega) < \frac{\bar{n}+1}{i}\} \in F_t$$

where $\bar{n} = \sup\{n: \frac{n+1}{i} < t\}$.

The result is straightforward when ξ is simple, and therefore for ξ^i replacing . Since ξ^i decreases to ξ it follows for ξ after taking limits using the theorem and the right continuity of the sample paths of I. //

The proof of the Corollary yields another useful result, and one which we shall use shortly, for it we allow $d = b = \infty$, so that ξ_1 can be infinite:

Lemma 7A *Let* M *be a metric space and*

$$x:[a,\xi_1) \times \Omega \to M$$

a right continuous F_**-process.*

Then for any stopping time ξ with $a \leq \xi < \xi_1$ the process $\hat{x}$ defined by $\hat{x}(t) = x(t\wedge\xi)$ is also an F_*-process.

Proof. Take a sequence $\{\xi^i\}$ of simple stopping times decreasing to ξ, as in the proof of the Corollary. For each i the restriction of $x(t\wedge\xi^i)$ to Ω_t^1 is easily seen to be F_t-measurable. Therefore so is its limit $\hat{x}(t)$. //

(B) There is the following similar but more elementary result than Corollary 7A. This time set

$$I(t) = \int_c^{t\wedge\xi_1} B(s)(dz^1(s),\ldots,dz^q(s)).$$

Proposition 7B If $t \in [c,d]$ then for almost all $\omega \in \Omega_t$

$$I(t\wedge\xi) = \int_c^{t\wedge\xi} B(s)(dz^1(s),\ldots,dz^q(s))$$

whenever both integrals exist.

Proof. This follows immediately from the definition of the integrals in terms of belated Riemann sums, and from the fact that convergence in measure implies the existence of a subsequence converging almost surely. //

Remark 7B. We can extend Corollary 7A to the case of a general stopping time $\xi_1:\Omega \to [c,d]$. For this suppose B satisfies the conditions of the theorem and set

$$I(t) = \int_c^{t\wedge\xi_1} B(s)(dz^1(s),\ldots,dz^q(s)) \qquad c \leq t \leq d.$$

Assume that $\{I(t):c \leq t \leq d\}$ has a version with right continuous sample paths. Then for that version, if $c \leq t \leq d$,

$$\int_c^{t\wedge\xi} B(s)(dz^1(s),\ldots,dz^q(s)) = I(t\wedge\xi) \quad \text{a.s.}$$

Proof. Take $\xi^i:\Omega \to [c,d]$ as in the proof of Corollary 7A. Since ξ^1 is simple it follows from the definition that

$$I(t\wedge\xi^i\wedge\xi_1) = \int_c^{t\wedge\xi^i\wedge\xi_1} B(dz^1,\ldots,dz^q) \quad \text{a.s.}$$

As before $t\wedge\xi^i\wedge\xi_1$ decreases to $t\wedge\xi$ as $i\to\infty$ and the result follows using

Theorem 7A and the right sample continuity of I. //

(C) The next proposition will be particularly useful when z is a Brownian motion. Suppose that q = 1 and set $z = z^1$. The stopping times ξ, ξ_1 are as in section A.

Proposition 7C Suppose B is adapted and that z is an F_*-*martingale* i.e. $\mathbb{E}_s(z(t)) = z(s)$, s < t. Assume $\int_c^{t\wedge\xi} B(s)dz(s)$ exists as a limit in $\mathcal{L}^1(\Omega, F_t, \mu; H)$ of belated Riemann sums, for each $t \in [c,d]$. Then

$$\mathbb{E}_s \int_c^{t\wedge\xi} B(r)dz(r) = \int_c^{s\wedge\xi} B(r)dz(r) \qquad c \leq s \leq t \leq d$$

Proof. Note first that $\int_c^{t\wedge\xi} Bdz = \int_c^t \bar{B}dz$. Next if $c \leq s \leq t$ then $\int_c^t \bar{B}dz$ is a limit of Riemann sums $S(\Pi)$ with one of the partition points of Π being equal to s.
Say

$$S(\Pi) = \sum_j \bar{B}(\tau_j)\Delta_j z$$

with $t_k = s$.

$$\text{Then} \qquad \mathbb{E}_s(S(\Pi)) = \sum_{j<k} \bar{B}(\Pi_j)\Delta_j z + \mathbb{E}_s \sum_{k\leq j} \mathbb{E}_{t_j} \bar{B}(\tau_j)\Delta_j z$$

$$= \sum_{j<k} \bar{B}(\tau_j)\Delta_j z$$

which is a typical Riemann sum for $\int_c^s \bar{B}dz$. The result follows by the continuity of $\mathbb{E}_s$ acting on $\mathcal{L}^1$. //

Corollary 7C Assume that B satisfies the conditions of Theorem 7A, (with m not necessarily bounded on Ω) and that B has continuous sample paths.
Set

$$I(t) = \int_c^{t\wedge\xi} B(r)dz(r) \qquad c \leq t \leq d$$

and assume that $I:[c,d] \times \Omega \to H$ is bounded. Then if z is a martingale

$$\mathbb{E}_c(I(t)) = 0 \qquad c \leq t \leq d.$$

Proof. Let ξ_R be the first exit time after c of $|B(r)|$ from $[0,R)$. Set $\tau_R = t\wedge\xi_R\wedge\xi$. By the dominated convergence theorem and Theorem 7A

$$\mathbb{E}_c(I(t)) = \lim_{R\to\infty} \mathbb{E}_c\left(\int_c^{\tau_R} B(r)dz(r)\right)$$

But

$$\int_c^{\tau_R} B(r)dz(r) = 0$$

by the Proposition. //

(D) In order to deal with the case where each $B(-,\omega)$ is continuous on $[a,\xi(\omega))$ but not bounded we need some sort of extension of Theorem 7A. For this we continue with our notation but allow $d = b = \infty$, so ξ may also be infinite. We also take $\xi = \xi_1$. Suppose that for all $c\leq t\leq d$ and all stopping times η with $c\leq\eta<\xi$ on $\{\omega: c < \xi(\omega)\}$ the integrals

$$I(\eta)(t) = \int_c^{t\wedge\eta} B(s)(dz^1(s),\ldots,dz^q(s))$$

are defined and F_t-measurable, thereby defining F_*-processes

$$I(\eta) : [c,\eta) \times \Omega \to H.$$

By Proposition 7B the family $\{I(\eta) : c \leq \eta < \xi$ when $c < \xi(\omega)$, η a stopping time$\}$ satisfy the conditions of Lemma 6B and so determines an essentially unique F_*-process.

$$\bar{I} : [c,\xi) \times \Omega \to G,$$

provided ξ is *predictable* i.e. $\xi = \sup\ \{\eta: c \leq \eta < \xi, \eta$ a stopping time$\}$ on $\{\omega: c < \xi(\omega)\}$.

By the following lemma we can indicate the relationship between $\bar{I}$ and the integrals by writing

$$\bar{I}(t) = \int_c^{t\wedge\xi} B(s)(dz^1(s),\ldots,dz^q(s)) \qquad \text{a.s. on } \Omega_t$$

thus giving some meaning to the integral on the right hand side, as a function of t, even if it does not exist in the sense of section A. By Proposition 7B this agrees with the usual meaning when the integrals do exist.

Lemma 7D The process $\bar{I}$ satisfies

$$\bar{I}(t) = \int_c^{t\wedge\xi} B(s)(dz^1(s),\dots,dz^q(s)) \qquad \text{a.s. on } \Omega_t$$

in the sense that for each t the right hand side exists on Ω_t as a limit in measure of Riemann sums $S(\Pi):\Omega_t \to H$ for belated partitions Π of $[c,t]$ and equality holds almost surely on Ω_t. (We could equally well write $\int_c^t$ instead of $\int_c^{t\wedge\xi}$).

Proof. Suppose $\varepsilon > 0$. Since ξ is predictable we can find a stopping time η with $c \leq \eta < \xi$ on $\{\omega : c < \xi(\omega)\}$ such that $\mu\{\omega \in \Omega_t : \eta(\omega) \leq t\} < \varepsilon/2$. Also for $\alpha > 0$ we can find $\delta > 0$ such that if Π is a belated partition of $[a,t]$ with mesh $\Pi < \delta$ then

$$\mu\{\omega \in \Omega: \left| \int_c^{t\wedge\eta} B dz^1 \dots dz^q - S(\Pi,\bar{B})\right| > \alpha\} < \varepsilon/2$$

where $\bar{B}(s) = \chi_{[0,\eta)}(s)\, B(s)$.

But then $\mu\{\omega \in \Omega_t : |\bar{I}(t) - S(\Pi,B)| > \alpha\}$

$$\leq \mu\{\omega \in \Omega_t : |\bar{I}(t) - S(\Pi,\bar{B})| > \alpha\}$$

$$+ \mu\{\omega \in \Omega_t : \eta(\omega) \leq t\}$$

$$< \varepsilon$$

since $\bar{I}(t) = I(\eta)(t) = \int_c^{t\wedge\eta} B dz^1 \dots dz^q$ on $\{\omega \in \Omega : t < \eta(\omega)\}$ by definition of $\bar{I}$. //

When an admissible version of $\bar{I}$ exists we will take it. In fact:

Proposition 7D Suppose $q = 1,2$ and z^i, $i = 1$ to q satisfy $A(q)$ and have continuous sample paths. Assume B is F_τ-random on Ω_τ, and $B|[a,\tau] \times \Omega_\tau$ has almost surely continuous sample paths for $a \leq \tau < b$. Then if ξ is predictable there is an admissible process $\bar{I} : [c,\xi) \times \Omega \to H$ uniquely defined up to equivalence such that

$$\bar{I}(t) = \int_c^{t\wedge\xi} B(s)(dz^1(s),\ dz^2(s)) \qquad \text{a.s. on } \Omega_t \quad c \leq t < d$$

or

$$\bar{I}(t) = \int_c^{t\wedge\xi} B(s)dz^1(s) \qquad \text{a.s. on } \Omega_t \quad c \leq t < d$$

when $q = 1$.

Proof. We shall take $q = 1$, $z = z^1$. It suffices to show that for any stopping time with $c \leq \eta < \xi$ the integrals $I(\eta)(t) = \int_c^{t\wedge\eta} B(s)(dz(s))$ exist and determine admissible processes. For this we will have to use one of the results to be proved in the next chapter. In fact for such η define

$$\hat{B}(\tau,\omega) = B(\tau\wedge\eta(\omega),\omega).$$

Then $\hat{B}$ satisfies condition B_0 of Chapter III by Lemma 7A and the sample continuity of B and so Theorem 5C of Chapter IV will show that

$$J(t) = \int_c^t \hat{B}(s,\omega)\ dz(s)$$

exists as an admissible process $J : [c,d) \times \Omega \to G$.

Now $B|[c,\eta) \times \Omega$ satisfies the conditions of Theorem 7A and so $I(\eta)(t)$ exists for each t. Moreover, since $\hat{B}$ and B agree on $[c,\eta) \times \Omega$, applying Theorem 7A, Corollary 7A and Proposition 7B we see that $I(\eta)(t) = J(t\wedge\eta)$ almost surely, proving that $I(\eta)$ has an admissible version. //

Remark 7D *Under the conditions and notation of the Proposition, suppose* η *is a stopping time with* $c \leq \eta < \xi$ *almost surely on* $\{\omega : c < \xi(\omega)\}$. *Then defining*

$\bar{I}(\eta)(\omega) = I(\eta(\omega))$ *we have*

$$\bar{I}(\eta) = \int_c^{\eta} B(s)dz^1(s) \qquad \text{a.s.}$$

(*or if* $q = 2$: $\quad \bar{I}(\eta) = \int_c^{\eta} B(s)dz^1(s)dz^2(s) \quad$ a.s.)

Proof. (For $q = 1$ and $z = z^1$). By definition of $\bar{I}$, we have

$$\bar{I}(t\wedge\eta) = \int_c^{t\wedge\eta} B(s)dz(s). \qquad \text{a.s.} \quad c < t < d.$$

By Theorem 7A as $t \to d$ so

$$\int_c^{t\wedge\eta} Bdz \to \int_c^{\eta} Bdz$$

in measure. There is therefore a sequence $\{t_i\}$ in $[c,d]$ converging to d for which the corresponding integrals converge almost surely. Since $\bar{I}$ is admissible $\bar{I}(t_i\wedge\eta) \to \bar{I}(\eta)$ a.s. and the result follows. //

§8 NON-BELATED SECOND ORDER INTEGRALS

The following result is not surprising after the discussion at the end of §4. As a particular case we can take $K = T = [a,b]$ and $\phi(\omega,t) = t$ for all $(\omega,t) \in \Omega \times T$. Under the given continuity assumptions the lemma then shows that we could equally well take Riemann partitions in the definition of second order integrals i.e. partitions with $t_j \leq \tau_j \leq t_{j+1}$, each j (or more precisely that they are as effective as Cauchy partitions). The lemma will play an important role in the proof of the Itô formula in Chapter V, and is also used in Chapter IV in the proof that certain third order integrals vanish.

Lemma 8 Let G_1, G_2, H be Banach spaces and K a closed convex subset of a Banach space.

Assume

$$B:\Omega \times K \to \mathbb{L}(G_1,G_2;H) \text{ is } \mathcal{F}\text{-random},$$

$$\phi:\Omega \times T \to K \quad \text{is measurable},$$

and

$$u^i:[a,b] \to \mathcal{L}^2(\Omega,\mu;G_i) \text{ is } \tfrac{1}{2}\text{-Hölder continuous}, \; i = 1,2.$$

Also suppose that almost all sample paths of B and ϕ are continuous. Then, for belated partitions Π, as mesh $\Pi \to 0$

$$\sup_{0\leq s_j\leq 1} |S(\Pi;B\circ(\text{id} \times \phi);u^1,u^2) - \sum_j B(\phi(t_j)+s_j\Delta_j\phi)(\Delta_j u^1,\Delta_j u^2)| \to 0$$

in measure.

[Here $\Pi = (t_1,\ldots,t_{m+1};\tau_1,\ldots,\tau_m), \Delta_j\phi = \phi(t_{j+1})-\phi(t_j)$, etc.]

Proof. We can, and will, assume that T is compact. Given $\delta > 0$, $\varepsilon > 0$ we must show that there exists $\eta > 0$ such that if mesh $\Pi < \eta$ then,

$$\mu\{\omega\in\Omega: |S(\Pi;B\circ(id_\Omega\times\phi);u^1,u^2) - \sum_j B(\phi(t_j)+s_j\Delta_j\phi)(\Delta_j u^1,\Delta_j u^2)| \leq \delta \text{ all } 0\leq s_j\leq 1\} \geq 1-\varepsilon$$

Let M be a common $\frac{1}{2}$-Hölder bound for u^1,u^2. Choose measurable maps $\rho(\gamma):\Omega \to R(\geq 0)$, for $\gamma > 0$ such that for almost all $\omega \in \Omega$

$$\rho(\gamma)(\omega) = \sup \{|B(\phi(\tau_1)) - B(s_j\phi(\tau_2) + (1-s_j)\phi(\tau_3))|\}$$

where the supremum is taken over all $s_j \in [0,1]$ and all $\tau_1,\tau_2,\tau_3 \in T$ with

$$\max \{|\tau_2-\tau_3|, |\tau_3-\tau_1|, |\tau_1-\tau_2|\} \leq \gamma.$$

Since for almost all ω the closed convex hull $\overline{co}\,\{\phi(\omega,\tau): \tau \in T\}$ is compact we can choose such $\rho(\gamma)$ to be finite everywhere, and since, almost surely, $B(-,\omega)$ is uniformly continuous on compact sets, we see $\rho(\gamma) \to 0$ a.s. as $\gamma \to 0$. Thus there exists $\Omega_1 \in F$ and $\eta > 0$ such that $\mu(\Omega_1) > 1 - \varepsilon/2$ and if $\omega \in \Omega_1$ and $\gamma \leq \eta$ then $\rho(\gamma)(\omega) < \frac{1}{2}\,\delta\varepsilon\, M^{-2}(b-a)^{-1}$.

Now

$$\Big\| \sum_j (|\Delta_j u^1|\ |\Delta_j u^2|) \Big\|_{\mathcal{L}^1} \leq \sum_j \|\Delta_j u^1\|\ \|\Delta_j u^2\|$$

$$\leq M^2 \sum_j \Delta_j t = M^2(b-a)$$

whence, if $\Omega_2 = \{\omega \in \Omega;\ \sum_j |\Delta_j u^1|\ |\Delta_j u^2| < 2M^2(b-a)\varepsilon^{-1}\}$

then

$$\mu(\Omega_2) \geq 1 - \frac{\varepsilon}{2}.$$

If $\omega \in \Omega_1 \cap \Omega_2$ and mesh $\Pi < \eta$ then, for $0 \leq s_j \leq 1$,

$$|S(\Pi;B\circ(id_\Omega\times\phi);u^1,u^2) - \sum_j B(\phi(t_j) + s_j\Delta_j\phi)(\Delta_j u^1,\Delta_j u^2)|$$

$$\leq \sum_j |B(\phi(\tau_j))-B(\phi(t_j)+s_j\Delta_j\phi)|\ |\Delta_j u^1|\ |\Delta_j u^2|$$

$$< \frac{1}{2} \frac{\delta\varepsilon}{M^2(b-a)} \frac{2M^2(b-a)}{\varepsilon} = \delta.$$

Since $\mu(\Omega_1 \cap \Omega_2) > 1-\varepsilon$ the proof is complete. //

CHAPTER IV : SPECIAL CASES

§1 INTEGRALS OF ORDER GREATER THAN TWO

Theorem 1 Suppose $q \geq 3$ and that the z^ρ satisfy $A(q)$. Assume B satisfies condition B_σ. Then if for some k_0 almost all the sample paths of z^{k_0} are continuous on $[a,b]$ it follows that

$$\int_a^b B(dz^1,\ldots,dz^q) = 0 \qquad \text{a.e.}$$

Proof. We can assume $k_0 = 1$. If it exists define

$$v:[a,b] \times \Omega \to G_1 \hat{\otimes} G_2 \hat{\otimes} G_3$$

by

$$v(t) = \int_a^t dz_1 \otimes dz_2 \otimes dz_3.$$

Then, according to Proposition 4B of Chapter III, we would have

$$\int_a^b B(dz^1,\ldots,dz^q) = \int_a^b \bar{B}(dv,dz^4,\ldots,dz^q) \qquad \text{a.s.}$$

for some $\bar{B}$. It therefore suffices to prove that v exists and $v(t) = 0$, a.s., for $a \leq t \leq b$.

For this define

$$A:[a,b] \times \Omega \to \mathbb{L}(G_2,G_3;G_1 \hat{\otimes} G_2 \hat{\otimes} G_3)$$

by

$$A(t,\omega)(g_2,g_3) = z^1(t) \otimes g_2 \otimes g_3.$$

For a partition Π of $[a,b]$ we have

$$\sum_j \Delta_j z^1 \otimes \Delta_j z^2 \otimes \Delta_j z^3 = \sum_j A(t_{j+1})(\Delta_j z^2,\Delta_j z^3) - \sum_j A(t_j)(\Delta_j z^2,\Delta_j z^3)$$

which converges to zero in measure as mesh $\Pi \to 0$ by the special case of Lemma 8 of Chapter III obtained by setting $K = T = [a,b]$ and $\phi(\omega,t) = t$ with $B(\omega,t) = A(t,\omega)$ and $s_j \equiv 1$ for all j. //

Remarks 1(i) The proof in (McShane 1974) yields the same result when B satisfies B(1).

(ii) Our proof also yields an alternative proof of the existence of the integrals under consideration; at least when combined with the proof of the substitution theorem Proposition 4B of Chapter III. This is because the second part of that proof actually yields the result that I_1 and I_2 both exist if either one of them does. For our case $I_2 = \int_a^b \bar{B}(dv,dz^4,\ldots,dz^q)$ which certainly exists since we have shown that $v \equiv 0$.

§2 LIPSCHITZ SAMPLE PATHS AND SAMPLE PATHS OF BOUNDED VARIATION

(A) The most useful consequence, for our purposes, of the theorems below is the easily proved special case that under general conditions

$$\int_a^b B(dz^1,dt) = 0$$

(i.e. we take $z^2(t,\omega) = t$, all ω). Theorem 2B is the simpler of the two and will be strong enough for what we want. Theorem 2A does not use any of our existence theorems.

Theorem 2A Suppose $q = 2$ and that B satisfies either condition B(1) or condition B_0, with H a Hilbert space. Assume that z^1 satisfies A(1) and that z^2 has sample paths satisfying the Hölder condition: for almost all $\omega \in \Omega$

$$|z^2(t',\omega) - z^2(t'',\omega)| = o(|t'-t''|^{\frac{1}{2}}) \text{ as } |t'-t''| \to 0$$

where t' and t" are in [a,b]. Then

$$\int_a^b B(dz^1,dz^2) = 0 \qquad \text{a.e.}$$

Proof. By Lemmas 4A(i) and (ii) of Chapter III we need only consider the case where B satisfies B(1). We will prove that the Riemann sums converge to zero in measure: i.e. given $\alpha > 0$ and $\varepsilon_1 > 0$

$$\mu\{\omega : |S(\Pi)| > \alpha\} < \varepsilon_1$$

for all belated partitions Π of [a,b] with sufficiently small mesh. For this choose $\varepsilon > 0$ with $\varepsilon(1 + \alpha^{-2}\lambda) \leqslant \varepsilon_1$ where

$$\lambda = K^{\frac{1}{2}}(\sup\{\|B(t)\| : a \leqslant t \leqslant b\} + 1)(2(b-a) + 1)$$

and K is the constant of Condition A(2).

For $\gamma > 0$ choose $\sigma(\gamma) \in \mathcal{L}^0(\Omega,F;R)$ with

$$\sigma(\gamma) = \sup\left\{\left|\frac{z^2(t')-z^2(t")}{\sqrt{|t' - t"|}}\right| : a\leqslant t'<t"\leqslant b,\ |t'-t"| < \gamma\right\} \quad \text{a.e.}$$

Then $\sigma(\gamma) \to 0$ in measure as $\gamma \to 0$, so we can choose $\gamma > 0$ and $\Omega_\gamma \in F$ with $\mu(\Omega_\gamma) < \varepsilon$ such that $\sigma(\gamma)(\omega) < \varepsilon$ if $\omega \notin \Omega_\gamma$. Now

$$\mathbb{E}_{t_j}(|\Delta_j z^1|(\Delta_j t)^{\frac{1}{2}}) \leqslant (\Delta_j t)^{\frac{1}{2}}\sqrt{\mathbb{E}_{t_j}(|\Delta_j z^1|^2)} \leqslant K^{\frac{1}{2}}\ \Delta_j t$$

and

$$\mathbb{E}_{t_j}(|\Delta_j z^1|^2 \Delta_j t) \leqslant K(\Delta_j t)^2$$

for Π as usual, with mesh $\Pi < \delta$. If also mesh $\Pi < \gamma$ then using the fundamental estimate: (Lemma 2 of Chapter III)

$$\begin{aligned}
\int_{\Omega-\Omega_\gamma} |S(\Pi)|^2\, d\mu &\leqslant \int_{\Omega-\Omega_\gamma} \left(\sum_j |B(\tau_j)|\ |\Delta_j z^1|\ \ |\Delta_j z^2|\right)^2 d\mu \\
&\leqslant \varepsilon \int_{\Omega-\Omega_\gamma} \left(\sum_j |B(\tau_j)|\ \ |\Delta_j z^1|(\Delta_j t)^{\frac{1}{2}}\right)^2 d\mu \\
&\leqslant \varepsilon \int_{\Omega} \left(\sum_j |B(\tau_j)|\ |\Delta_j z^1|(\Delta_j t)^{\frac{1}{2}}\right)^2 d\mu \\
&\leqslant \varepsilon\, 2\{\sum_{j=1}^m K^{\frac{1}{2}}\ \Delta_j t\, \|B(\tau_j)\| + \\
&\quad + \varepsilon \sum_{j=1}^m K(\Delta_j t)^2\, \|B(\tau_j)\|^2\}^{\frac{1}{2}} \\
&\leqslant \varepsilon\, K^{\frac{1}{2}}(\sup\{\|B(t)\| : a\leqslant t\leqslant b\} + 1)\Big(2(b-a) \\
&\quad + \mu(\Pi)^{\frac{1}{2}}(b-a)^{\frac{1}{2}}\Big) \\
&\leqslant \varepsilon\, \lambda
\end{aligned}$$

for all Π with sufficiently small mesh. Therefore

$$\mu\{\omega : |S(\Pi)| > \alpha\} < \mu\{\omega \in \Omega - \Omega_\gamma : |S(\Pi)|^2 > \alpha^2\}$$
$$+ \mu(\Omega_\gamma)$$
$$< \varepsilon\lambda\alpha^{-2} + \varepsilon < \varepsilon_1. \qquad //$$

(B) Theorem 2B Suppose that z^1 has continuous sample paths and that almost all sample paths of z^2 are of bounded variation. Then for $a < t < b$

$$\int_a^t dz^1 \otimes dz^2 = 0 \qquad \text{a.s.}$$

(in particular it exists). Consequently whenever B satisfies B_0 with H a Hilbert space and z^1 and z^2 satisfy $A(2)$

$$\int_a^b B(dz^1, dz^2) = 0 \qquad \text{a.s.}$$

Proof. Define

$$A : [a,b] \times \Omega \to \mathbb{L}(G_2; G_1 \hat{\otimes} G_2)$$

by

$$A(t,\omega)g = z^1(t,\omega) \otimes g.$$

Then for a partition Π of $[a,b]$ the Riemann sum for our first integral is

$$S(\Pi) = \sum_j \Delta_j z^1 \otimes \Delta_j z^2 = \sum_j A(t_{j+1})\Delta_j z^2 - \sum_j A(t_j)\Delta_j z^2$$
$$\to \int_a^b A\, dz^2 - \int_a^b A\, dz^2 = 0$$

as mesh $\Pi \to 0$, for almost all ω, since we can interpret the integrals as classical Riemann-Stieltjes integrals defined for each ω (almost surely). Thus

$$\int_a^t dz^1 \otimes dz^2 = 0,$$

and so we have

$$\int_a^b B(dz^1, dz^2) = \int_a^b \tilde{B}\, dz^1 \otimes dz^2 = 0$$

by the substitution theorem, Proposition 4B of Chapter III. //

§3 INDEPENDENT Z's

Theorem 3 Let $q = 2$ and assume the conditions of one of the existence theorems. Suppose, for $a \leq s \leq t \leq b$

$$|\mathbb{E}_s[(z_t^1-z_s^1) \otimes (z_t^2-z_s^2)]| \leq |\mathbb{E}_s(z_t^1-z_s^1) \otimes \mathbb{E}_s(z_t^2-z_s^2)|$$

and

$$\mathbb{E}_s(|z_t^1-z_s^1|^2|z_t^2-z_s^2|^2) \leq \mathbb{E}_s(|z_t^1-z_s^1|^2)\mathbb{E}_s(|z_t^2-z_s^2|^2) \quad \text{a.e.}$$

e.g. assume $z^1(t)-z^1(s)$ and $z^2(t)-z^2(s)$ are conditionally independent as conditioned by F_s. Then

$$\int_a^b B(dz^1,dz^2) = 0 \qquad \text{a.e.}$$

Proof. Let $\Pi = t_1,\dots,t_{m+1};\tau_1,\dots,\tau_m)$ be a belated partition of $[a,b]$, with mesh $\Pi < \delta$. Then

$$\begin{aligned} |\mathbb{E}_{t_j}(\Delta_j z^1 \otimes \Delta_j z^2)| &\leq |\mathbb{E}_{t_j}(\Delta_j z^1) \otimes \mathbb{E}_{t_j}(\Delta_j z^2)| \\ &\leq |\mathbb{E}_{t_j}(\Delta_j z^1)|\ |\mathbb{E}_{t_j}(\Delta_j z^2)| \\ &\leq K^2(\Delta_j t)^2 \leq K^2\ \Delta_j t \text{ mesh } \Pi \end{aligned}$$

and

$$\begin{aligned} \mathbb{E}_{t_j}(|\Delta_j z^1 \otimes \Delta_j z^2|^2) &\leq \mathbb{E}_{t_j}(|\Delta_j z^1|^2)E_{t_j}(|\Delta_j z^2|^2) \\ &\leq K^2(\Delta_j t)^2 \leq K^2\ \Delta_j t \text{ mesh } \Pi. \end{aligned}$$

In case B satisfies B(1) we can apply the fundamental estimate to see $\|S(\Pi)\| \to 0$ as mesh $\Pi \to 0$. When B satisfies condition B_0 we can use Lemmas 4A(i) and (ii) of Chapter III as usual. //

§4 BROWNIAN MOTION

In this section the filtration is taken to be the *natural filtration* of the Brownian motion z, i.e.

$$F_t = \sigma\{z(s):0 \leq s \leq t\}.$$

In §5 of Chapter V we show that the results remain valid for suitable enlargements of that filtration (in fact those for which z is an F_*-martingale). First observe that Brownian motions satisfy Conditions A(p) for all p = 1,2,... by Proposition 3C of Chapter 2.

When $B(t,\omega) \in \mathbb{L}(\mathbb{R}^n,\mathbb{R}^n;H)$ we will let $trB(t,\omega)$ denote the trace

$$\mathrm{tr}\, B(t,\omega) = \sum_i B(t,\omega)(e_i,e_i)$$

where $e_1,\dots,e_n$ is an orthonormal base for $\mathbb{R}^n$.

Theorem 4 Let z be a Brownian motion on $\mathbb{R}^n$. Suppose

$$B: T \times \Omega \to \mathbb{L}(\mathbb{R}^n,\mathbb{R}^n;H)$$

satisfies either Condition B(1) or Condition B_0. Then

$$\int_a^b B(t,\omega)(dz(t,\omega),dz(t,\omega)) = \int_a^b \mathrm{tr}\, B(t,\omega)dt$$

where the right hand side denotes the vector valued Riemann integral of

$$\mathrm{tr}\, B:[a,b] \to \mathcal{L}^2(\Omega,F,\mu;H)$$

in Case B(1), or

$$\mathrm{tr}\, B:[a,b] \to \mathcal{L}^0(\Omega,F;H)$$

for B_0.

Proof. We can reduce to assuming that B satisfies B(1) by the standard method. Let B be represented by the matrix (B_{ij}), so $B_{ij}: T \times \Omega \to \mathbb{L}(\mathbb{R},\mathbb{R};H)$ satisfies B(1). By Proposition 3D(ii) of Chapter II, and Theorem 3

$$\int_a^b B_{ij}\, dz^i\, dz^j = 0 \qquad i \neq j.$$

whence

$$\int_a^b B(dz,dz) = \sum_i \int_a^b B_{ii}\, dz^i\, dz^i$$

and we can assume from now on that n = 1.

For Π a belated partition of [a,b] as usual, set

$$\Delta_j = (\Delta_j z)^2 - \Delta_j t \quad j = 1,\ldots,m.$$

Then $$\mathbb{E}_{t_j}(\Delta_j) = 0$$

and $$\mathbb{E}_{t_j}(|\Delta_j|^2) = 2\,(\Delta_j t)^2,$$

by Proposition 3C of Chapter II.

By the fundamental estimate

$$\left\| \sum_{j=1}^{m} B(\tau_j)\Delta_j \right\| \leqslant \left\{ \sum_{j=1}^{m} 2(t_{j+1}-t_j)^2 \;\|B(\tau_j)\|^2 \right\}^{\frac{1}{2}}$$

which converges to zero as mesh $\Pi \to 0$. Thus, taking Cauchy partitions Π, since the above shows

$$\|S(\Pi;B;z,z) - S(\Pi;B;dt)\| \to 0 \text{ as mesh } \Pi \to 0$$

and since

$$S(\Pi;B;t) \to \int_a^b B(t)dt \text{ in } \mathcal{L}^2(\Omega,\mu;R),$$

where the last integral is the Riemann integral for $\mathcal{L}^2$-valued functions, we see

$$\int_a^b B(dz,dz) = \int_a^b B(t)dt \quad \text{as required.} \quad //$$

<u>Corollary 4A</u> <u>Let</u> z <u>be a 1-dimensional Brownian motion and suppose</u> $0 \leqslant a = t_1 \leqslant t_2 \leqslant \ldots \leqslant t_{m+1} = b$. <u>Then</u>

$$\sum_{j=1}^{m} (\Delta_j z)^2 \to b-a \text{ in } \mathcal{L}^2 \text{ as } \max \Delta_j t \to 0.$$

<u>Proof</u>. Take $B(t,w) \equiv 1$ in the theorem. //

<u>Corollary 4B</u> <u>Let</u> z <u>be a 1-dimensional Brownian motion. Then almost all the sample paths of</u> z <u>are not of bounded variation and do not satisfy a Hölder condition of order</u> α <u>if</u> $\frac{1}{2} < \alpha < 1$, <u>on any</u> [a,b] <u>in</u> T.

<u>Proof</u>. Since convergence in $\mathcal{L}^2$ implies that a subsequence converges almost everywhere, by Corollary 4A there is a sequence of partitions $\{\Pi_i\}_{i=1}^{\infty}$ of [a,b] with partition points $t_1^i,\ldots,t_{m_i+1}^i$, say, and an $\tilde{\Omega} \in F$

with $\mu(\tilde{\Omega}) = 1$ such that

$$\lim_{i\to\infty} \sum_{j=1}^{m_i} |z(t_{j+1},\omega)-z(t_j,\omega)|^2 = b-a$$

for all ω in $\tilde{\Omega}$. Then, for $\omega \in \tilde{\Omega}$, the paths $t \mapsto z(t,\omega)$ cannot satisfy either of the regularity conditions mentioned. //

Remark. Theorems 2 and 3 can be written as the "Itô rules"

"$dw\ dw = dt$"

"$dt\ dw = 0$"

for 1-dimensional Brownian motions w. Knowing that there are versions of Brownian motion with continuous sample paths we also have

"$dw\ dw\ dw = 0$".

Alternatively this follows using Proposition 4 of Chapter III by

"$dw\ dw\ dw = dw\ dt = 0$"

for integrals $\int_a^b B\ dw\ dw\ dw$ when B satisfies B_0. However it is a straightforward exercise in the use of the estimates to give a direct proof of this rule in the case of Brownian motion.

§5 CONTINUITY OF SAMPLE PATHS

The proof of the most important special case of Proposition 5B below: the case $q = 1$ with z^1 a martingale, is a simple consequence of our estimates and the martingale inequality

$$\mu\{\omega : \sup_{a\leq t\leq b} |C(t,\omega)| > \delta\} \leq \frac{1}{\delta^2} \|C(b)\|^2$$

for any $\delta > 0$ and square integrable martingale C on [a,b] with continuous sample paths (e.g. see (Friedman, 1975), Vol. 1, Corollary 3.3). We give an alternative proof for this special case. However it will be very convenient to have estimates which are valid for more general processes z^1.

(A) We start with some uniform analogues of the estimates in §2 of Chapter III.

Lemma 5A Given: (a) a probability space (Ω, F, μ) with σ-algebras $F_1 \subset F_2 \subset \ldots \subset F_m \subset F_{m+1} = F$

(b) A Banach space G, and a Hilbert space H, both separable

(c) F_j-random maps

$$u_j : \Omega \to \mathbb{L}(G;H) \qquad j = 1,\ldots,m$$

(d) $\Delta_j : \Omega \to G$ each F_{j+1}-measurable, $j = 1,\ldots,m$, with constants C_j, D_j such that

$$|\mathbb{E}_j(\Delta_j)| \leq C_j$$

$$\mathbb{E}_j(|\Delta_j|^2) \leq D_j \qquad j = 1,\ldots,m.$$

Then, for $\alpha > 0$,

$$\mu\{\omega: \sup_{1 \leq k \leq m} |\sum_{j=1}^{k} u_j(\Delta_j)| > \alpha\} \leq \frac{1}{\alpha} \sum_{k=1}^{m} C_k \|u_k\| + \frac{1}{\alpha^2} \sum_{k=1}^{m} D_k \|u_k\|^2$$

Proof. Set

$$S_k = \sum_{j=1}^{k} u_j(\Delta_j) \text{ and } \xi_k = |S_k|^2.$$

For $k = 1,\ldots,m$ and fixed $\alpha > 0$ define

$$\chi_k : \Omega \to \{0,1\} \text{ by}$$

$$\chi_k(\omega) = 1 \text{ iff } \xi_1(\omega) \leq \alpha^2, \ldots, \xi_{k-1}(\omega) \leq \alpha^2, \xi_k(\omega) > \alpha^2,$$

(or iff $\xi_1(\omega) > \alpha^2$ when $k = 1$).

Then (i) χ_k is F_{k+1}-measurable

(ii) $\alpha^2 \chi_k(\omega) < \xi_k(\omega)\chi_k(\omega)$

(iii) $\mu\{\omega: \sup_{1 \leq k \leq m} |S_k| > \alpha\} = \sum_{k=1}^{m} \int_\Omega \chi_k \, d\mu$

(iv) if $\ell_k(\omega) = 1 - \chi_1(\omega) - \ldots - \chi_k(\omega)$ then

$$\ell_k(\omega)|S_k(\omega)|^2 = |\ell_k(\omega)S_k(\omega)|^2 \leq \alpha^2$$

Trivially

$$\sum_{k=1}^{m} \xi_k \chi_k = \xi_1 + (\xi_2-\xi_1)\,\ell_1 + \ldots + (\xi_m-\xi_{m-1})\,\ell_{m-1} - \xi_m\ell_m$$

while

$$\left|\int_\Omega (\xi_k-\xi_{k-1})\ell_{k-1}\, d\mu\right| \leq 2\left|\int_\Omega \langle u_k(\Delta_k), S_{k-1}\rangle \ell_{k-1}\, d\mu\right|$$

$$+ \int_\Omega u_k(\Delta_k)|^2\, \ell_{k-1}\, d\mu\,.$$

By the proof of Lemma 2, Chapter III

$$\int_\Omega u_k(\Delta_k)|^2\, \ell_{k-1}\, d\mu \leq \int_\Omega u_k(\Delta_k)|^2\, d\mu$$

$$\leq D_k\, \|u_k\|^2$$

and also, since ℓ_{k-1} is F_k-measurable,

$$\left|\int_\Omega \langle S_{k-1}, u_k(\Delta_k)\rangle\, \ell_{k-1}\, d\mu\right| = \left|\int_\Omega \mathbb{E}_k\, \langle \ell_{k-1}\, S_{k-1}, u_k(\Delta_k)\rangle\, d\mu\right|$$

$$= \left|\int_\Omega \langle \ell_{k-1}\, S_{k-1}, u_k(\mathbb{E}_k(\Delta_k))\rangle\, d\mu\right|$$

$$\leq \|\ell_{k-1}\, S_{k-1}\|\; \|u_k(\mathbb{E}_k(\Delta_k))\|$$

$$\leq \|\ell_{k-1}\, S_{k-1}\|\, C_k\, \|u_k\| \qquad \ldots (*)$$

$$\leq \alpha C_k\, \|u_k\|$$

using (iv). Thus

$$\int_\Omega \left(\sum_{k=1}^{m} \xi_k \chi_k\right) d\mu \leq \sum_{k=1}^{m} \left(\alpha\, C_k\, \|u_k\| + D_k\, \|u_k\|^2\right).$$

Application of (iii) and (ii) completes the proof. //

Remark 5A We can obtain a result which is stronger for large α: In the inequality (*) of the proof, instead of majorizing $\|\ell_{k-1}\, S_{k-1}\|$ by α we can apply Lemma 2 of Chapter 3 to replace it by

$$\|S_{k-1}\| \leq 2 \sum_{j=1}^{m} C_j \|u_j\| + \{\sum_{j=1}^{m} D_j \|u_j\|^2\}^{\frac{1}{2}}.$$

This leads to the estimate

$$\mu\{\omega: \sup_{1\leq k\leq m} |\sum_{j=1}^{k} u_j(\Delta_j)| > \alpha\} \leq \frac{M}{\alpha^2}$$

where

$$M = \left(2 \sum_{j=1}^{m} C_j \|u_j\| + \{\sum_{j=1}^{m} D_j \|u_j\|^2\}^{\frac{1}{2}}\right) \sum_{k=1}^{m} C_k \|u_k\| + \sum_{k=1}^{m} D_k \|u_k\|^2 .$$

(B) For a partition $\Pi = \{t_1,\ldots,t_{m+1};\tau_1,\ldots,\tau_m\}$ of $[a,b]$, if $s \in [a,b]$ let Π^s be the partition $\{t_1,\ldots,t_{r(s)},s;\ \tau_1,\ldots,\tau_{r(s)}\}$ of $[a,s]$ where $t_{r(s)} \leq s < t_{r(s)+1}$. First we give a "continuous time" version of Lemma 5A.

Lemma 5B Suppose $z: [a,b] \to \mathcal{L}^0(\Omega,\mathcal{F};G)$ satisfies A(1) and has continuous sample paths. Let Π be a partition of $[a,b]$ with mesh $\Pi < \delta$ and suppose given $\mathcal{F}_{t_j}$-random maps

$$u_j:\Omega \to \mathbb{L}(G;H) \qquad j = 1 \text{ to } m$$

with $\|u_j\| < \infty$ for each j, where H is a Hilbert space. Define

$$S:[a,b] \times \Omega \to H$$

by

$$S(t,\omega) = \sum_{j=1}^{r(s)} u_j(\Delta_j z) + u_{r(s)}(z(s) - z(t_{r(s)})).$$

Then, if $\alpha > 0$,

$$\mu\{\omega: \sup_{a\leq t\leq b} |S(t,\omega)| > \alpha\} \leq \frac{1}{\alpha} \sum_{j=1}^{m} K\Delta_j t \|u_j\| + \frac{1}{\alpha^2} \sum_{j=1}^{m} K\Delta_j t \|u_j\|^2 .$$

Proof. For a positive integer n let $\{s_j^k: j = 1 \text{ to } m,\ k = 1 \text{ to } n\}$ satisfy

$$t_j = s_j^1 < s_j^2 < \ldots < s_j^{n+1} = t_{j+1}$$

with

$$s_j^{k+1} - s_j^k = \frac{1}{n} \Delta_j t$$

Then

$$|S(s_{j_0}^{k_0})| = |\sum_{j=1}^{j_0-1} \sum_{k=1}^{n} u_j(z(s_j^{k+1}) - z(s_j^k)) + \sum_{k=1}^{k_0-1} u_{j_0}(z(s_{j_0}^{k+1}) - z(s_{j_0}^k))|$$

for $1 < k_0 \leq n+1$, $1 \leq j_0 \leq m$. By Lemma 5A

$$\mu\{\omega: \sup_{j,k} |S(s_j^k)| > \alpha\} \leq \frac{1}{\alpha} \sum_{j=1}^{m} K\, \Delta_j t\, \|u_j\| + \frac{1}{\alpha^2} \sum_{j=1}^{m} K\Delta_j t\, \|u_j\|^2 .$$

The result follows by taking the limit as $n \to \infty$, because of the sample continuity of z. //

Proposition 5B For $q = 1$ or 2 let

$$z^\rho : [a,b] \to \mathcal{L}^0(\Omega, F; G_\rho) \qquad \rho = 1,\ldots,q$$

and

$$B : T \times \Omega \to \mathbb{L}(G_1,\ldots,G_q;H)$$

satisfy the conditions of the main existence theorem: Theorem 3 of Chapter III. Assume also that all the sample paths of each z^ρ are continuous on $[a,b]$. Then for each Π we have a measurable map

$$C(\Pi) : \Omega \to C([a,b];H)$$

$$C(\Pi)(\omega)(s) = S(\Pi^s)(\omega).$$

As mesh $\Pi \to 0$ through belated partitions Π so $\{C(\Pi)\}$ converges in probability to an element $I \in \mathcal{L}^0(\Omega, F_b; C(a,b ;H))$ with $I(-,t)$ a version of $\int_a^t B\,dz^1 \ldots dz^q$ for $t \in [a,b]$.

Moreover if $\alpha > 0$

$$\mu\{\omega: \sup_{a \leq t \leq b} |I(\omega)(t)| > \alpha\} \leq \frac{K}{\alpha} \int_a^b \|B(t)\|\, dt + \frac{K}{\alpha^2} \int_a^b \|B(t)\|^2\, dt$$

Proof. In general the proof is very similar to that of the main existence theorem, but with Lemma 5A used instead of Lemma 2 of Chapter III. Since the only second order integrals we actually use will reduce to first order integrals (e.g. via Theorem 4) we will only give the proof for $q = 1$. However first observe that the measurability of $C(\Pi)$ is immediate from the fact that the Borel σ-algebra of $C([a,b];H)$ is generated by the algebra C of cylinder sets described in §1 Chapter II.

Set $z = z^1$. Suppose

$$\Pi = (t_1,\dots,t_{m+1};\tau_1,\dots,\tau_m)$$

and

$$\Pi' = (t_1,\dots,t_{m+1};\tau_1',\dots,\tau_m'),$$

and both are belated with mesh less than δ.

Taking $u_j = B(\tau_j) - B(\tau_j)$ in Lemma 5B we obtain, for $\alpha > 0$,

$$\mu\{\omega: \sup_{a\leq t\leq b} |C(\Pi)(t) - C(\Pi')(t)| > \alpha\}$$

$$\leq K(b-a)\{\frac{1}{\alpha} \sup_j \|B(\tau_j)-B(\tau_j')\| + \frac{1}{\alpha^2} \sup_j \|B(\tau_j)-B(\tau_j')\|^2\}$$

which converges to 0 with mesh Π and mesh Π'. On the other hand suppose we have two partitions Π_1, Π_2, say

$$\Pi_i = \{t_1^i,\dots,t_{m(i)+1}^i,\tau_1^i,\dots,\tau_{m(i)}^i\},\ i = 1,2.$$

Let Π_i'', $i = 1,2$ both have subdivision points the union of the set of subdivision points of Π_1 and Π_2 with Π_i'' a subdivision of Π_i, each i (i.e. the same τ-points: see the proof of Theorem 3 of Chapter III). Then

$$C(\Pi_i) = C(\Pi_i'') \qquad i = 1,2$$

so that

$$C(\Pi_1) - C(\Pi_2) = C(\Pi_1'') - C(\Pi_2'')$$

and we have already made a uniform estimate about such a difference.

Thus $\{C(\Pi)\}$ is Cauchy in probability and converges to an element I as required. Also Lemma 5B applied to $C(\Pi)$ yields

$$\mu\{\omega: \sup_{a\leq t\leq b} |C(\Pi)(t)| > \alpha\} \leq \frac{1}{\alpha} \Sigma \|B(\tau_j)\| K\Delta_j t + \frac{1}{\alpha^2} \Sigma \|B(\tau_j)\|^2 K\Delta_j t,$$

and the estimate follows letting mesh $\Pi \to 0$. //

Alternative Proof of existence when $q = 1$ and z is also a martingale using the martingale inequality stated before Lemma 5A. Keeping the same notation $\{C(\Pi)(t): a\leq t\leq b\}$ and $\{C(\Pi')(t): a\leq t\leq b\}$ are martingales, both square integrable, by the same argument as in the proof of Proposition 7C of

Chapter III. Therefore $C(\Pi) - C(\Pi')$ determines a square integrable martingale; whence if $\delta > 0$

$$\mu\{\omega: \sup_{a \leq t \leq b} |C(\Pi)(t) - C(\Pi')(t)| > \delta\} \leq \frac{1}{\delta^2} \|S(\Pi) - S(\Pi')\|$$

since $C(\Pi)(b) = S(\Pi)$ and similarly for Π'. However by the main existence theorem $\{S(\Pi)$: belated $\Pi\}$ is Cauchy in L^2 as mesh $\Pi \to 0$. Thus $\{C(\Pi)$: belated $\Pi\}$is Cauchy in measure as required. //

<u>Remark 5B</u> (i) Using Remark 5A and the Cauchy-Schwarz inequality, as in the proof of Corollary 2 of Chapter III, we obtain the alternative estimate, for $\varepsilon > 0$:

$$\mu\{\omega: \sup_{a \leq t \leq b} |I(\omega)t| > \varepsilon\} \leq \frac{J(a,b)}{\varepsilon^2}$$

where $J(a,b) = K(2K(b-a)^{\frac{1}{2}} + K^{\frac{1}{2}}) \left(\int_a^b \|B(t)\|^2 dt\right)^{\frac{1}{2}} \int_a^b \|B(t)\| dt + K \int_a^b \|B(t)\|^2 dt$

$$\leq K(2K(b-a) + K^{\frac{1}{2}} (b-a)^{\frac{1}{2}} + 1) \int_a^b \|B(t)\|^2 dt.$$

(ii) If also $E_{t_j} \Delta_j z^\rho = 0$, $\rho = 1,2$, all j (i.e. the martingale case) a more careful use of Lemma 5A would yield the estimate

$$\mu\{\omega: \sup_{a \leq t \leq b} |I(\omega)t| > \varepsilon\} \leq K/_{\varepsilon^2} \int_a^b \|B(t)\|^2 dt.$$

(iii) By Lemma 5A(ii) the Proposition (without the estimate) still holds when B only satisfies B_0.

(C) Since the limit in probability I of Proposition 5B is only defined up to sets of measure zero in F_b we cannot immediately make claims about the F_t-measurability of $I(-)(t)$. This is because we have not assumed that F_t contains all sets of measure zero in F. For example if z is a Brownian motion and $F_t = \sigma\{z(s): 0 \leq s \leq t\}$ then such an assumption would not hold. However Lemma 3D of Chapter I shows that we could always enlarge F_t to a σ-algebra $\tilde{F}_t$ containing all such sets of measure zero without changing any of the estimates on conditional expectations on which our work has been based. This simplifies the statements and proofs of a lot of the results without changing anything of the essence. *We will therefore, from now on, make the following assumption:*

Assumption 5C: Each σ-algebra F_τ for $\tau \in T$ contains all sets in F which have measure zero.

Now any two versions of $\int_b^t B\,dz^1 \ldots dz^q$ differ only on sets of measure zero, for each t. Also under our conditions there exists an F_t-measurable version. Consequently, under our new assumption, all versions are F_t-measurable. This, together with Remark 5B(iii) proves the theorem:

Theorem 5C. Let q = 1 or 2. Suppose B satisfies B(1) or B_0 and $z^1,\ldots,z^q$ have continuous sample paths and satisfy A(q). Let

$$u(t) = \int_a^t B\,dz^1 \ldots dz^q.$$

Then u(t) is F_t-measurable for each t and u may be chosen to have continuous sample paths. //

In future we shall always choose versions with continuous sample paths when they exist. Therefore we have the following from Proposition 5B and Remarks 5B:

Estimates 5C. Under the above conditions, if B satisfies B(1) and $\alpha > 0$,

$$\mu\{\omega: \sup_{a \leq t \leq b} \left|\int_a^t B\,dz^1 \ldots dz^q\right| > \alpha\} \leq \frac{K}{\alpha}\int_a^b \|B(t)\|\,dt + \frac{K}{\alpha^2}\int_a^b \|B(t)\|^2\,dt$$

and

$$\mu\{\omega: \sup_{a \leq t \leq b} \left|\int_a^t B\,dz^1 \ldots dz^1\right| > \alpha\} \leq \frac{\gamma}{\alpha^2}\int_a^b \|B(t)\|^2\,dt$$

where

$$\gamma = K(2K(b-a) + K^{\frac{1}{2}}(b-a)^{\frac{1}{2}} + 1).$$

Also if q = 1 and $\mathbb{E}_s\, z^1(t) = z^1(s)$ a.e. for $a \leq s \leq t \leq b$ then γ above can be replaced by K. //

CHAPTER V : THE CHANGE OF VARIABLE FORMULA

§1 THE FUNDAMENTAL THEOREM OF STOCHASTIC CALCULUS

Heuristically the Itô formula says that in stochastic (Itô) calculus third order terms may be neglected but second order terms must be kept. This is also a well known heuristic rule for physicists working with Feynman path integrals (Feynman & Hibbs 1965). For a discussion of a particular example from a physicists' viewpoint see (Edwards & Gulyaev 1964).

Proposition 1 (Ito's Formula) Let G,H be Banach spaces and $\theta: G \to H$ a C^2 map. Suppose $u:[a,b] \to \mathcal{L}^2(\Omega,\mu;G)$ is $\frac{1}{2}$-Hölder continuous and almost all its sample paths are continuous. Then, if

$$\int_a^b D\theta(u_t)\, du_t$$

exists as a limit in measure of Riemann sums using Cauchy partitions then

$$\int_a^b D^2\theta(u_t)(du_t,du_t)$$

exists, and conversely. If so

$$\theta(u_t) = \theta(u_a) + \int_a^b D\theta(u_t)du_t + \tfrac{1}{2}\int_a^b D^2\theta(u_t)(du_t,du_t) \quad \text{a.e.}(1)$$

(For the notation for derivatives used here see Appendix A.)

Proof. Let $\Pi = (t_1,\ldots,t_{m+1};t_1,\ldots,t_m)$ be a Cauchy partition of [a,b]. Then

$$\theta(u(b)) - \theta(u(a)) = \sum_j \theta(u(t_{j+1})) - \theta(u(t_j))$$

$$= \sum_j \{D\theta(u(t_j))\,\Delta_j u + \int_0^1 (1-s)D^2\theta(u(t_j)+s\Delta_j u)(\Delta_j u,\Delta_j u)ds\}$$

by Taylor's theorem with integral remainder.

The result follows from Lemma 8 of Chapter III by taking K = G and $\phi = u$. //

Examples 1: Let u denote a Brownian motion on R. Then, using §4 of Chapter IV:

(i) taking $\theta(u) = \exp u$,

$$\exp u(t) = 1 + \int_0^t \exp u(s)\, du(s) + \tfrac{1}{2}\int_0^t \exp u(s)\, du(s)\, du(s)$$

$$= 1 + \int_0^t \exp u(s)\, du(s) + \tfrac{1}{2}\int_0^t \exp u(s)\, ds$$

Note that the first of these two equalities is true for a wide class of processes u, including those with smooth sample paths, whereas the second is definitely false in the smooth case.

(ii) taking $\theta(u) = u^2$,

$$u(t)^2 = \int_0^t 2u(s)\, du(s) + \int_0^t du(s)\, du(s)$$

$$= 2\int_0^t u(s)\, du(s) + t.$$

Thus $$\int_0^t u(s)\, du(s) = \tfrac{1}{2}\, u(t)^2 - \tfrac{1}{2}t.$$

This last equality was often considered rather unfortunate, especially in engineering circles. The Stratonovich Integral discussed in the next section does not lead to such 'anomalous' results.

Our third example relates to the tensor quadratic variation discussed in §4C of Chapter III. Let z^1 and z^2 be sample continuous processes with values in G_1 and G_2. Set $G = G_1 \times G_2$, $u_t = (z_t^1, z_t^2)$ and $H = G_1 \hat{\otimes} G_2$. Define θ by

$$\theta(g_1, g_2) = g_1 \otimes g_2.$$

Then under suitable conditions (in particular assuming either the first order integrals, or the second order integral, exists)

(iii) $$z_b^1 \otimes z_b^2 = z_a^1 \otimes z_a^2 + \int_a^b z_t^1 \otimes dz_t^2 + \int_a^b dz_t^1 \otimes z_t^2 + \int_a^b dz_t^1 \otimes dz_t^2.$$

§2 STRATONOVICH INTEGRALS

(A) For suitable $B:T \times E \to \mathbb{L}(G,H)$ and $u:[a,b] \to \mathcal{L}^0(\Omega,F;E)$ the *Stratonovich integral* (S) $\int_a^b B(t,u_t)dz_t$, for a G-valued process z, was defined to be the limit in measure of

$$\sum_j B(\tfrac{1}{2}[t_j + t_{j+1}],\tfrac{1}{2}[u(t_j) + u(t_{j+1})])\Delta_j z$$

taken over partitions Π of [a,b], with mesh $\Pi \to 0$. This is related to the *mid-point rule* used by physicists heuristically in Feynman path integration: see for example (Dowker 1975) and the references therein, (Schulman 1975), and (Mizrahi 1979). It has also been called the *Feynman-Stratonovich integral* and the *symmetric integral.*

With the notation of Proposition 1, if $x,y \in G$ and $w = \tfrac{1}{2}(x+y)$ then

$$\theta(x) - \theta(y) = (\theta(x) - \theta(w)) - (\theta(y) - \theta(w))$$

$$= D\theta(z)(x-y) + \int_0^1 (1-s)\Big(D^2\theta(w+s(x-w))-D^2\theta(w+s(y-w))\Big)(\tfrac{x-y}{2},\tfrac{x-y}{2})ds$$

by Taylor's theorem applied to $\theta(x)-\theta(w)$ and $\theta(y)-\theta(w)$. Taking $x = u(t_j)$, $y = u(t_{j+1})$, the method of Proposition 1 yields

$$\theta(u_a) - \theta(u_b) = (S)\int_a^b D\theta(u_t)du_t \qquad \text{a.e.} \qquad S(1)$$

provided the integral exists. (Essentially: $\theta(x) - \theta(y) = D\theta(w)(x-y)+3^{rd}$ order, and we can ignore third order terms, given continuous sample paths.) In terms of these integrals Examples I(i) and (ii) become

$$\exp u(t) = 1 + (S)\int_0^t \exp u(s)\, du(s),$$

and

$$(S)\int_0^t u(s)\, du(s) = \tfrac{1}{2}u(t)^2.$$

In general, computations using the Stratonovich integral give rise to equations having the same form as those of the classical integral or differential calculus e.g. S(1) as compared to (1). However it is not so tractable mathematically (for example to prove existence or give estimates).

(B) Suppose that B is C^1 on $[a,b] \times E$. Let D_1 and D_2 denote the partial derivatives of B with respect to the first and second variables respectively. Then for our partition Π

$$B(\tfrac{1}{2}[t_j + t_{j+1}], \tfrac{1}{2}[u(t_j) + u(t_{j+1})])$$

$$= B(t_j + \tfrac{1}{2}\Delta_j t, u(t_j) + \tfrac{1}{2}\Delta_j u)$$

$$= B(t_j, u(t_j)) + \tfrac{1}{2}\int_0^1 D_1 B(t_j + \tfrac{s}{2}\Delta_j t, u(t_j) + \tfrac{s}{2}\Delta_j u)(\Delta_j t)ds$$

$$+ \tfrac{1}{2}\int_0^1 D_2 B(t_j + \tfrac{s}{2}\Delta_j t, u(t_j) + \tfrac{s}{2}\Delta_j u)(\Delta_j u)ds.$$

Arguing as in the proof of Proposition 1, and using Theorem 2B of Chapter IV to dispose of the terms involving $\Delta_j t\ \Delta_j z$, we obtain:

<u>Proposition 2B</u> <u>Suppose that</u> B <u>is</u> C^1 <u>on</u> $[a,b] \times E$ <u>and that</u> u <u>and</u> z <u>are sample continuous, adapted, and</u> $\tfrac{1}{2}$-<u>Hölder continuous into</u> $\mathcal{L}^2$ <u>on</u> [a,b]. <u>Then if</u> z <u>satisfies</u> A(2)

$$(S) \int_a^b B(t,u_t)dz_t = \int_a^b B(t,u_t)dz_t + \tfrac{1}{2}\int_a^b D_2 B(t,u_t)(du_t)(dz_t)$$

<u>in the sense that whenever any two of the integrals exist then so does the third and there is equality a.e.</u> //

Proposition 2B shows that under reasonable conditions one can pass from Stratonovich integrals to the sort of integrals we have been considering (Itô integrals) and vice versa with no difficulty. It is only necessary to take into account the *Stratonovich correction term*

$$\tfrac{1}{2}\int_0^b D_2 B(t,u_t)(du_t)(dz_t)$$

(C) There is a more modern definition of Stratonovich integrals, which allows more general integrands. For $A:[a,b] \times \Omega \to \mathbb{L}(G;H)$ define

$$\int_a^b A(t)\circ dz_t = \int_a^b A(t)dz_t + \tfrac{1}{2}\int_a^b dA_t\, dz_t$$

whenever the integrals on the right hand side exist. When $A(t,\omega) = B(t,u_t(\omega))$ for u and B as in Proposition 2B and with B of class C^2 (in fact C^1, and C^2 in the second variable would suffice) Proposition 1 yields for $a < t < b$

$$A(t) = A(a) + \int_a^t D_1B(s,u_s)ds + \int_a^t D_2B(s,u_s)du_s$$
$$+ \tfrac{1}{2}\int_a^t D_2D_2B(s,u_s)(du_s,du_s)$$

whenever the integrals exist. It follows from the substitution theorem, Proposition 4B of Chapter III, and Theorems 1 and 2A of Chapter IV that at least when u and z also satisfy A(3) we have

$$\int_a^b B(t,u_t)\circ dz_t = \int_a^b A(t)dz_t + \tfrac{1}{2}\int_a^b D_2B(s,u_s)(du_s)dz_s.$$

Comparing with Proposition 2B we obtain

$$\int_a^b B(t,u_t)\circ dz_t = (S)\int_a^b B(t,u_t)dz_t.$$

This integral is also called the *Fisk-Stratonovich integral* after (Fisk 1963). See also (Ikeda & Watanabe 1981) Chapter III, §1.

(D) We could define "anticipating integrals", for example if u is 1-dimensional

$$(A)\int_u^b u(s)du(s) = \lim \sum_j u(t_{j+1})\Delta_j u, \quad \text{as mesh } \Pi \to 0,$$

Then

$$(A)\int_0^t u(s)du(s) = 2(S)\int_0^t u(s)du(s) - \int_0^t u(s)du(s)$$
$$= \tfrac{1}{2}u(t)^2 + \tfrac{1}{2}t,$$

when u is a Brownian motion.

In particular we see that $\{S(\Pi;u;u)\}$ does not converge as mesh $\Pi \to 0$ through arbitrary Riemann partitions Π. Thus we cannot use Riemann partitions to define first order integrals.

§3. TRANSFORMATION OF STOCHASTIC INTEGRALS

Next we have the version of Itô's formula when our process u is given by integrals.

Theorem 3 Let G be a separable Banach space and H,F separable Hilbert spaces. Assume $z:[a,b] \times \Omega \to G$ satisfies condition A(4) and almost all of its sample paths are continuous. Suppose

$$B:T \times \Omega \to \mathbb{L}(G;H)$$

and

$$C:T \times \Omega \to \mathbb{L}(G,G;H)$$

satisfy condition B_0. Let $u:[a,b] \times \Omega \to H$ be defined by

$$u(t) = u(a) + \int_a^t B\,dz + \int_a^t C(dz,dz)$$

where $u(a)$ is F_a-measurable. Then if $\theta:H \to F$ is C^2 we have

$$\theta(u_t) = \theta(u_a) + \int_a^t D\theta(u_s)B_s\,dz_s$$
$$+ \int_a^t D\theta(u_s)C_s(dz_s,dz_s)$$
$$+ \tfrac{1}{2}\int_a^t D^2\,\theta(u_s)(B_s dz_s, B_s dz_s).$$

Proof. The integrals all exist by Proposition 4A of Chapter III. When B and C both satisfy condition B(1), the map $t \mapsto u(t) - u(a)$ is $\frac{1}{2}$-Hölder continuous in $\mathcal{L}^2$-norm by Corollary 3 of Chapter III. We first prove the theorem under that additional hypothesis.

Set

$$u_1(t) = u(a) + \int_a^t Bdz$$

$$u_2(t) = \int_a^t C(dz,dz).$$

By Theorem 5C of Chapter IV there is a version of u with almost all its sample paths continuous. Hence by Proposition 1

$$\theta(u(t)) = \theta(u(a)) + \int_a^t D\theta(u(s))du(s) + \tfrac{1}{2}\int_a^t D^2\theta(u(s))(du(s),du(s))$$

$$= \theta(u(a)) + \int_a^t D\theta(u(s))du_1(s) + \int_a^t D\theta(u(s))du_2(s)$$

+ *(next page)*

$$+ \tfrac{1}{2} \int_a^t D^2\theta(u(s))(du_1(s),du_1(s))$$

$$+ \int_a^t D^2\theta(u(s))(du_1(s),du_2(s))$$

$$+ \tfrac{1}{2} \int_a^t D^2\theta(u(s))(du_2(s),du_2(s))$$

by the obvious linearity and bilinearity of the integrals. Applying Proposition 4B of Chapter III it remains only to show that the following integrals vanish:

$$I_1 \equiv \int_a^t D^2\theta(u)\cdot(C,B)((dz,dz),dz)$$

$$I_2 \equiv \int_a^t D^2\theta(u)\cdot(C,C)((dz,dz),(dz,dz)).$$

But this is true by Theorem 1 of Chapter IV.

When B_0 but not B(1) is satisfied we retract B and C to suitable maps B_N, C_N as in Lemma 4A(iii) of Chapter III, and work with the resulting modification u_N of u. The result will be true for u_N and hence for u except on a subset of Ω of arbitrarily small measure. Thus the theorem holds in general. //

Remark 3. We can also consider the more general case of $\theta:[a,b] \times H \to F$. Then it is only necessary to assume that θ is C^1 with a continuous second partial derivative in the second variable. The formula remains the same except for an additional term $\int_a^t D_1\theta(s,u(s))ds$ where D_1 denotes partial differentiation with respect to the first variable.

Corollary 3.1 Under the conditions of the theorem for

$$u(t) = u(a) + \int_a^t B\,dz + \int_a^t C(dz,dz)$$

and p = 1 or p $\geqslant$ 2 we have

$$|u_t|^{2p} = |u_a|^{2p} + 2p \int_a^t |u_s|^{2(p-1)} \langle u_s, B_s dz_s\rangle$$

$$+ 2p \int_a^t |u_s|^{2(p-1)} \langle u_s, C_s(dz_s,dz_s)\rangle \quad + (\textit{next page})$$

$$+ p \int_a^t |u_s|^{2(p-1)} \langle B_s dz_s, B_s dz_s \rangle$$

$$+ 2p(p-1) \int_a^t |u_s|^{2(p-2)} \langle u_s, B_s dz_s \rangle \langle u_s, B_s dz_s \rangle .$$

Proof. Define $\theta : H \to \mathbb{R}$ by $\theta(x) = |x|^{2p}$. Then for x,u,v in H:

$$D\theta(x)v = 2p\langle x,v\rangle |x|^{2(p-1)}$$

and

$$D^2\theta(x)(u,v) = 2p\langle u,v\rangle |x|^{2(p-1)} + 4p(p-1)\langle x,u\rangle\langle x,v\rangle |x|^{2(p-2)} .$$

The result follows from the therorem. //

§4 INTEGRATION BY PARTS

Let $u:[a,b] \times \Omega \to H$ and $B:[a,b] \times \Omega \to \mathbb{L}(H;F)$ be sample continuous and adapted where F is a separable Hilbert space and H need only be a separable Banach space. It will be convenient to suppose that B has image in some separable subspace E furnished with a norm $|\ |_E$ such that the induced maps

$$(E, |\ |_E) \xrightarrow{\subset} \mathbb{L}(H;F)$$

and

$$B:[a,b] \times \Omega \to E, |\ |_E$$

are continuous and sample continuous respectively.

Proposition 4 For u and B as above, both $\frac{1}{2}$-Hölder continuous into $\mathcal{L}^2$, if $a < t < b$

$$B_t(u_t) = B_a(u_a) + \int_a^t B_s(du_s) + \int_a^t dB_s(u_s) + \tfrac{1}{2}\int_a^t dB_s(du_s) \quad \text{a.s.}$$

in the sense that if the 2 first order integrals exist (using Cauchy partitions) then the second order integral exists and conversely, and if so the equation is true.

Proof. Immediate from Proposition 1 using the map

$$\theta : E \times H \to F$$
$$\theta(B,u) = B(u). \qquad //$$

Note that the first order integrals exist under Conditions P2 or P3 below, and the second order integral under Condition P1 (see the proof of Theorem 2B of Chapter IV) under which condition it vanishes. Observe that A(1) implies ½-Hölder continuity into $\mathcal{L}^2$. These conditions all refer to B as a map into E, $|\ |_E$, and u and B are assumed sample continuous:

Condition P1   One of B and u has sample paths of bounded variation.

Condition P2 E, $|\ |_E$ is a Hilbert space and B has a representation as a sum of stochastic integrals as for the process u of Proposition 3, and z satisfies A(1); OR H is a Hilbert space and u has such a representation while B satisfies A(1); OR both spaces are Hilbert and both u and B have such a representation.

Condition P3   Both u and B satisfy A(1).

§5 REMARKS ABOUT THE FILTRATIONS. A CHARACTERISATION OF BROWNIAN MOTION

(A) When we come to consider stochastic differential equations we will have a 'noise' process $z:[a,\infty) \times \Omega \to G$ which satisfies the relevant conditions A(q) for some filtration, and so in particular for the filtration $\{\tilde{F}_\tau : \tau \geq a\}$ where $\tilde{F}_\tau = \sigma\{z(s): a \leq s \leq \tau\}$. We will often wish to have a given initial function $u_a:\Omega \to H$ which is independent of the noise; and also not necessarily constant, as it would be if it were $\tilde{F}_0$-measurable for $\tilde{F}_0$ as above with z a Brownian motion. To keep all our processes adapted we may therefore have to enlarge the σ-algebras of our original filtration. For example we can replace $\tilde{F}_\tau$ by F_τ where $F_\tau = \tilde{F}_\tau \vee \sigma\{u_a\}$, using the notation $A \vee B$ for the smallest σ-algebra containing given σ-algebras A and B. We see from the following lemma that the relevant properties of z (e.g. Condition A(q), or the martingale property) are not disturbed by such an enlargement.

Lemma 5A   Suppose $f \in \mathcal{L}^1(\Omega,F,\mu;G)$, and A and B are σ-algebras in F with both f and A independent of B. Then

$$\mathbb{E}(f|A\vee B) = \mathbb{E}(f|A) \qquad \text{a.e.}$$

Proof. Certainly $\mathbb{E}(f|A)$ is $A\vee B$-measurable. Also if $B \in B$

$$\int_B \mathbb{E}(f|A)d\mu = \mathbb{E}(\chi_B \mathbb{E}(f|A)) = \mu(B)\mathbb{E}(f)$$

by independence of A and $\mathcal{B}$. But by independence of f and $\mathcal{B}$

$$\mu(B)\mathbb{E}(f) = \mathbb{E}(\chi_B f) = \int_B f\, d\mu.$$

Thus

$$\int_B \mathbb{E}(f|A)d\mu = \int_B f\, d\mu$$

for all B in $\mathcal{B}$, as well as for all B in A, and therefore for all B in $A \vee \mathcal{B}$ as required. //

(B) In general, for an arbitrary filtration $\{F_t : 0 \leq t < \infty\}$ we will say that (B, F_*) *is a Brownian motion* or B is an F_*-*Brownian motion* if B is a Brownian motion and an F_*-martingale. The following lemma shows that the results of §4 Chapter IV still hold when such filtrations are used. In fact we will only need to use filtrations obtained as described above in §A, and for these the lemma follows immediately from Lemma 5A. However, it is a nice exercise in the use of the Itô formula and martingale property.

Lemma 5B Let B be an n-dimensional F_*-Brownian motion with components $B^1,\ldots,B^n$. Then for $0 \leq s \leq t$ and $p = 1,2,\ldots$

(i) $$\mathbb{E}_s(|B_t - B_s|^{2p}) = \mathbb{E}(|B_t - B_s|^{2p} | \sigma\{B_r : 0 \leq r \leq s\}) \qquad \text{a.s.}$$

(ii) for $i \neq j$ $$\mathbb{E}_s((B_t^i - B_s^i)(B_t^j - B_s^j)) = 0 \qquad \text{a.s.}$$

and $$\mathbb{E}_s((B_t^i - B_s^i)^2(B_t^j - B_s^j)^2) = (t-s)^2 \qquad \text{a.s.}$$

Proof. Fix $s \geq 0$ and set $u_t = B_t - B_s$ for $t \geq s$. Then $u_t = \int_s^t dB_r$ and so by Corollary 3.1

$$|u_t|^{2p} = 2p \int_s^t |u_t|^{2(p-1)} \langle u_r, dB_r\rangle + p \int_s^t |u_r|^{2(p-1)} \langle dB_r, dB_r\rangle$$

$$+ 2p(p-1) \int_s^t |u_r|^{2(p-2)} \langle u_r, dB_r\rangle\langle u_r, dB_r\rangle.$$

Since Brownian motion increments are in $\mathcal{L}^q$ for all q by Proposition 3C of Chapter II, the first order integral is an F_*-martingale by Proposition 7C of Chapter III. Using Theorem 4 of Chapter IV we therefore have

$$\mathbb{E}_s|u_t|^{2p} = np \int_s^t \mathbb{E}_s|u_r|^{2(p-1)} dr + 2p(p-1) \int_s^t \mathbb{E}_s|u_r|^{2(p-1)} dr$$

This gives the correct answer when p = 1, and for p = 2,3,... it is the same recurrence relation for p = 2,3,... as for $\mathbb{E}(|B_t - B_s|^{2p} | \sigma\{B_r : 0 \leq r \leq s\})$. This proves (i).

For (ii) let u^i and u^j be distinct components of u. Applying the Itô formula as before we see

$$u_t^i \, u_t^j = M_t$$

and

$$(u_t^i)^2 \, (u_t^j)^2 = M_t' + \int_s^t 2\Big((u_r^i)^2 + (u_r^j)^2\Big)dr$$

where M and M' are F_*-martingales vanishing at time s. Thus

$$\mathbb{E}_s(u_t^i \, u_t^j) = 0$$

and, using (i),

$$\mathbb{E}_s\Big((u_t^i)^2 (u_t^j)^2\Big) = 2\int_s^t (r-s)\,dr = (t-s)^2. \quad //$$

(C) The temptation to give the Kunita-Watanabe proof of Lévy's characterisation of Brownian motion is now too great, even though we will make little use of it. It is discussed in the context of the "martingale problem" in (Williams 1980a).

Theorem 5C Let $z : [0,\infty) \times \Omega \to \mathbb{R}$ be sample continuous and such that, for some filtration $\{F_t : 0 \leq t < \infty\}$,

(i) z is an F_*-martingale, and

(ii) $\{z_t^2 - t : t \geq 0\}$ is an F_*-martingale.

Then z is a Brownian motion if $z_0 \equiv 0$.

Proof. For $\xi \in \mathbb{R}$ set $y_t = \exp(i\xi z_t + \frac{1}{2}\xi^2 t) = \theta(z_t, t)$ for $\theta : \mathbb{R} \times [0,\infty) \to \mathbb{C} \simeq \mathbb{R}^2$ given by $\theta(x,t) = \exp(i\xi x + \frac{1}{2}\xi^2 t)$. Now by (i) and (ii) if $0 \leq s \leq t$ and $z_0 \equiv 0$

$$\mathbb{E}_s(z_t - z_s)^2 = \mathbb{E}_s z_t^2 - 2\mathbb{E}_s z_s z_t$$

$$= z_s^2 + t-s + z_s^2 - 2(z_s \mathbb{E}_s z_t) = t-s,$$

and so $\quad \mathbb{E}(z_t - z_s)^2 = t-s.$

Therefore z is ½-Hölder continuous into $\mathcal{L}^2$ and satisfies A(1) so we can apply Proposition 1 to obtain

$$y_t = 1 + \int_0^t i\xi\, y_s\, dz_s + \int_0^t \tfrac{1}{2}\,\xi^2\, y_s\, ds - \tfrac{1}{2}\int_0^t \xi^2\, y_s\, dz_s dz_s.$$

However exactly the same proof as that of Theorem 4 Chapter IV shows that the last integral is equal to the one before. Thus

$$y_t = 1 + \int_0^t i\xi\, y_s\, dz_s.$$

Now y_s is bounded on [0,t] and sample continuous. It therefore satisfies B(1) by the dominated convergence theorem. Consequently it is an F_*-martingale by **Proposition 7C of Chapter III.**

Thus, for all $\xi \in \mathbb{R}$,

$$\mathbb{E}_s \exp(i\xi(z_t - z_s) + \tfrac{1}{2}\xi^2(t-s)) = 1 \qquad \text{a.s.}$$

giving

$$\mathbb{E}_s \exp i\xi(z_t - z_s) = \exp(-\tfrac{1}{2}\xi^2(t-s))$$

Let $0 = t_0 < t_1 < \ldots < t_m$, and now take

$\xi = (\xi_1, \ldots, \xi_m) \in \mathbb{R}^m = \mathbb{R}^{\underline{t}}$ for $\underline{t} = \{t_1, \ldots, t_m\}$.

Then

$$\begin{aligned}\mathbb{E}\exp\Big(\sum_{j=1}^{m} i\,\xi_j\Delta_j z\Big) &= \mathbb{E}\Big\{\exp\Big(\sum_{j=1}^{m-1} i\,\xi_j\Delta_j z\Big)\mathbb{E}_{t_{m-1}}(\exp i\,\xi_m\Delta_m z)\Big\}\\ &= \exp(-\tfrac{1}{2}\,\xi_m^2\Delta_m t)\,\mathbb{E}\Big\{\exp\Big(\sum_{j=1}^{m-1} i\,\xi_j\Delta_j z\Big)\Big\}\\ &= \ldots = \exp\Big(-\tfrac{1}{2}\sum_{j=1}^{m}\xi_j^2\,\Delta_j t\Big).\end{aligned}$$

Thus the Fourier transform of the distribution (image measure) of $(\Delta_1 z, \ldots, \Delta_m z):\Omega \to \mathbb{R}^{\underline{t}}$ is uniquely determined and therefore so is the distribution itself. We can either recognise it as being the same as that which would occur if z was a Brownian motion B, or realise that it must be the same since such a B will satisfy analogues of (i) and (ii). In any case it follows that the finite dimensional distribution $z_{\underline{t}}$ is the same as that of $B_{\underline{t}}$ since $z(t_j) = \Delta_1 z + \ldots + \Delta_{j-1} z$ and therefore $z_{\underline{t}}(\mu)$ is determined by the distributions of $(\Delta_1 z, \ldots, \Delta_j z)$, $j = 1$ to m, and similarly for $B_{\underline{t}}(\mu)$. //

For generalisations of the following corollary, which show that a wide class of martingales are just time changed Brownian motions, see (Ikeda & Watanabe 1981) §7 Chapter II.

Corollary 5C *Suppose* B *is an* n*-dimensional* F_**-Brownian motion and* $A:[0,\infty) \times \Omega \to \mathbb{L}(\mathbb{R}^n;\mathbb{R})$ *satisfies* B_0 *and has Euclidean norm* $A(s,\omega)\, A(s,\omega)^* = 1$ *almost surely, all* $0 \leq s < \infty$. *Set*

$$u(t) = \int_0^t A(s)dB(s).$$

Then u *is a 1-dimensional* F_**-Brownian motion.*

Proof. By Corollary 3.1

$$u(t)^2 = 2\int_0^t u(s)A(s)dB(s) + \int_0^t \operatorname{tr} A^*(s)A(s)ds,$$

giving

$$u(t)^2 - t = 2\int_0^t u(s)A(s)dB(s)$$

since $\operatorname{tr} A^*(s)A(s) = A(s)A^*(s) = 1$. Now since A is bounded and satisfies B_0 it satisfies $B(1)$ by the dominated convergence theorem. Therefore u_t is in $\mathcal{L}^2$ for each t and uA satisfies $B(1)$. Proposition 7C of Chapter III now shows that $\{u_t\}$ and $\{u_t^2 - t\}$ are both F_*-martingales, and so the result follows from the theorem. //

§6 APPLICATION: BROWNIAN MOTION IN POLAR COORDINATES

(A) As usual let $H, \langle\ ,\ \rangle$ be a separable Hilbert space and let $u:[a,b] \to \mathcal{L}^2(\Omega,F,\mu;H)$ be $\frac{1}{2}$-Hölder continuous and sample continuous. Set $r(x) = |x|$ for x in H. Then r is C^2 on $H-\{0\}$ with derivatives

$$Dr(x)v = r(x)^{-1} \langle x,v\rangle$$

and

$$D^2r(x)(v',v) = r(x)^{-1} \langle v',v\rangle - r(x)^{-2} \langle x,v\rangle \langle x,v'\rangle.$$

Suppose $u(t,\omega) \in H - \{0\}$ for all t, almost all ω. Then whenever the integrals exist Proposition 1 shows that

$$r(u_t) = r(u_a) + \int_a^t r(u_s)^{-1} \langle u_s, du_s\rangle$$

$$+ \tfrac{1}{2}\int_a^t r(u_s)^{-1} \langle du_s, du_s\rangle - \tfrac{1}{2}\int_a^t r(u_s)^{-2}\langle u_s, du_s\rangle^2$$

a.s.

(B) Now let $B:[0,\infty) \times \Omega \to \mathbb{R}^n$ be an n-dimensional Brownian motion with $n \geqslant 2$. Let $x_0 \in \mathbb{R}^n$ with $x_0 \neq 0$. Then for almost all $\omega \in \Omega$ the sample path $B(-,\omega)$ never hits the point $-x_0$; and so if we set $u_t = B_t + x_0$ then we can apply the above formula to u. (Proofs of this fact may be found in (Ito & McKean 1965) 32.7 and (Knight 1981) page 27; for generalizations see (Friedman 1975) Volume II, Chapter 11.) Setting $r_t = r(u_t)$, we obtain by §4 of Chapter IV

$$r_t = r_0 + \int_0^t r_s^{-1} \langle u_s, dB_s\rangle + \tfrac{1}{2}\int_0^t r_s^{-1}\, n\, ds - \tfrac{1}{2}\int_0^t r_s^{-2}\, r_s\, ds,$$

giving

$$r_t = r_0 + \tilde{B}_t + \tfrac{1}{2}(n-1)\int_0^t r_s^{-1}\, ds \qquad (1)$$

where $\tilde{B}_t = \int_0^t r_s^{-1} \langle u, dB_s\rangle$ is a 1-dimensional Brownian motion by Corollary 5C.

It will follow from the results of Chapter VII that for any 1-dimensional Brownian motion $\tilde{B}$ existence and uniqueness holds for (1), and moreover the isonomy class of the solution $\{r_t : t \geqslant 0\}$ does not depend on what model is used for the Brownian motion $\tilde{B}$. These solutions are called *Bessel processes*, sometimes written *Bes(n)*, and have been extensively studied: see (Pitman & Yor 1981). For a discussion of the case when $x_0 = 0$ see (McKean 1960). The case n = 1 (*reflecting* Brownian motion) needs more sophisticated treatment: $r(u_t)$ will vanish infinitely often and it is necessary to examine the corresponding *sojourn time density* or *local time*, see (Ikeda & Watanabe 1981) §4 Chapter III.

(C) In order to consider the angular behaviour of Brownian motion we must take n = 2 for the moment. With this condition let x_0, B_t, u_t, and r_t be as above. We will seek a sample continuous

$$\theta:[0,\infty) \times \Omega \to \mathbb{R}$$

with

$$u_t = (r_t \cos \theta_t, r_t \sin \theta_t) \qquad \text{a.s.}$$

and $0 \leqslant \theta_0 < 2\pi$.

Such exists and is unique since $\mathbb{R} \to S^1$, $\theta \mapsto (\cos \theta, \sin \theta)$ is a covering, see Appendix A). The easiest way to obtain an equation for θ_t is to formally set $\theta = \tan^{-1} (\frac{u^2}{u^1})$ where $u = (u^1, u^2)$ and write down the Itô formula, forgetting any difficulties. This leads to

$$\theta_t = \theta_0 + \int_0^t r_s^{-2} (u_s^1 \, du_s^2 - u_s^2 \, du_s^1). \tag{2}$$

Using

$$r_t = r_0 + \int_0^t r_s^{-1}(u_s^1 \, du_s^1 + u_s^2 \, du_s^2) + \tfrac{1}{2} \int_0^t r_s^{-1} \, ds \tag{3}$$

we can *define* θ_t by (2) and use the Itô formula to check that $r_t e^{i\theta_t}$ yields the original Brownian motion. This is left as an exercise: note that '$dr_t \, d\theta_t$' = 0. Physicists particularly may like the discussion in (Edwards & Gulyaev 1964).

§7 APPLICATION: COMPUTATION OF COVARIANCES

Suppose $u:[a,b] \times \Omega \to H$ is defined by $u_t = u_a + \int_a^t A_s \, dB_s$ where $A:[a,b] \times \Omega \to \mathbb{L}(\mathbb{R}^n;H)$ satisfies B(1), for H a separable Hilbert space, and where B is an n-dimensional F_*-Brownian motion, and $u_a \in \mathcal{L}^2(\Omega,F_a;\mu;H)$. The *covariance* Cov (u_s,u_t) is by definition

$$\text{Cov}(u_s,u_t) = \mathbb{E}\,\{(u_s - \mathbb{E}u_s) \otimes (u_t - \mathbb{E}u_t)\} \in H \otimes H.$$

For simplicity take $H = \mathbb{R}$. Then the covariance is real valued and if $a \leqslant s \leqslant t \leqslant b$, using Corollary 3B of Chapter I and the martingale property of u_t (Proposition 7C of Chapter III):

$$\begin{aligned}
\text{Cov}(u_s,u_t) &= \mathbb{E}\{(u_s - \mathbb{E}u_s)(u_t - \mathbb{E}u_t)\} \\
&= \mathbb{E}\,\mathbb{E}_s\{(u_s - \mathbb{E}u_s)(u_t - \mathbb{E}u_t)\} \\
&= \mathbb{E}\{(u_s - \mathbb{E}u_s)\mathbb{E}_s(u_t - \mathbb{E}u_t)\} \\
&= \mathbb{E}\{(u_s - \mathbb{E}u_s)(u_s - \mathbb{E}u_t)\}
\end{aligned}$$

$$= \mathbb{E}\,(u_s - \mathbb{E}u_a)^2 = \mathbb{E}(u_s)^2 - (\mathbb{E}u_a)^2.$$

By Corollary 3.1

$$\mathbb{E}(u_s)^2 = \mathbb{E}\{(u_a)^2 + 2\int_a^s u_r\, A_r(dB_r) + \int_a^s A_r(dB_r)A_r(dB_r)\}$$

$$= \mathbb{E}(u_a)^2 + \mathbb{E}\int_a^s \text{trace } A_r^* \, A_r \, dr + I(s)$$

where $I(s) = 2\mathbb{E}\int_a^s u_r A_r(dB_r)$ and $A_r^*(\omega)$ denotes the adjoint (or transpose) of $A_r(\omega)$. By the martingale property we can expect $I(s)$ to vanish: we know this happens if $u_r A_r$ satisfies B(1), and in particular therefore if A is bounded on $[a,b] \times \Omega$. If so we obtain by Fubini's theorem

$$\text{Cov}(u_s,u_t) = \int_a^{t\wedge s} \|A_r\|^2 \, dr + \text{Cov}(u_a,u_a) \qquad (1)$$

since the *Hilbert-Schmidt*, or *Euclidean*, norm $|T|_2 = \sqrt{\text{trace } T^*T}$ on $\mathbb{L}(\mathbb{R}^n;\mathbb{R}^p)$ agrees with the operator norm when $p = 1$.

Exercise 7 Let v be the 1-dimensional Ornstein-Uhlenbeck velocity process:

$$v_t = e^{-\beta t}\, v_0 + \gamma e^{-\beta t}\int_0^t e^{\beta s}\, dB_s \qquad 0 < t < \infty$$

where β and γ are constnts and $v_0:\Omega \to \mathbb{R}$ is in $\mathcal{L}^2$ and is independent of the 1-dimensional Brownian motion B. Prove that

$$\text{Cov}(v_s,v_t) = e^{-\beta(t+s)}(\text{Cov}(v_0,v_0) - \gamma^2/2\beta) + e^{-\beta|t-s|}\gamma^2/2\beta$$

and

$$\mathbb{E}v_t = e^{-\beta t}\,\mathbb{E}v_0.$$

CHAPTER VI : STOCHASTIC INTEGRAL EQUATIONS

1 EXISTENCE

(A) Again, let (Ω,F,μ) be a probability space, H a separable Hilbert space, and $\mathcal{L}^2(\Omega,F;H)$ the totality of maps $x:\Omega \to H$ of class L^2, with its semi-norm $\| \ \|$. The corresponding Hilbert space of equivalence classes is $L^2(\Omega,F;H)$. For our fixed subset T of $\mathbb{R}$ we have the increasing family $\{F_\tau\}_{\tau\in T}$ of σ-subalgebras of F.

If $T_0 \subset T$ set

$$S(T_0) = \{x:T_0 \to \mathcal{L}^2(\Omega,F;H): x \text{ is continuous and each } x(t) \text{ is } F_t\text{-measurable}\}$$

and

$$S(T_0) = \{x: T_0 \to L^2(\Omega,F;H): x \text{ is continuous and each } x(t) \text{ can be represented by an } F_t\text{-measurable map}\}.$$

<u>Lemma 1A</u> <u>When T_0 is compact</u>, $S(T_0)$ <u>with norm</u> $|x| = \sup_{t\in T_0} \|x(t)\|$ <u>is a Banach space.</u>

<u>Proof.</u> It suffices to show that $S(T_0)$ is a closed subspace of the usual Banach space $C(T_0;L^2(\Omega,F;H))$. To do this note that, for each $t \in T$, the space $L^2(\Omega,F_t;H)$ is included as a closed subspace of $L^2(\Omega,F;H)$ and then

$$S(T_0) = \{x \in C(T_0;L^2(\Omega,F;H)): x(t) \in L^2(\Omega,F_t;H) \text{ each } t \in T_0\}.$$

The result follows since for a sequence $\{x_n\}_{n=1}^\infty$ in $S(T_0)$, convergence in $C(T_0;L^2(\Omega,F;H))$ implies convergence of $\{x_n(t)\}_{n=1}^\infty$ in $L^2(\Omega,F_t;H)$ for each $t \in T_0$. //

A process $y:T_0 \to \mathcal{L}^0(\Omega,F;H)$ will be called *regular* if it is sample continuous and belongs to $S(T_0)$.

(B) Let G be a separable Banach space. Write $\mathbb{L}^s(G,\ldots,G;H)$ for the subspace of $\mathbb{L}(G,\ldots,G;H)$ consisting of symmetric maps and denote by $\mathbb{P}_q(G;H)$ the Banach space of polynomials of degree $q \geq 1$ with zero constant term:

$$\mathbb{P}_q\,(G;H) = \mathbb{L}(G,H) \times \mathbb{L}^s\,(G,G;H) \times \ldots \times \mathbb{L}^s\,(G,\ldots,G;H),$$

together with the norm

$$|(A_1,\ldots,A_q)| = q \max_{1\leq\rho\leq q} |A_\rho|.$$

For suitable $p:T \times \Omega \to \mathbb{P}_q(G:H)$ and $z:[a,b] \times \Omega \to G$ the obvious notation

$$\int_a^b p(t,\omega)\; dz(t,\omega)$$

will be used for the corresponding sum of stochastic integrals of order ρ, $1 \leq \rho \leq q$. For example suppose

$$X:[a,b] \times \Omega \to \mathbb{L}(\mathbb{R}^n;\mathbb{R}^m),\; Y:[a,b] \times \Omega \to \mathbb{L}(\mathbb{R}^n,\mathbb{R}^n;\mathbb{R}^m)$$

and

$$A:[a,b] \times \Omega \to \mathbb{R}^m.$$

Take $G = \mathbb{R}^n \times \mathbb{R}$ and $H = \mathbb{R}^m$. Define

$$p^1(t,\omega)(v,s) = X(t,\omega)v + A(t,\omega)s$$

$$p^2(t,\omega)((v,s),(v',s')) = Y(t,\omega)(v,v')$$

for (v,s) and (v',s') in $\mathbb{R}^n \times \mathbb{R}$. Set $p(t,\omega) = (p^1(t,\omega),\, p^2(t,\omega)) \in \mathbb{P}_2(G;H)$ and take $z(t,\omega) = (B(t,\omega),t)$ where B is an n-dimensional Brownian motion. Then by definition

$$\int_a^b p(t)dz(t) = \int_a^b X(t)dB(t) + \int_a^b A(t)dt + \int_a^b Y(t)(dB(t),dB(t)).$$

Our basic existence theorem for stochastic integral equations will involve 'kernels' $P:T \times H \times \Omega \to \mathbb{P}_q(G;H)$ which satisfy the following *regularity and global Lipschitz conditions:*

Conditions 1B (a) For all $\tau \in T$ the map $(y,\omega) \mapsto P(\tau,y,\omega)$ is Borel $(H)*F_\tau$-random where Borel $(H)*F_\tau$ denotes the product σ-algebra.

(b) For each $y \in H$ almost all the sample paths $\tau \mapsto P(\tau,y,\omega)$ are continuous on T.

(c) There is a constant M such that

$$\sup_{\tau \in T} |P(\tau,0,\omega)| \leqslant M \quad \text{a.s.}$$

(d) There is a constant L (the *Lipschitz constant*) such that

$$|P(\tau,y_1,\omega) - P(\tau,y_2,\omega)| \leqslant L|y_1 - y_2|$$

for all y_1, $y_2 \in H$ and $\tau \in T$, almost surely.

(C) We now have our basic existence thoerem for stochastic integral equations:

Theorem 1C Let G be a separable Banach space, and H a separable Hilbert space. Suppose $P: T \times H \times \Omega \to \mathbb{P}(G;H)$ satisfies the regularity and global Lipschitz conditions of Conditions 1B. Let $[a,b] \subset T$, and suppose $z:[a,b] \to \mathcal{L}^0(\Omega,F;G)$ satisfies condition A(q). Then, for given $u \in S([a,b])$, there is a solution $x \in S([a,b])$ to the equation

$$x(t)(\omega) = u(t)(\omega) + \int_a^t P(s,x(s)(\omega),\omega)dz(s,\omega) \quad a \leqslant t \leqslant b.$$

Any two such solutions x,x' have x(t) = x'(t) almost surely for each $t \in [a,b]$.

If also z is sample continuous and u is regular then x may be chosen to be regular. All regular solutions are equivalent.

Remark To be completely strict in the interpretation of the belated integrals in the equation our solution x should be defined on $(-\infty,b] \cap T$. However x can always be given such an extension to make the integrals exist (e.g. $x(\tau) = x(a)$ if $\tau < a$), and by using Cauchy partitions the actual values of the extension are clearly irrelevant. We can therefore safely ignore such points here and in future.

Proof. For $x \in S([a,b])$ and $s,t \in [a,b]$ define

$$P(s,x_t):\Omega \to \mathbb{P}_q(G;H)$$

by

$$P(s,x_t)(\omega) = P(s,x_t(\omega),\omega).$$

Then, almost surely,

$$|P(s,x_s)(\omega) - P(t,x_t)(\omega)| \leq |P(s,x_s)(\omega) - P(s,x_t)(\omega)|$$
$$+ \; P(t,x_t)(\omega) - P(s,x_t)(\omega)|$$
$$\leq A(s,t)(\omega) + B(s,t)(\omega)$$

where

$$A(s,t)(\omega) = L|x_s(\omega) - x_t(\omega)|$$

and

$$B(s,t)(\omega) = |P(t,x_t)(\omega) - P(s,x_t)(\omega)|.$$

Now $\lim_{s\to t} A(s,t) = 0$ in $\mathcal{L}^2$, because $x \in S([a,b])$, and $\lim_{s\to t} B(s,t)(\omega) = 0$ almost surely by Condition 1B(b). Moreover, by Condition 1B(c), (d), for almost all ω

$$|B(s,t)(\omega)| \leq 2L|x_t(\omega)| + 2M;$$

whence by the dominated convergence theorem $\lim_{s\to t} B(s,t) = 0$ in $\mathcal{L}^2$.

Thus the map $s \mapsto P(s,x_s)$ is continuous in $\mathcal{L}^2$-norm, and so satisfies Condition B(1) of Chapter III. Consequently, by our main existence theorem, Theorem 3 of Chapter III, for every x in $S([a,b])$ and every t in $[a,b]$, the integral $\int_a^t P(s,x_s)dz_s$ exists as an element of $\mathcal{L}^2(\Omega,F_t;H)$.

By definition of the integral, $\int_a^t P(s,x_s)dz_s$ determines an element of $L^2(\Omega,F_t;H)$ depending only on the class of x in $S([a,b])$. Moreover the integral is continuous in t as a map into $\mathcal{L}^2(\Omega,F;H)$ by Corollary 3 of Chapter III. Thus we obtain a map $\theta:S([a,b]) \to S([a,b])$ by the equation

$$\theta(x)(t) = u(t) + \int_a^t P(s,x(s))dz(s)$$

interpreted in terms of equivalence classes.

Let $m = L^2\beta^2$ where $\beta = 2K(b-a)^{\frac{1}{2}} + K^{\frac{1}{2}}$ as in Corollary 3 of Chapter III.

We will show by induction on n that for $x,y \in S([a,b])$

$$\| \theta^n(x)(t)-\theta^n(y)(t)\|^2 < \frac{1}{n!} \quad m^n \; |x-y|^2 \; (t-a)^n \qquad a<t<b.$$

In fact

$$\| \theta^{n+1}(x)(t) - \theta^{n+1}(y)(t)\|^2 = \|\int_a^t [P(s,\theta^n(x)(s) -$$

$$- P(s,\theta^n(y)(s))]dz(s)\|^2$$

$$< \beta^2 \int_a^t \|P(s,\theta^n(x)(s)) - P(s,\theta^n(y)(s))\|^2 \; ds$$

$$< L^2 \; \beta^2 \int_a^t \|\theta^n(x)(s) - \theta^n(y)(s)\|^2 \; ds$$

$$< m \int_a^t \frac{1}{n!} \; m^n \; |x-y|^2 \; (s-a)^n \quad ds$$

$$= m^{n+1} \frac{1}{(n+1)!} |x-y|^2 \; (t-a)^{n+1}$$

assuming the result is true for the integer n.

Since $\sum_n \left(\frac{1}{n!} \quad m^n(t-a)^n \quad |x-y|^2\right)^{\frac{1}{2}} < \infty$

a very slight modification of the usual contraction mapping principle [for example as in (Jameson)] shows that θ has a unique fixed point: the class $\bar{x}$ of the required solution x; and moreover $\bar{x} = \lim_{n\to\infty} \theta^n(x_0)$ for any $x_0 \in S([a,b])$. This last fact will be used later.

The continuity of sample paths for some solution when z and u are sample continuous follows from Theorem 5C of Chapter IV and the first part of this proof: these imply that if $x_1 \in S([a,b])$ is any solution there is a version x of $t \mapsto u(t) + \int P(s,x_1(s))dz(s)$ which is sample continuous; but this is a version of x_1 and every version of x_1 is a solution. //

Corollary 1C Let G, H, and P be as in the theorem. Suppose $a < b \leq \infty$ and the $\frac{1}{2}$-open interval $[a,b)$ is contained in T. Let $u \in S([a,b))$. Then the results of the theorem still hold for $a \leq t < b$ and for processes $z:[a,b) \to \mathcal{L}^0(\Omega,F;G)$ which satisfy condition A(q) on each compact subset of $[a,b)$.

Proof. Choose an increasing sequence $\{c_i\}_{i=1}^{\infty}$ in $[a,b)$ with $c_1 = a$ and $c_i \to b$ as $i \to \infty$. Choose a solution $x_i \in S([a,c_i])$ on $[a,c_i]$, for each i, by the theorem. Define a solution x on $[a,b)$ by $x(t) = x_i(t)$ if $c_i \leq t < c_{i+1}$.

When u and z are sample continuous, choose each x_i to be regular. Then $x_j|[a,c_i] \sim x_i$ if $i < j$, and so x can be taken to be regular.

Remark. In his book (McShane 1974) Chapter V, Theorem 1.5, McShane proves an existence theorem when the constants L, M of Conditions 1B are allowed to be random variables.

§2 DEPENDENCE ON INITIAL DISTRIBUTION (GLOBALLY LIPSCHITZ CASE)

(A) Throughout this section we assume the notation and conditions of Theorem 1C. In particular $P:T \times H \times \Omega \to \mathbb{P}_q(G:H)$ satisfies the regularity and global Lipschitz conditions of Condition 1B. Consideration of more general situations is postponed until later chapters, as is a discussion of higher differentiability. The standard reference is (Gikhman & Skorohod 1972). The results given here and in §3 below will not play a fundamental role later, although Theorem 2B will prove useful.

(B) Theorem 2B Let $F: S([a,b]) \to S([a,b]$ be defined so that F(u) is a solution to the stochastic integral equation

$$x(t)(\omega) = u(t)(\omega) + \int_a^t P(s,x(s)(\omega),\omega)\, dz(s,\omega) \quad a \leq t \leq b. \quad (*)$$

Then F is Lipschitz on $S([a,b])$; in fact if L is a Lipschitz constant for P, and $\beta = 2K\sqrt{(b-a)} + \sqrt{K}$ is the constant of Corollary 3, Chapter III, associated with z, then $\sqrt{2}\exp[\beta^2 L^2(b-a)]$ serves as a Lipschitz constant for F.

Proof. For $u, v \in S([a,b])$ set $x_u = F(u)$, $x_v = F(v)$. With β and L as stated, using Corollary 3 of Chapter III:

$$\begin{aligned}\|F_t(u) - F_t(v)\| &\leq \|u_t - v_t\| + \left\|\int_a^t (P(s,F_s(u)) - P(s,F_s(v)))dz_s\right\| \\ &\leq |u-v| + \left\{\int_a^t \|P(s,F_s(u)) - P(s,F_s(v))\|^2 ds\right\}^{\frac{1}{2}} \\ &\leq |u-v| + \beta L \left\{\int_a^t \|F_s(u) - F_s(v)\|^2 ds\right\}^{\frac{1}{2}}.\end{aligned}$$

Thus

$$\|F_t(u) - F_t(v)\|^2 \leq 2\|u-v\|^2 + 2\beta^2L^2 \int_a^t \|F_s(u) - F_s(v)\|^2 ds.$$

Applying Gronwall's lemma (Chapter I, §5):

$$\sup_{a\leq t\leq b} \|F_t(u) - F_t(v)\|^2 \leq 2|u-v|^2\left(1+2\beta^2L^2 \int_a^b \exp[2\beta^2L^2(b-s)]ds\right)$$

$$\leq 2|u-v|^2 \exp\,[2\beta^2L^2(b-a)],$$

as required. //

Corollary 2B If also z is sample continuous, for each u which is regular on [a,b] let $\bar{F}(u)$ be a solution of (*) in $\mathcal{L}^0(\Omega,F;C([a,b];H))$. Then

$$\bar{F}:\mathcal{L}^2(\Omega,F_*,\mu;C([a,b];H)) \to \mathcal{L}^0(\Omega,F,\mu;C([a,b];H))$$

is continuous. (The F_* in $\mathcal{L}^2(\Omega,F_*,\mu;\ C([a,b];H))$ indicates that only adapted processes are considered).

Proof. For $\varepsilon > 0$, and for γ as in Estimate 5C of Chapter IV

$$\mu\{\omega:\ \sup_{a\leq t\leq b}\ |\bar{F}_t(u)(\omega) - \bar{F}_t(v)(\omega)| > \varepsilon\}$$

$$\leq \mu\{\omega:\ \sup_{a\leq t\leq b}\ |u_t(\omega) - v_t(\omega)| > \varepsilon/2\}$$

$$+\ \mu\{\omega:\ \sup_{a\leq t\leq b}\ \left|\int_a^t \left(P(s,\bar{F}_s(u)) - P(s,\bar{F}_s(v))\right) dz_2\right| > \varepsilon/2\}$$

$$\leq \mu\{\omega:\ \sup_{a\leq t\leq b}\ |u_t(\omega) - v_t(\omega)| > \varepsilon/2\}$$

$$+\ 4\gamma\,\varepsilon^{-2} \int_a^b \|P(s,\bar{F}_s(u)) - P(s,\bar{F}_s(v))\|^2\ ds$$

$$\leq 4\ \varepsilon^{-2}\,\|u-v\|^2 + 4\gamma\,\varepsilon^{-2}\,L^2(b-a)\ \sup_{a\leq t\leq b}\|\bar{F}_t(u) - \bar{F}_t(v)\|^2.$$

The corollary now follows immediately from the theorem. //

For the sort of estimates needed to get continuity of $\bar{F}$ when its target space is taken as $\mathcal{L}^2$ not $\mathcal{L}^0$ see (Friedman 1975) Theorem 3.7 Chapter 4 or (Ikeda & Watanabe 1981) equation 6.7 Chapter I.

(C) We want to discuss the differentiability of the map F of Proposition 2B, (with $\mathcal{L}^2$ replaced by L^2). This will be done using a version of the implicit function theorem, and we need some preliminary lemmas first.

For a function $f:U \to V$ between open subsets of real Hausdorff topological vector spaces E and F we use the notation $GDf(x)v$ for the *Gâteaux*, or *directional*, *derivative* of f at x in the direction v:

$$GDf(x)v = \lim_{t \downarrow 0} \frac{f(x+tv) - f(x)}{t} \qquad x \in U, \quad v \in E.$$

For the relation between Gâteaux differentiability and ordinary (i.e. Fréchet) differentiability see Appendix A.

<u>Definition</u> A map $P:T \times H \times \Omega \to \mathbb{P}_q(G;H)$ will be said to be $\tilde{C}^1$ if it satisfies Conditions 1B and if also the partial derivative

$$D_2P:T \times H \times \Omega \to \mathbb{L}(H;\mathbb{P}_q(G;H))$$

exists and satisfies

(a) for all $\tau \in T$ the map $(y,\omega) \mapsto D_2 P(\tau,y,\omega)$ is [Borel(H) * $\mathcal{F}$]-random as a map into

$$\mathbb{L}(H,G;H) \times \ldots \times \mathbb{L}(H,G,\ldots,G;H);$$

and (b) for almost all $\omega \in \Omega$ the map $(\tau,y) \mapsto D_2P(\tau,y,\omega)$ is continuous on $T \times H$.

Note that if P is independent of Ω and satisfies Conditions 1B then it is $\tilde{C}^1$ if and only if the partial derivative D_2P exists and is continuous.

<u>Lemma 2C</u> <u>Suppose</u> $P:T \times H \times \Omega \to \mathbb{P}_q(G;H)$ <u>is</u> $\tilde{C}^1$ <u>on</u> $T_0 \times H \times \Omega$ <u>where</u> $T_0 \subset T$. <u>For regular</u> $x:T_0 \to \mathcal{L}^0(\Omega,\mathcal{F};H)$ <u>define</u>

$$Q^x:T_0 \times H \times \Omega \to \mathbb{P}_q(G;H)$$

by

$$Q^x(t,v,\omega) = D_2P(t,x_t(\omega),\omega)v.$$

<u>Then</u> Q^x <u>satisfies Conditions</u> 1B <u>on</u> $T_0 \times H \times \Omega$ <u>with the same Lipschitz constants as</u> P.

Proof. If L is a Lipschitz constant for P then certainly for almost all ω,

$$\sup \{|D_2P(t,y,\omega)| : (t,y) \in T_0 \times H\} < L$$

and so Q^x satisfies 1B(d) with Lipschitz constant L. Condition 1B(c) is trivially true since $Q^x(t,0,\omega) = 0$ all t,ω. The other conditions come directly from the definitions. //

(D) If E is a topological vector space and f is an E valued map defined on an interval $[0,\delta)$ we shall write

$$\text{"}f(t) = o(t^+)\text{" if } \lim_{t\downarrow 0} t^{-1}f(t) = 0.$$

A map $h:U \to V$ of open subsets of topological vector spaces will be said to be *somewhat Lipschitz at the point* x_0 of U if whenever $a,b:[0,\delta) \to U$ are maps of some interval $[0,\delta)$ into U with $a(0) = b(0) = x_0$, both continuous at 0, such that $a(t) - b(t) = o(t^+)$ then $h(a(t)) - h(b(t)) = o(t^+)$.

Lemma 2D Let U, V be open subsets of real Hausdorff topological vector spaces E, F. Suppose $f: U \to V$ is a homeomorphism with inverse g. Assume

(i) f is Gâteaux differentiable at the point x_0 of U and $GDf(x_0):E \to F$ is surjective

(ii) g is somewhat Lipschitz, at $y_0 = f(x_0)$.

Then g is Gâteaux differentiable at y_0 and

$$GDf(x_0) \circ GDg(y_0) = id_F.$$

Moreover $GDf(x_0)$ is bijective.

Proof. For $w \in F-\{0\}$ choose $v \in E$ with $GDf(x_0)(v) = w$. Then

$$f(x_0 + tv) = f(x_0) + tw + o(t^+)$$

whence

$$g(y_0 + tw) = g(f(x_0 + tv) + o(t^+))$$

$$= x_0 + tv + \Big(g(f(x_0 + tv) + o(t^+)) - g(f(x_0 + tv))\Big)$$

$$= g(y_0) + tv + o(t^+)$$

by the Lipschitz condition.

Thus g is differentiable at y_0 in the direction w with $GDg(y_0)(w) = v$, as required. Bijectivity of $GDf(x_0)$ follows since v was an arbitrary element of $GDf(x_0)^{-1}w$. //

Let Reg ([a,b]) denote the subspace of S([a,b]) consisting of those classes which have regular versions, and give Reg ([a,b]) the norm induced on it from S([a,b]).

__Theorem 2D__ __Assume that__ z __is sample continuous and satisfies__ A(q). __Assume that__ P __is__ $\tilde{C}^1$ __and let__ F:Reg([a,b]) → Reg ([a,b]) __be defined as in Theorem 2B__, (__but now considered as a map on equivalence classes__). __Then__ F __is Gâteaux differentiable on__ Reg ([a,b]) __and for each__ u, v __in__ Reg ([a,b]), y = GDF(u)v __is the solution to__

$$y_t(\omega) = v_t(\omega) + \int_a^t D_2P(s,F_s(u)(\omega),\omega)(y_s(\omega))dz_s(\omega)$$

__Proof__. Define

$$\phi:\text{Reg}([a,b]) \to \text{Reg}([a,b]) \text{ by } \phi(x)_t = x_t - \int_a^t P(s,x_s)dz_s.$$

(We shall persistently confuse functions with their equivalence classes).

By Theorem 2B, ϕ is a homeomorphism with Lipschitz inverse F. For x, v ∈ Reg ([a,b]) and $r > 0$

$$\phi(x + rv)_t - \phi(x)_t = rv_t - \int_a^t \Big(D_2P(s,x_s)rv_s + \theta(s,x,r,v)\Big)dz_s$$

where $\theta(s,x,r,v):\Omega \to \mathbb{P}_q(G;H)$ is o(r) almost surely for each s,x,v. To show that ϕ is differentiable at x in the direction v it suffices to show that $\sqrt{\int_a^b \|\theta(s,x,r,v)\|^2 ds}$ is o(r), by Corollary 3 of Chapter III. However

$$\frac{1}{r}|\theta(s,x,r,v)| \leq \frac{1}{r}|P(s,x_s + rv_s) - P(s,x_s)| + |D_2P(s,x_s)v_s|$$
$$\leq 2L|v|$$

where L is a Lipschitz constant for P, and so we can apply the dominated convergence theorem. Thus $GD\phi(x)v$ exists and

$$[GD\phi(x)v]_t = v_t - \int_a^t D_2P(s,x_s)v_s\, dz_s.$$

By Lemma 2C and Theorem 1C, the map $GD\phi(x)$: Reg ([a,b]) → Reg ([a,b]) is surjective for each x. Thus we can apply Lemma 2D to complete the proof. //

Remark. By weakening Condition 1B(b) we could prove the corresponding result for F:S([a,b]) → S([a,b]) without assuming z is sample continuous provided we assume additional uniform continuity conditions on D_2P.

§3 ALMOST SURE CONTINUITY WITH RESPECT TO THE INITIAL POINT

We will now prove a theorem on the sample continuity of solutions with respect to their initial conditions. It seems to have first appeared in (Blagoveščenskii & Freidlin 1961) and was apparently known independently to I.V. Girsanov. I learnt of the theorem from the work of Baxendale, who also furnished the reference to Totoki's proof of the generalization of Kolmogorov's theorem, and from the work of Malliavin who used a quite different method: see Chapter VIII. *We will assume that* z *has continuous sample paths and that* q = 1, *i.e. that the equations involve only first order integrals.*

(A) First we need some $\mathcal{L}^p$ estimates. Again the standard reference is (Gikhman & Skorohod 1972)

Theorem 3A With the assumptions and notation of Theorem 1C suppose also that u is regular and that for p = 2 or some $p \geq 4$ it has $\sup\{|u(t)|^p : a \leq t \leq b\}$ in $\mathcal{L}^2$. Then the solution $x = F(u) \in S([a,b])$ to

$$x_t = u_t + \int_a^t P(s,x_s)dz_s$$

has $x(t) \in \mathcal{L}^{2p}$ for each $t \in [a,b]$.

Moreover if u,v ∈ Reg ([a,b]) with both

$$\sup\{|u_t|^p : a \leq t \leq b\} \text{ and } \sup\{|v_t|^p : a \leq t \leq b\} \text{ in } \mathcal{L}^2$$

we have

$$\sup \|F_t(u)-F_t(v)\|_{\mathcal{L}^{2p}} : a\leq t\leq b\} \leq 2^{1-\frac{1}{2}p^{-1}} e^{(b-a)\Lambda} \sup_{a\leq t\leq b} \{\|u_t - v_t\|_{\mathcal{L}^{2p}}\} \tag{1}$$

where $\Lambda = 2^{2p-4}[(p-1)L^2 + 2L]^2\, p\beta^2$.

Proof. We can assume that $x \in \text{Reg}([a,b])$. Take $0 < R < \infty$ and let τ be the first exit time of $|x(t)|$ from the interval $[0,R)$. By Proposition 5C of Chapter III the map τ is a stopping time. Set

$$x'(t) = x(t \wedge \tau)$$

$$u'(t) = u(t \wedge \tau) + \int_a^{t\wedge\tau} P(s,0)dz(s).$$

and

$$P'(s,y,\omega) = P(s,y,\omega) - P(s,0,\omega).$$

By Corollary 3.1 of Chapter V and Corollary 7A of Chapter III, setting $p = 2q$,

$$|x_t' - u_t'|^{2q} = 2q \int_a^{t\wedge\tau} |x_s|^{2(q-1)} \langle x_s, P'(s,x_s)dz_s\rangle$$

$$+ \ q \int_a^{t\wedge\tau} |x_s|^{2(q-1)} \langle P'(s,x_s)dz_s, P'(s,x_s)dz_s\rangle$$

$$+ \ 2q(q-1) \int_a^{t\wedge\tau} |x_s|^{2(q-2)} \langle x_s, P'(s,x_s)dz_s\rangle^2 \tag{2}$$

and we can equally well replace each x by x' in the right hand side. Now $|P'(s,x'(s))| \leq L\,|x'(s)|$ almost surely. Consequently using Theorem 7A of Chapter III and the inequality

$$(a + b)^p \leq 2^{p-1}(a^p + b^p) \qquad a,b \geq 0, \ \ p \in \mathbb{R}$$

if we set $y(t) = |x'(t)|^{2q}$ we obtain

$$2^{1-p}\|y_t\| \leq \|(|u_t'|^p)\| + \beta(pL+qL^2+p(q-1)L^2)\{\int_a^t \|y_s\|^2 ds\}^{\frac{1}{2}}$$

giving

$$\|y_t\|^2 \leq 2^{2p-1}\|(|u_t'|^p)\|^2 + 2^{2p-1}q^2\beta^2[(p-1)L^2+2L]^2 \int_a^t \|y_s\|^2 ds.$$

Applying Gronwall's lemma we get

$$\|y_t\|^2 \leq 2^{2p-1}\|(|u_t'|^p)\|^2 + 2^{2p-1}\lambda\int_a^t e^{\lambda(t-s)}\|(|u_s'|^p)\|^2 ds \qquad \text{for } a \leq t \leq b \tag{3}$$

where

$$\lambda = 2^{2p-3}[(p-1)L^2 + 2L]^2 p^2\beta^2,$$

Using our hypotheses on u and P together with the fact that $\tau \to b$ as $R \to \infty$ we see that x(t) is in $\mathcal{L}^{2p}$.

For the Lipschitz condition (1) we can now use the method and notation of Theorem 2B together with some of the arguments above to see that (3) holds with y(t) replaced by $|F_t(u) - F_t(v)|^{2p}$ and u' replaced by u-v. Equation (1) then follows directly. //

When z is a Brownian motion better estimates can be obtained by integrating equation (2) in the proof directly using the martingale property of the first order integral and the fact that the second order integral reduces to a deterministic integral. For stronger results see (Stroock & Varadhan 1979).

(B) Now let $H = \mathbb{R}^n$. For fixed $t \in [a,b]$ consider the 'flow map'

$$F_t : \mathbb{R}^n \to \mathcal{L}^0(\Omega, F, \mathbb{R}^n)$$

where $F_t(u) = F(u)(t)$ is the solution as before of

$$x(t) = u + \int_a^b P(s,x(s))dz(s) \qquad a \le t \le b$$

(but now u is constant).

Theorem 3B (Blagoveščenskii & Freidlin) Under the conditions of Theorem 1C with $\dim H < \infty$ there is a sample continuous version of F_t.

Proof. By (1) of Theorem 3A, if $\delta > 0$ and $x_1, x_2 \in \mathbb{R}^n$

$$\mu\{\omega: |F_t(x_1)-F_t(x_2)| > \delta\} \le \delta^{-2n} \int_\Omega |F_t(x_1)-F_t(x_2)|^{2n} d\mu$$
$$\le \delta^{-2n}\, 2^{2n}\, e^{2n(b-a)\Lambda} |x_1-x_2|^{2n}.$$

Thus the finite dimensional distributions of F_t satisfy Condition C of Totoki's theorem: Theorem 2C of Chapter II. For any compact subset K of R^n there will therefore exist a process parametrized by K which is sample continuous and isonomous to $F_t|K$. In fact, as shown in Appendix C, there is a sample continuous version of $F_t|K$. Since $\mathbb{R}^n$ is a countable union of compact subsets the result follows. //

Remarks 3B

(i) For dim $H = \infty$ our arguments show that for each finite dimensional subspace H_0 of H there is a version of F_t such that $F_t|H_0$ is sample continuous.

(ii) Results about differentiability and surjectivity of F_t will be given in Chapter VIII.

(iii) With more care we could get sample continuity in $[a,b] \times \mathbb{R}^n$ of F: see (Kunita 1980).

§4 PIECEWISE LINEAR APPROXIMATION

For $X : H \to \mathbb{L}(G;H)$ differentiable define

$$M(X) : H \to \mathbb{P}_2(G;H)$$

by

$$M(X)(x)(g) = X(x)(g) + \tfrac{1}{2} DX(x)(X(x)g)(g) \quad x \in H, \quad g \in G$$

noting that $DX(x) \in \mathbb{L}(H;\mathbb{L}(G;H))$. Note further that $M(X)$ is Lipschitz if X is bounded and both X and DX are Lipschitz.

Also, given a process $z:[a,b] \to \mathcal{L}^0(\Omega,F;G)$ and a (Cauchy) partition Π of $[a,b]$, $\Pi = (t_1,\ldots,t_{m+1})$, define the *piecewise linear approximation* z_π to z,

$$z_\pi:[a,b] \times \Omega \to G$$

by $z_\pi(s,\omega) = (t_{j+1} - t_j)^{-1} \left\{(t_{j+1}-s)z(t_j,\omega)+(s-t_j)z(t_{j+1},\omega)\right\}$ $t_j \leq s < t_{j+1}$.

The following is a special case of a slightly more general theorem by McShane: see (McShane 1974) Chapter 6 Theorem 3-62. In his book McShane gives a general discussion of approximation theorems of this type. See also (Ikeda et al. 1977) and (Dowell 1979). The latter also discusses the behaviour of solutions when Brownian (i.e. 'white') noise is approximated by Ornstein-Uhlenbeck (or 'coloured') noise. Malliavin gives corresponding results when z is approximated by its *regularizations* $\{z_\varepsilon:\varepsilon > 0$ where

$$z_\varepsilon(t) = \int_0^\varepsilon \frac{1}{\varepsilon} z(t+s)\phi(\frac{s}{\varepsilon})ds$$

for a suitable bump function $\phi:\mathbb{R} \to \mathbb{R}$. See (Malliavin 1977a) or (Malliavin 1978a). The proof we give was extracted from McShane's by R.M. Dowell.

Theorem 4 Let $X:H \to \mathbb{L}(G;H)$ be C^1 and bounded with both X and DX globally Lipschitz. Let $[a,b] \subset T$ and suppose $z:[a,b] \to \mathcal{L}^0(\Omega,\mathcal{F};G)$ satisfies Condition A(2). Suppose $x \in \mathcal{S}([a,b])$ is a solution to the stochastic integral equation

$$x(t) = x_a + \int_a^t M(X)(x(s))dz(s) \qquad a \leqslant t \leqslant b$$

and for each partition Π of $[a,b]$ let

$$x_\pi:[a,b] \times \Omega \to H$$

be the solution to the family of ordinary differential equations

$$\frac{d}{dt} x_\pi(t,\omega) = X(x_\pi(t,\omega)) \frac{dz_\pi}{dt}(t,\omega) \qquad a \leqslant t \leqslant b$$

$$x_\pi(a,\omega) = x_a(\omega)$$

Then $\{x_\pi\}_\Pi$ converges to x in $\mathcal{L}^2$, uniformly on $[a,b]$, as mesh $\Pi \to 0$.

Proof. For a fixed partition Π set $y = x_\pi$, $x_j = x(t_j)$, and $y_j = y(t_j)$. Then

$$\Delta_j y = y'(t_j)\,\Delta_j t + \int_0^1 (1-s)y''(t_j + s\,\Delta_j t)\,(\Delta_j t)^2 ds$$

$$= X(y_j)\Delta_j z + \int_0^1 (1-s)DX\Big(y(t_j+s\Delta_j t)\Big)\Big(X(y(t_j+s\Delta_j t))\Delta_j z\Big)(\Delta_j z)ds.$$

Thus

$$y_k - y_1 = \sum_{j=1}^{k-1} \Delta_j y$$

$$= \sum_{j=1}^{k-1} M(X)(y_j)(\Delta_j z) + A_k$$

where

$$A_k = \sum_{j=1}^{k-1} \int_0^1 (1-s)\Big\{DX(y(t_j+s\Delta_j t))\circ X(y(t_j+s\Delta_j t)) - $$

$$- DX(y_j)\circ X(y_j)\Big\}\,(\Delta_j z)(\Delta_j z)ds.$$

Consequently:

$$y_k - x_k = A_k + \sum_{j=1}^{k-1} \{M(X)(y_j) - M(X)(x_j)\}\Delta_j z + S_k$$

where

$$S_k = \sum_{j=1}^{k-1} M(X)(x_j)\Delta_j z - \int_a^{t_k} M(X)(x_s)dz_s.$$

By the definition of stochastic integrals, and the main existence theorem, Theorem 3 of Chapter III, $S_k \to 0$ in $\mathcal{L}^2$ as mesh $\Pi \to 0$, uniformly in k. Such a term we call $\varepsilon(\Pi)$.

Next we show that A_k is $\varepsilon(\Pi)$: let K, δ be the constants of condition A(2), and suppose Mesh $\Pi < \delta$: For $0 \leqslant s \leqslant 1$

$$\|y(t_j + s\,\Delta_j t) - y(t_j)\| = \left\|\int_0^s X\Big(y(t_j + r\Delta_j t)\Big)\Delta_j z dr\right\|$$

$$\leqslant B\,\|\Delta_j z\|$$

$$\leqslant B(K\Delta_j t)^{\frac{1}{2}} \qquad (2)$$

where B is a bound for X. Moreover, if β is the constant of Corollary 2 of Chapter III, by that corollary:

$$\|A_k\|^2 \leqslant \beta^2 \sum_{j=1}^{k-1} \left\|\int_0^1 \{DX(y(t_j + s\Delta_j t)) \circ X(y(t_j + s\Delta_j t)) - \right.$$

$$\left. - DX(y_j) \circ X(y_j)\} ds\right\|^2 \Delta_j t$$

$$\leqslant \beta^2 \sum_{j=1}^{k-1} \int_0^1 L\,\|y(t_j + s\Delta_j t) - y_t\|^2 \,ds\,\Delta_j t$$

where L is a Lipschitz constant for $DX \circ X$. Consequently:

$$\|A_k\|^2 \leqslant \beta^2 \sum_{j=1}^{k-1} L\,B^2 K(\Delta_j t)^2$$

$$\leqslant \beta^2 L\,B^2 K\,(b-a)\text{ Mesh }\Pi,$$

showing that A_k is $\varepsilon(\Pi)$.

Next we consider the term $\sum_{j=1}^{k-1} \{M(X)(y_j) - M(X)(x_j)\}\Delta_j z$: let L_1 be a Lipschitz constant for M(X), and set

$$N(s) = \sup_{a \leqslant r \leqslant s} \|y(r) - x(s)\| \qquad a \leqslant s \leqslant b.$$

Then

$$\|\sum_{j=0}^{k-1}\{M(X)(y_j)-M(X)(x_j)\}\Delta_j z\|^2 \leqslant \beta^2 \sum_{j=1}^{k-1} \|M(X)(y_j)-M(X)(x_j)\|^2 \Delta_j t$$

$$\leqslant \beta^2 L_1^2 \sum_{j=1}^{k-1} N(t_j)^2 \Delta_j t$$

$$\leqslant \beta^2 L_1^2 \int_a^{t_k} N(s)^2 ds$$

since N(s) is non-decreasing. Going back to (1) we now have

$$\|y_k - x_k\| \leqslant \|\varepsilon(\Pi)\| + \beta L_1 \surd\left\{\int_a^{t_k} N(s)^2 ds\right\}.$$

On the other hand if $t_k \leqslant t \leqslant t_{k+1}$, then $x(t) - x_k$ is $\varepsilon(\Pi)$ by the uniform continuity of $x:[a,b] \to \mathcal{L}^2(\Omega,F;H)$ while $y(t) - y_k$ is $\varepsilon(\Pi)$ by (2). Thus

$$\|y(t) - x(t)\| \leqslant \|\varepsilon(\Pi)\| + \beta L_1 \surd\left\{\int_a^{t_k} N(s)^2 ds\right\}$$

giving

$$N(t) \leqslant \|\varepsilon(\Pi)\| + \beta L_1 \surd\left\{\int_a^{t} N(s)^2 ds\right\} \qquad a \leqslant t \leqslant b.$$

Whence

$$N(t)^2 \leqslant 2 \|\varepsilon(\Pi)\|^2 + 2\beta^2 L_1^2 \int_a^t N(s)^2 ds.$$

By Gronwall's Lemma, §5 of Chapter I:

$$N(t)^2 \leqslant 2 \|\varepsilon(\Pi)\|^2 + C \int_a^t e^{C(t-s)} \|\varepsilon(\Pi)\|^2 ds \quad a \leqslant t \leqslant b$$

where $C = 2\beta^2 L_1^2$. Thus $N(t) \to 0$ uniformly in [a,b] as required. //

<u>Addendum 4.1</u> <u>Suppose in addition that</u> z <u>is sample continuous and</u> x <u>is regular</u>. <u>Then</u> $\{x_\pi\}_\Pi$ <u>converges uniformly on</u> [a,b] <u>to</u> x <u>in measure: for each</u> $\delta > 0$

$$\mu\{\omega: \sup_{a\leqslant t\leqslant b} |x_\pi(t) - x(t)| > \delta\} \to 0 \text{ as mesh } \Pi \to 0.$$

<u>Proof</u>. From formula (1) of the theorem, with $\Pi = (t_1,\ldots,t_{m+1})$,

$$y_k - x_k = A_k + J_k + S_k$$

where

$$J_k = \sum_{j=1}^{k-1} \{M(X)(y_j) - M(X)(x_j)\}\Delta_j z.$$

By Lemma 5A of Chapter IV

$$\mu\{\omega: \sup_{1\le k\le m} |J_k| > \delta\} \le \frac{1}{\delta} K \sum_{k=1}^{m} \Delta_k t \left(\|u_k\| + \frac{1}{\delta}\|u_k\|^2\right)$$

where

$$\|u_k\| = \|M(X)(y_k) - M(X)(x_k)\| \le L_1 \; \|y_k - x_k\| \to 0$$

uniformly in k as mesh $\Pi \to 0$ by the theorem. Similarly, since $\|y(t_k + s\,\Delta_k t) - y_k\| \to 0$ uniformly in s,k as mesh $\Pi \to 0$, we have $\sup_{1\le k\le m} |A_k| \to 0$ in measure as mesh $\Pi \to 0$.

For S_k, let $S_k(\Pi)$ denote its dependence on Π. From Proposition 5B of Chapter IV we can choose a refinement Π' of Π, $\Pi' = (t_1',\ldots,t_m')$, to make $\sup_{1\le k\le m} |S_k(\Pi')|$ arbitrarily small in measure. However Lemma 5A of Chapter IV, together with the uniform $\mathcal{L}^2$ continuity of x on [a,b], ensures that if mesh Π is small so is $\sup_{1\le k\le m} |S_k(\Pi)-S_{k'}(\Pi')|$ in measure, (where $t_{k'}' = t_k$).

Finally for $\gamma > 0$ set $\sigma_\gamma(z)(\omega) = \sup\{|z(t,\omega)-z(s,\omega)|:|t-s|<\gamma\}$. Then $\sigma_\gamma(z) \to 0$ in measure as $\gamma \to 0$, as does $\sigma_\gamma(x)$. It follows that given $\gamma_1 > 0$, $\gamma_2 > 0$ there exists $\Omega_0 \in F$ and $\varepsilon > 0$ with

$$\mu(\Omega_0) > 1 - \gamma_1$$

and such that for mesh $\Pi < \varepsilon$, $\omega \in \Omega_0$, $k = 1,\ldots,m$:

$$|\Delta_k z| < \gamma_2/B$$

and

$$|x(t) - x_k| < \gamma_2 \qquad t_k \le t \le t_{k+1}.$$

Consequently by the mean value theorem

$$\omega \in \Omega_0,\ \text{mesh}\ \Pi < \varepsilon \Rightarrow |y(t) - y_k| < \gamma_2 \qquad t_k \le t \le t_{k+1}$$

which is enough to finish the proof. //

§5 LOCAL UNIQUENESS

Next we have a key result. The proof is essentially that of (Gikhman & Skorohod 1972). It is annoying that there is not a simpler proof of such an obvious looking statement (especially in the case where P has a Lipschitz extension over all of H, from which the general case easily follows). However in the most important cases for us: when x_1 and x_2 are solutions of suitable integral equations which agree over U, the results will follow using §4 by taking approximating ordinary differential equations.

Theorem 5 Let U be an open subset of H and $P:[a,b)\times U \to \mathbb{P}_q(G;H)$ a continuous bounded map which is Lipschitz on U uniformly in t. Let $x_i:[a,b)\times\Omega\to H$, $i = 1,2$, be admissible processes with first exit times $\tau_i(U):\Omega\to[a,b)$ from U. Assume they both satisfy, a.s.,

$$x_i(t) = x_i(a) + \int_a^t P(s,x_i(s))dz(s) \qquad (t,\omega)\in[a,\tau_i(U))\times\Omega$$

and

$$x_1(a)|\Omega_0 = x_2(a)|\Omega_0$$

for a certain $\Omega_0 \in F_a$. Then

$$\tau_1(U)|\Omega_0 = \tau_2(U)|\Omega_0 \qquad \text{a.s.}$$

and

$$x_1|[a,\tau_1(U))\times\Omega_0 \sim x_2|[a,\tau_2(U))\times\Omega_0.$$

Proof. We can clearly normalize the measure of Ω_0 if $\mu(\Omega_0) > 0$ to assume $\Omega_0 = \Omega$. Set

$$\tau(\omega) = \min\{\tau_1(U)(\omega),\tau_2(U)(\omega)\}$$

and define

$$\alpha:[a,b)\times\Omega\to\{0,1\} \text{ by } \alpha(t)(\omega) = 1 \text{ if and only if } t<\tau(\omega).$$

Write $P(s,x_1(s)(\omega)) - P(s,x_2(s)(\omega)) = Q(s,\omega)$ where $Q:[a,b)\times\Omega\to\mathbb{P}_q(G;H)$. Write $y(t) = x_1(t) - x_2(t)$. Then, by Remark 7A of Chapter III

$$\alpha(t)y(t) = \alpha(t)\int_a^{t\wedge\tau} Q(s)dz(s).$$

Applying Theorem 7A of Chapter III

$$\|\alpha(t)y(t)\| \leqslant \beta\{\int_a^t \|\alpha(s)Q(s)\|^2 ds\}^{\frac{1}{2}}.$$

Since P is Lipschitz on U there is a constant L with

$$|\alpha(s)Q(s)| \leqslant L|\alpha(s)y(s)| \qquad \text{a.s.}$$

Therefore

$$\|\alpha(t)y(t)\| \leqslant 2\beta L\{\int_a^t \|\alpha(s)y(s)\|^2 ds\}^{\frac{1}{2}}.$$

Squaring both sides and applying Gronwall's Lemma yields

$$\alpha(t)y(t) = 0 \qquad \text{a.s.} \quad \text{for } t \in [a,b).$$

Thus $x_1(t) = x_2(t)$ up to the time one of them exists from U. By continuity $x_1(\tau(\omega),\omega) = x_2(\tau(\omega),\omega)$ a.s. when $\tau(\omega) < b$. Therefore $\tau(\omega) = \tau_1(U)(\omega) = \tau_2(U)(\omega)$ a.s. since at these times the processes are in the boundary of U. This completes the proof. //

§6. ISONOMY INVARIANCE

(A) The basic ideas about isonomy were discussed in §4 of Chapter I. We first prove the isonomy invariance of stochastic integration.

Theorem 6A *Suppose* $p^i: T \to \mathcal{L}^0(\Omega,F;\mathbb{P}_q(G;H))$ *and*

$$z^i:[a,b] \to \mathcal{L}^0(\Omega,F;G) \qquad i = 1,2$$

are such that the stochastic integrals

$$I^i(t)(\omega) = \int_a^t p^i(s,\omega)dz^i(s,\omega) \qquad i = 1,2$$

both exist for $a \leqslant t \leqslant b$, *determining* $I^i:[a,b] \to \mathcal{L}^0(\Omega,F;H)$. *Then*

$$\{p^1, z^1\} \dot{\sim} \{p^2, z^2\}$$

implies

$$\{I^1,p^1,z^1\} \dot{\sim} \{I^2,p^2,z^2\}.$$

Proof. For i = 1,2 and Π a belated partition of [a,b] define

$$S^i(\Pi):[a,b] \to \mathcal{L}^0(\Omega,F;H)$$

by

$$S^i(\Pi)(t) = S(\Pi^t;p^i;z^i)$$

where Π^t is the partition induced by Π on $[a,b]$, as in §5B Chapter IV. Then for $\underline{s} = \{s_1,\ldots,s_k\} \subset T$, each $I^i_{\underline{s}}$ is the limit in measure of the Riemann sums $S^i(\Pi)_{\underline{s}}$.

By Lemma 4B of Chapter I, if $\{p^1,z^1\} \mathrel{\dot{\sim}} \{p^2,z^2\}$ we have

$$\{S^1(\Pi),p^1,z^1\} \mathrel{\dot{\sim}} \{S^2(\Pi),p^2,z^2\}.$$

Now convergence in measure implies convergence in distribution (see §6B of Chapter VIII below) and so the finite dimensional joint distributions of $\{I^i,p^i,z^i\}$ are limits of the corresponding finite dimensional distributions of $\{S^i(\Pi),p^i,z^i\}$, for $i = 1,2$. The result follows. //

(B) Next we have isonomy invariance for equations, sometimes called, *uniqueness in law;* see (Watanabe & Yamada 1971).

Theorem 6B Suppose $P:T \times H \to \mathbb{P}_q(G;H)$ together with z^i,u^i, $i = 1,2$, satisfy the conditions of the existence theorem, Theorem 1C, ensuring that there is a solution $x^i \in S([a,b])$ to

$$x^i_t = u^i_t + \int_a^t P(s,x^i_s)dz^i_s \qquad i = 1,2.$$

Assume

$$\{u^1,z^1\} \mathrel{\dot{\sim}} \{u^2,z^2\}.$$

Then

$$\{x^1,u^1,z^1\} \mathrel{\dot{\sim}} \{x^2,u^2,z^2\}.$$

Proof. According to the proof of Theorem 1C we can define

$$\theta^i:S([a,b]) \to S([a,b]) \qquad i = 1,2$$

by the equation

$$\theta^i(x)_t = u^i_t + \int_a^t P(s,x_s)dz^i_s,$$

and then

$$x^i = \lim_{n\to\infty} (\theta^i)^n(x^o)$$

where $x^o \in S([a,b])$ is arbitrary. In particular we can take $x^o \equiv 0$. Then, by induction, using Lemma 4B of Chapter I and Theorem 6A we see that

$$\{(\theta^1)^n(x^o),u^1,z^1\} \sim \{(\theta^2)^n(x^o),u^2,z^2\} \qquad n = 1,2,\dots .$$

The result follows: this time because convergence in $\mathcal{L}^2$ implies convergence in distribution. //

§7 EXAMPLES WITH EXPLICIT SOLUTIONS

(A) The simplest example is the *Ornstein-Uhlenbeck velocity process* $\{v_t:t \geq 0\}$ in $\mathbb{R}^n$. This is defined to be the solution of

$$v_t = v_0 + \gamma B_t - \beta \int_0^t v_s \, ds$$

for constants β and γ, where $v_0:\Omega \to \mathbb{R}^n$ is given and independent of the given n-dimensional Brownian motion $\{B_t:t \geq 0\}$. The solution is

$$v_t = e^{-\beta t} v_0 + \gamma e^{-\beta t} \int_0^t e^{\beta s} \, dB_s \qquad t \geq 0, \tag{1}$$

or equivalently, on integration by parts,

$$v_t = e^{-\beta t} v_0 + \gamma B_t - \gamma\beta \, e^{-\beta t} \int_0^t e^{\beta s} B_s \, ds. \tag{2}$$

The fact that it is a solution is readily verified by direct substitution of (2) into the equation using another integration by parts.

Equation (2) shows that $v_t - e^{-\beta t} v_0$ is linear in B. It follows that v is *Gaussian* and so determined by $\mathbb{E}v_t$ and Cov (v_s,v_t) which are given in Exercise 7 of Chapter V. For the definition and basic facts of Gaussian processes see any of the standard texts, e.g. (Kingman & Taylor 1966), or the 'Stepping Stone' in (DeWitt-Morette & Elworthy 1981).

This process is the derivative of the *Ornstein-Uhlenbeck position process* $\{x_t:t \geq 0\}$:

$$x_t = x_0 + \int_0^t v_s \, ds.$$

Together, we have the system

$$\left.\begin{aligned} v_t &= v_0 + \gamma B_t - \beta\int_0^t v_s\, ds \\ x_t &= x_0 + \int_0^t v_s\, ds \end{aligned}\right\} \qquad (3)$$

with v_0 and x_0 assumed given.

For a discussion of x as a model of physical Brownian motion see (Nelson 1967).

(B) When z is 1-dimensional and sample continuous the solution of the 1-dimensional problem

$$x_t = x_0 + \int_0^t x_s\, dz_s$$

is easily verified, using Itô's formula, to be

$$x_t = x_0 \exp\,(z_t - \tfrac{1}{2}\langle z,z\rangle_t) \qquad (4)$$

using the notation

$$\langle z,z\rangle_t = \int_0^t dz_s\, dz_s.$$

The solution of more general inhomogeneous linear equations can also be written down explicitly in the 1-dimensional case: see (McShane 1974) Chapter V. In higher dimensions there is a *stochastic product integral* representation for solutions of some linear equations, see (Ibero 1976).

Exercise 7B Let $a:[0,\infty) \times \Omega \to \mathbb{R}$ and $A:[0,\infty) \times \Omega \to \mathbb{L}(\mathbb{R}^n;\mathbb{R})$. When $\{B_t : t \geq 0\}$ is an n-dimensional Brownian motion show that under suitable conditions on x_0, a, and A, the equation (for 1-dimensional x):

$$x_t = x_0 + \int_0^t x_s\, a_s\, ds + \int_0^t x_s\, A_s\, dB_s \qquad 0 \leq t < \infty$$

has solution

$$x_t = x_0 \exp\left\{\int_0^t a_s ds + \int_0^t A_s dB_s - \tfrac{1}{2}\int_0^t \operatorname{trace} A_s^* A_s\, ds\right\}.$$

Hint: save work by setting $z_t = \int_0^t a_s ds + \int_0^t A_s dB_s$ and using (4)!

CHAPTER VII : STOCHASTIC DIFFERENTIAL EQUATIONS ON MANIFOLDS

We will define stochastic differential equations on manifolds in a way which will make them look as similar to ordinary differential equations as possible. It may be helpful to keep in mind the fact that ordinary differential equations will be a special case of our stochastic differential equations.

The notation in this chapter is the same as before, but now we assume that we have a filtration $\{F_t : a \leq t < b\}$ with b allowed to be infinite. In particular the σ-algebras F_t each contain all sets of measure zero in F. Throughout G and E will denote separable Banach spaces and H a separable Hilbert space, while

$$z:[a,b) \to \mathcal{L}^0(\Omega;E)$$

is sample continuous and satisfies condition A(q), $q \geq 4$, on each compact subset of [a,b). Some of the terminology from manifold theory and differential geometry is explained in Appendices A and B.

§1 STOCHASTIC DYNAMICAL SYSTEMS

(A) Let M be a separable metrizable C^3 manifold modelled on the Hilbert space H. Let $\underline{E}$ denote the trivial E-bundle, $E \times M \to M$, over M, and let TM be the tangent bundle of M. A *stochastic dynamical system, (S.D.S.)*, (X,z) on M consists of a section X of $\mathbb{L}(\underline{E},TM)$ for some E, together with a stochastic process z with values in E. In particular $X(m) \in \mathbb{L}(E,T_mM)$ for each $m \in M$. We will write $X \in \underset{\sim}{\mathbf{L}}(\underline{E},TM)$.

We shall only consider stochastic dynamical systems with X of class C^1 and having locally Lipschitz first derivatives (in coordinate charts) e.g. X of class C^2, and where z satisfies our standing hypotheses.

Examples 1A (i) Let V be a vector field on M, and define $X \in \tilde{\mathbb{L}}(\underline{R},TM)$ by $X(t,m) = tV(m)$. Define $z:(-\infty,\infty) \times \Omega \to \mathbb{R}$ by $z(t,w) = t$. Then (X,z) can be called an *ordinary dynamical system* or a *drift*.

(ii) Let (X_1,z_1), (X_2,z_2) be stochastic dynamical systems on M with $X_i \in \tilde{\mathbb{L}}(\underline{E}_i,TM)$, i = 1,2., and with z_1,z_2 having the same domain. Define $X \in \tilde{\mathbb{L}}(\underline{E}_1 \times \underline{E}_2,TM)$ by $X((v_1,v_2),m) = X_1(v_1,m) + X_2(v_2,m)$, and define z by $z(t,\omega) = (z_1(t,\omega), z_2(t,\omega))$. The resulting S.D.S. (X,z) will be called the *direct sum* of (X_1,z_1) and (X_2,z_2) and written $(X_1 \oplus X_2, z_1 \oplus z_2)$.

We shall be mainly interested in direct sums $(X_1 \oplus X_2, z_1 \oplus z_2)$ where $E_1 = R^n$, z_1 is a Brownian motion and (X_2,z_2) is a drift.

(iii) Let N^n be a finite dimensional Riemannian manifold with orthonormal frame bundle $\pi:O(N) \to N$. An element $u \in O(N)$ can be considered as an isometry $u:R^n \to T_{\pi(u)}N$. The Levi-Civita connection of N determines a splitting of the tangent bundle to O(N),

$$TO(N) = VTO(N) \oplus HTO(N)$$

into vertical and horizontal components: $VTO(N) = \ker T\pi$ and $T\pi|HTO(N)$ is an isomorphism on each fibre (see §1 Appendix B). Define $X \in \tilde{\mathbb{L}}(\underline{R}^n,TO(N))$ to be the trivialization of HTO(N) given by

$$X(u)e = (T\pi|HTO(N))^{-1}u(e),$$

(in other words X is the natural inverse of the canonical 1-form of O(N)). If we take z to be a Brownian motion on R^n then (X,z) will be called the *canonical S.D.S.* of the Riemannian manifold N, although it is an S.D.S. on O(N).

(iv) Let M be a submanifold of E, where E is now assumed to be a Hilbert space and define $X \in \tilde{\mathbb{L}}(\underline{E},TM)$ so that $X(m):E \to T_mM$ is the orthogonal projection at each point m of M.

(v) Let M be a Hilbert Lie group. Let $E = T_1M$, the tangent space at the identity of M and define $X \in \tilde{\mathbb{L}}(\underline{E},TM)$ by left translation (see Appendix A). An E-valued process z then determines a *left invariant* S.D.S. (X,z) on M.

(B) When (X,z) is a stochastic dynamical system on the Hilbert space H itself we will consider X as a map $X:H \to \mathbb{L}(E;H)$ and follow the notation of Chapter VI §4 to define $M(X):H \to \mathbb{P}_2(E;H)$ by $M(X)(x)e = X(x)e + \frac{1}{2}DX(x)(X(x)e)e$, remembering that $DX(x) \in \mathbb{L}(H;\mathbb{L}(E;H))$.

A *regular localization* (r.l.) for X at the point m of M is a triple $\Lambda = ((U,\phi),U_0,\lambda)$ where:

(i) (U,ϕ) is a C^2 chart about m, with $\phi(U)$ a bounded open subset of H, $\phi(U) = W$ say.

(ii) U_0 is an open neighbourhood of m in U with closure $\overline{\phi(U_0)} \subset W$. Set $W_0 = \phi(U_0)$.

(iii) $\lambda:H \to [0,1]$ is C^2 with supp $\lambda \subset W$ and $\lambda|W_0 \equiv 1$.

(iv) If $X_\Lambda:H \to \mathbb{L}(E;H)$ is defined by $X_\Lambda(h) = \lambda(h)\,\phi_*(X)(h)$, where $\phi_*(X)(h) = T_{\phi^{-1}(h)}\phi \circ X(\phi^{-1}(h))$, then $M(X_\Lambda):H \to \mathbb{P}_2(E;H)$ is globally Lipschitz.

Clearly a regular localization exists for each point of M, although we do have to use the theorem on the existence of smooth partitions of unity on a Hilbert space to obtain a suitable λ, see (Lang 1962), (however the insistence that λ is C^2 is only for technical simplification).

A family $\{\Lambda^\alpha: \alpha \in A\}$ of regular localizations will be said to *cover* a subset A of M if $A \subset \cup\{U_0^\alpha: \alpha \in A\}$.

Suppose $x:[a,\xi) \times \Omega \to M$ is admissible and $\Lambda = ((U,\phi),U_0,\lambda)$ is an r.l. for X. Then for $t_0 \in [a,b)$ set

$$\Omega_{t_0}^\Lambda = \{\omega \in \Omega: t_0 < \xi(\omega) \ \&\ x(\omega,t_0) \in U_0\},$$

so $\Omega_{t_0}^\Lambda \in F_{t_0}$.

Let $x_{t_0}^\Lambda:[t_0,b) \times \Omega_{t_0}^\Lambda \to H$ be the uniquely defined regular solution to the stochastic integral equation

$$y(t) = \phi \circ x(t_0) + \int_{t_0}^t M(X_\Lambda)(y)dz \qquad t \geq t_0$$

and let $\tau_{t_0}^\Lambda:\Omega_{t_0}^\Lambda \to [a,b]$ be its exit time from W_0. Finally let $\tau(U_0,t_0)$ be the first exit time of $x|[t_0,\xi) \times \Omega_{t_0}^\Lambda$ from U_0.

Let (X,z) be an S.D.S. on M, and Λ an r.l. for X. Suppose $x:[a,\xi) \times \Omega \to M$ is admissible. Then we say Λ *affirms* x (as a solution of the stochastic differential equation $dx = Xdz$) if for each $t_0 \in [a,b)$

$$\theta \circ x|[t_0,\tau(U_0,t_0)\wedge\xi) \times \Omega^{\Lambda}_{t_0} \sim x^{\Lambda}_{t_0}|[t_0,\tau(U_0,t_0)\wedge\xi) \times \Omega^{\Lambda}_{t_0}$$

where $\theta:M \to H$ is some, not necessarily continuous, extension of ϕ.

(C) Let $x:[a,\xi) \times \Omega \to M$ be admissible. Then x is said to be a *locally regular solution* (or just a *solution*) of the *stochastic differential equation* $dx = X\,dz$ if there exists a cover of M by local regularizations for X each of which affirms x.

When $M = H$ the above definition incorporates the main principle of McShane's work see (McShane 1976): a stochastic differential equation $dx = X\,dz$ must be interpreted as the integral equation

$$x_t = x_a + \int_a^t M(X)(x_s)dz_s \tag{M}$$

rather than as

$$x_t = x_a + \int_a^t X(x_s)dz_s. \tag{I}$$

Note that the difference between these equations is the term

$$\frac{1}{2}\int_a^t DX(x_s)(X(x_s)dz_s)dz_s$$

which is just the Stratonovich correction term (see §2 of Chapter V) and a regular solution to (M) will satisfy

$$x_t = x_a + \int_a^t X(x_s) \circ dz_s. \tag{S}$$

In fact most stochastic analysts would write

$$dx = X \circ dz$$

where we write $dx = X\,dz$ and keep our notation (with $M = H$) to mean (I). However we shall rarely use Itô equations like (I) and for these rare occasions, if we need a differential notation it will be

$$(\text{Itô})\quad dx = X\,dz.$$

There are three reasons for choosing this McShane/Stratonovich interpretation: it reduces to the standard definition when (X,z) is an ordinary dynamical system (or when z has smooth sample paths), it is invariantly defined, and it behaves well under the process of approximating z by other processes (c.f. Chapter IV §4).

The sort of approach we are using seems to have first appeared in (Clark 1973), but much of the track had already been laid by Itô (1950). Among articles in the intervening years are (Itô 1963, Dynkin 1968, and Daletskii & Shnaiderman 1969). However it was not until it was realized that the Stratonovich style interpretation was the one to use that it became clear how to combine the geometry with the analysis. This interpretation was certainly widespread in 1974 e.g. (Dowker 1975, Schulman 1975, and Elworthy 1975). The definition in the form just given appeared in (Eells & Elworthy 1976).

Before moving forward we should note that, with M = H and an S.D.S. (X,z), we have two notions: that of a solution to the integral equation $x(t) = x(s) + \int_a^t M(X)dz$ and that of a solution to the stochastic differential equation dx = Xdz with given initial condition. The first notion is defined globally, and clearly a regular solution to the integral equation is a solution to the differential equation. Nevertheless for general X, when the solutions to the integral equation may not exist for all time the converse does not seem so immediate and is left until §9D. In that section we also show the equivalence of our definition of a solution to (X,z) with that used by some other authors e.g. (Ikeda & Watanabe 1981).

(C) Our first lemmas are concerned with invariance properties.

Lemma 1D Let M = H with M(X) globally Lipschitz. Suppose $x:[a,b) \times \Omega \to H$ is a regular solution to the stochastic integral equation

$$x_t = x_a + \int_a^t M(X)(x_s)dz_s.$$

Then every regular localization Λ of X affirms x. In particular x is a solution of dx = X dz.

Proof. Let $\theta:H \to H$ be a C^2 extension of $\phi|\bar{U}_o$. (Such a θ can be obtained by using a suitable bump function on H, (Lang 1962).) Then

$$\theta_*(X)|W_o = X_\Lambda|W_o$$

whence

$$M(\theta_*X)|W_o = M(X_\Lambda)|W_o.$$

By the Itô formula, Theorem 3 of Chapter V,

$$\theta(x_t) = \theta(x_a) + \int_a^t D\theta(x)X(x)dz + \tfrac{1}{2}\int_a^t D\theta(x)DX(x)(X(x)dz)dz$$
$$+ \tfrac{1}{2}\int_a^t D^2\theta(x)(X(x)dz,X(x)dz) \qquad (1)$$

On the other hand, if $y \in U_0$, writing $\theta(y) = \tilde{y} \in W_0$ and $\theta_*X(\tilde{y}) = \tilde{X}(\tilde{y})$; for $e \in E$

$$\begin{aligned} M(\tilde{X})(\tilde{y})e &= \tilde{X}(\tilde{y})e + \tfrac{1}{2} D_y\Big(\tilde{X}(\tilde{y})\Big)(\tilde{X}(\tilde{y})e)(e) \\ &= \tilde{X}(\tilde{y})e + \tfrac{1}{2} D_y\Big(D\theta(\theta^{-1}\tilde{y})X(\theta^{-1}\tilde{y})\Big)(\tilde{X}(\tilde{y})e)(e) \\ &= \tilde{X}(\tilde{y})e + \tfrac{1}{2}D^2\theta(y)\Big(D\theta^{-1}(\tilde{y})\tilde{X}(\tilde{y})e\Big)X(y)e \\ &= \tfrac{1}{2} D\theta(y)DX(y)\Big(D\theta^{-1}(\tilde{y})(\tilde{X}(\tilde{y})e)\Big)e \\ &= D\theta(y)X(y)e + \tfrac{1}{2}D\theta(y)DX(y)(X(y)e)e \\ &+ \tfrac{1}{2} D^2\theta(y)(X(y)e,X(y)e). \end{aligned} \qquad (2)$$

Thus, for each $t_0 \in [a,b)$, the restriction $\theta o x|[t_0,\tau(U_0,t_0)) \times \Omega^{\Lambda}_{t_0}$ satisfies the stochastic integral equation

$$y(t) = \theta o x(t_0) + \int_{t_0}^t M(X_\Lambda)(y(s))dz(s).$$

Since $\theta o x$ is admissible and its first exit time from W_0 is $\tau(U_0,t_0)$ we can apply Theorem 5 of Chapter VI to conclude that it is equivalent to $x^{\Lambda}_{t_0}$ on $[t_0,\tau(U_0,t_0)) \times \Omega^{\Lambda}_{t_0}$. //

In the above proof we could have avoided the computation (2) of $M(\tilde{X})(\tilde{y})e$ by writing our equations in terms of Stratonovich integrals e.g. replacing $\int_a^t M(X)(x_s)dz_s$ by $\int_a^t X(x_s) \circ dz_s$. The change of variable then involves no second derivatives.

(E) Lemma 1E Let (X,z) be an S.D.S. on M as usual and $\Lambda = ((U,\phi), U_0,\lambda), \Lambda' = ((U',\phi'), U'_0,\lambda')$ regular localizations of X with

$$U' \subset U \quad U'_0 \subset U_0, \text{ and } \lambda o \phi | \operatorname{supp} \lambda' o \phi' \equiv 1.$$

Suppose Λ affirms the admissible process $x:[a,\xi) \times \Omega \to M$ as a solution of $dx = Xdz$; then so does Λ'.

Proof. If Λ affirms x so does $((U,\phi),U_0',\lambda)$. We can therefore assume $U_0' = U_0$. Factorize ϕ' by $\phi' = \psi o \phi$ for $\psi:\phi(U') \to W'$ where $W' = \phi'(U')$. Then $\tilde{\Lambda} = ((\phi(U'),\psi), W_0,\lambda')$ in an r.1. of X_Λ. In fact $(X_\Lambda)^{\sim}_{\tilde{\Lambda}} = X_{\Lambda'}$ because $\lambda' \cdot (\lambda o \psi^{-1}) = \lambda'$. By Lemma 1D, $\tilde{\Lambda}$ affirms $x^{\Lambda}_{t_0}$ for each $t_0 \in [a,b)$. Thus

$$\phi' o x|[t_0,\xi \wedge \tau(U_0,\tau_0)) \times \Omega^{\Lambda}_{t_0} \sim \psi o x^{\Lambda}_{t_0}|[t_0, \xi \wedge \tau^{\Lambda}_{t_0}) \times \Omega^{\Lambda}_{t_0}$$

$$\sim (x^{\Lambda}_{t_0})^{\tilde{\lambda}}_{t_0}|[t_0, \xi \wedge \tau^{\tilde{\Lambda}}_{t_0}) \times \Omega^{\Lambda}_{t_0}$$

$$= x^{\Lambda'}_{t_0}|[t_0, \xi \wedge \tau^{\Lambda'}_{t_0}) \times \Omega^{\Lambda'}_{t_0},$$

as required. //

(F) Theorem 1F Let (X,z) be an S.D.S. on M and $x:[a,\xi)\times\Omega\to M$ a locally regular solution of $dx = Xdz$. Suppose $h:M \to N$ is a C^3 diffeomorphism of M onto a manifold N. Then $y = hox$ is a locally regular solution to $dy = Ydz$ where $Y = h_*X$.

Proof. Immediate from the definitions. //

§2 UNIQUENESS AND EXISTENCE OF SOLUTIONS

(A) Lemma 2A For $i = 1,2$ let $x_1:[a,\xi_1) \times \Omega \to M$ be locally regular solutions of $dx = Xdz$ with $x_1(a) = x_2(a)$, almost everywhere. Then

$$x_1|[a,\xi_1 \wedge \xi_2) \times \Omega \sim x_2|[a,\xi_1 \wedge \xi_2) \times \Omega.$$

Moreover if $t \in (a,b)$ and $\xi_1(\omega) < t < \xi_2(\omega)$ then almost surely

$$x_2(\omega)|[a,\xi_1(\omega)) = x_1(\omega).$$

Proof. Set $\xi = \xi_1 \wedge \xi_2$ and $\Omega_t = \{\omega:t < \xi(\omega)\}$. Choose any countable dense subset T' of $[a,b)$ with $a \in T'$.

By Lemma 1E there is a cover of M by localy regularizations $\{\Lambda^j\}_{j=1}^{\infty}$ of X which affirm both x_1 and x_2. Set

$$E(t) = \{\omega \in \Omega_t : x_1(t,\omega) = x_2(t,\omega)\}$$

and

$$\Lambda^j[t_0,t_1] = \{\omega \in E(t_0) : x_1(t_0,\omega) \in U_0^j \ \&$$
$$t_1 < \xi \wedge \tau_1(U_0^j,t_0) \wedge \tau_2(U_0^j,t_0)\}$$

where τ_1,τ_2 denote exit times for x_1, x_2. Thus $E(t) \in F_t$ and $\Lambda^j[t_0,t_1] \in F_{t_1}$.

For $j = 1,2,\ldots$ let $\theta^j:M \to H$ be extensions of ϕ^j as usual. Then by hypothesis there exists $Z^j[t_0,t_1] \in F_{t_1}$ of measure zero such that if $\omega \in \Lambda^j[t_0,t_1] - Z^j[t_0,t_1]$ then

$$\theta^j x_1(\omega)|[t_0,t_1] = \tilde{x}_j(\omega)|[t_0,t_1] = \theta^j x_2(\omega)|[t_0,t_1]$$

where $\tilde{x}_j:[t_0,b) \times E(t_0) \to H$ is a regular solution of

$$y(t) = x_1(t_0) + \int_{t_0}^{t} M(X_{\Lambda^j}(y))dz.$$

Suppose $\omega \in \Omega_t$ and for $i = 1,2$, $x_i(\omega)|[a,t]$ is continuous. Let $s \in [a,t)$ and assume $x_1(\omega)\,[a,s) = x_2(\omega)|[a,s)$. Then there exists j with $\omega \in \Lambda^j[t_0,t_1]$ for some $t_0,t_1 \in T'$ satisfying $a \leqslant t_0 \leqslant s < t_1 \leqslant t$. Then if $\omega \notin Z^j[t_0,t_1]$ we have

$$x_1(\omega)|[a,t_1] = x_2(\omega)|[a,t_1].$$

Noting that t_1 is strictly greater than s and using the sample continuity and the fact that $\cup\{Z^j[t_0,t_1]: j = 1 \text{ to } \infty,\ t_0 < t_1 \leqslant t \ \&\ t_0,t_1 \in T\}$ has measure zero we see that $x_1(\omega)|[a,t] = x_2(\omega)|[a,t]$ almost surely for $\omega \in \Omega_t$, as required.

For the last part set $D_t = \{\omega:\xi_1(\omega) \leqslant t < \xi_2(\omega)\}$. Then

$$D_t = \cup\{D_t \cap \Omega_{t'} : t' \leqslant t \ \&\ t' \in T'\}.$$

The result follows since we have shown that if $t' \in T'$ with $t' \leqslant t$ then, almost surely for $\omega \in \Omega_{t'}$ we have $x_1(\omega)|[a,t'] = x_2(\omega)\,[a,t']$. //

(B) <u>Lemma 2B</u> <u>Let</u> $x_1:[a,\xi_1)\times \Omega \to M$ <u>and</u> $x_2:[c,\xi_2) \times \tilde{\Omega} \to M$ <u>be locally regular solutions of</u> $dx = Xdz$ <u>such that</u> $x_2(c,\omega) = x_1(c,\omega)$ <u>almost</u>

surely on $\tilde{\Omega}$, where $c \in (a,b)$, $\tilde{\Omega} \in F_t$ and $\tilde{\Omega} \subset \Omega_c^1 = \{\omega : c < \xi_1(\omega)\}$.

Then there exists a locally regular solution $x:[a,\xi) \times \Omega \to M$ where

$$\xi(\omega) = \begin{cases} \max \{\xi_1(\omega), \xi_2(\omega)\} & \text{if } \omega \in \tilde{\Omega} \\ \xi_1(\omega) & \text{otherwise.} \end{cases}$$

Proof. Set $\Omega_t^2 = \{\omega \in \tilde{\Omega} : t < \xi_2(\omega)\}$. Then $\Omega_t = \Omega_t^1 \cup (\Omega_t^2 - \Omega_t^1) \cap \tilde{\Omega}$. Note that for $a \leqslant t \leqslant c$, $\Omega_t = \Omega_t^1$ since $\tilde{\Omega} \subset \Omega_c^1$. Define $x:[a,\xi) \times \Omega \to M$ by

$$x(t,\omega) = \begin{cases} x_1(t,\omega) & \text{if } \omega \in \Omega_t^1 \\ x_2(t,\omega) & \text{if } \omega \in \Omega_t^2 - \Omega_t^1 \end{cases} .$$

Then, for $\omega \in \Omega_t$,

$$\begin{array}{ll} \text{if } a \leqslant t \leqslant c & x(\omega)|[a,t] = x_1(\omega)|[a,t] \\ \text{if } c \leqslant t < b \text{ and } \omega \in \Omega_t^1 & x(\omega)|[c,t] = x_1(\omega)|[c,t] \\ \text{if } c \leqslant t < b \text{ and } \omega \notin \Omega_t^1 & x(\omega)|[c,t] = x_2(\omega)|[c,t] \end{array}$$

almost surely; the last assertion by Lemma 2A. Thus x is an admissible process.

To check that x is a solution let Λ be a regular localization which affirms both x_1, x_2 (recall that Lemma 1E ensures that M is covered by such Λ). For $a \leqslant t_0 < t < b$ set

$$C[t_0,t] = \{\omega \in \Omega_{t_0}^{\Lambda} : t < \xi(\omega) \wedge \tau(U_0,t_0)\}.$$

Then $C[t_0,t]$ is the disjoint union

$$C[t_0,t] = C_1[t_0,t] \cup C_2[t_0,t]$$

where

$$C_1[t_0,t] = \Omega_t^1 \cap C[t_0,t]$$

$$C_2[t_0,t] = (\Omega_t^2 - \Omega_t^1) \cap C[t_0,t].$$

But if $\omega \in C_1[t_0,t]$ then $x(\omega)|[t_0,t] = x_1(\omega)|[t_0,t]$ and if $\omega \in C_2[t_0,t]$ then $x(\omega)|[t_0,t] = (x_1(\omega)|[t_0,c]) \cup (x_2(\omega)|[c,t])$ almost surely. The result follows from the local uniqueness theorem, Theorem 5 of Chapter VI. //

Remark 2B The corresponding result holds in the same way with $c = a$ and $\tilde{\Omega} \in F_a$.

(C) Proposition 2C If $x:[a,\xi) \times \Omega \to M$ is a locally regular solution of $dx = Xdz$ then every regular localization of X affirms x.

Proof. Let Λ be a regular localization of X and suppose $t_o \in [a,b)$. Set

$$\xi' = \tau(U_o,t_o) \wedge \xi\colon \Omega_{t_o}^{\Lambda} \to [a,b]$$

and

$$x' = \theta \circ x|[t_o,\xi') \times \Omega_{t_o}^{\Lambda} \to W_o.$$

Then, using Lemma 1E, we see that x' is a locally regular solution to $dx' = (X_\Lambda|U_o)dz$. However, by Lemma 1D, so is $x_{t_o}^{\Lambda}|[t_o, \tau_{t_o}^{\Lambda}) \times \Omega_{t_o}^{\Lambda}$. Lemma 2A therefore implies that x' and $x_{t_o}^{\Lambda}$ are equivalent when restricted to $[t_o,\tau_{t_o}^{\Lambda} \wedge \xi') \times \Omega_{t_o}^{\Lambda}$. Now for almost all ω in $\Omega_{t_o}^{\Lambda}$ with $\tau(U_o,t_o)(\omega) < \xi(\omega)$ we have $\theta \circ x(\tau(U_o,t_o)(\omega),\omega) \in \partial W_o$, the boundary of W_o.

Using this and applying the last part of Lemma 2A to $x_{t_o}^{\Lambda}$ and x' we see that $\xi' \prec \tau_{t_o}^{\Lambda}$ almost surely, and so x is affirmed by Λ. //

(D) The next proposition will be used to inject solutions of the local integral equations on H into our manifold M.

Proposition 2D (Open embeddings) Let (X,z) and (Y,z) be stochastic dynamical systems on the Hilbert manifolds M and N respectively. Suppose $h:N \to M$ is a C^3 diffeomorphism of N onto an open subset of M such that $h_*Y = X|h(N)$. Then if $y:[a,\xi) \times \Omega \to N$ is a locally regular solution of $dy = Ydz$, the process $x = h \circ y : [a,\xi) \times \Omega \to M$ is a locally regular solution of $dx = Xdz$.

Proof. By Theorem 1F we can assume that N is an open subset of M and h is the inclusion, so that $Y = X|N$. The only problem is with regular localizations of X around points on the topological boundary ∂M of N in M. To deal with these choose an ascending sequence of open subsets $\{N_i\}_{i=1}^{\infty}$ of M with

$$\bar{N}_1 \subset N_{i+1} \subset \ldots \subset N = \bigcup_i N_i .$$

Let τ_i be the first exit time of x from N_i and set $\xi_i = \tau_i \wedge \xi$, with $x_i = |[a,\xi_i) \times \Omega$. Then each x_i is a l.r. solution to dx = X dz since it is affirmed by all regular localizations whose charts lie in $N - \bar{N}_i$ or in N_{i+1}.

Let Λ be a regular localization for X. Suppose $t_o \in [a,b)$. For $\theta:M \to H$ as usual, we know by the previous proposition that for i = 1,2,...

$$\theta o x_i|[t_o,\xi_i \wedge \tau(U_o,t_o)) \times \Omega^{\Lambda}_{t_o} \sim x^{\Lambda}_{t_o}|[t_o,\xi_i \wedge \tau(U_o,t_o)) \times \Omega^{\Lambda}_{t_o}$$

where $\tau(U_o,t_o)$, $x^{\Lambda}_{t_o}$ and $\Omega^{\Lambda}_{t_o}$ are defined in terms of x. In other words $\theta o x$ and $x^{\Lambda}_{t_o}$ are equivalent when restricted to $[t_o,\xi_i \wedge \tau(U_o,t_o)) \times \Omega^{\Lambda}_{t_o}$ for any i = 1,2,... . Since $\xi = \sup_i \xi_i$ it follows (e.g. by the uniqueness part of Lemma 6B of Chapter III) that the processes are equivalent when restricted to $[t_o,\xi \wedge \tau(U_o,t_o)) \times \Omega^{\Lambda}_{t_o}$. This means that Λ affirms x. //

(E) The following is our main existence and uniqueness theorem.

<u>Theorem 2E</u> <u>Let</u> $x_a:\Omega \to M$ <u>be</u> F_a<u>-measurable</u>. <u>Then there exists a locally regular solution</u> $x: [a,\xi) \times \Omega \to M$ <u>of</u> dx = Xdz <u>with</u> $x(a) = x_a$ a.e. <u>such that if</u> $x_1:[a,\xi_1) \times \Omega \to M$ <u>is any other locally regular solution with</u> $x_1(a) = x_a$ a.e. <u>then</u> $\xi_1 \leq \xi$ a.e. <u>and</u> $x|[a,\xi_1) \times \Omega \sim x_1$.

<u>Moreover</u> $\xi > a$ <u>almost everywhere</u>.

<u>Proof</u>. Let $A \subset \mathcal{L}^o(\Omega,F;R^*)$ be the set of all ξ_α where $x_\alpha:[a,\xi_\alpha) \times \Omega \to M$ is a locally regular solution of dx = Xdz with $x_\alpha(a) = x_a$ almost everywhere. Choose a 'least upper bound' ξ of A as given by Lemma 6A of Chapter III, and an admissible map $x:[a,\xi) \times \Omega \to M$ as in Lemma 6B of Chapter III, with

$$x(t,\omega) = x_\alpha(t,\omega) \qquad \text{a.s. on } \{\omega: t < \alpha(\omega)\}$$

for each $t \in [a,b)$ and each $\alpha \in A$. Since the uniqueness is immediate from the definition of x it is enough to check (i) that x is a l.r. solution, and (ii) that $\xi > a$ almost surely.

For (i) suppose $t_0 \in [a,b)$ and Λ is a regular localization of X. Then $\phi^{-1} x^{\Lambda}_{t_0} | [t_0, \tau^{\Lambda}_{t_0}) \times \Omega^{\Lambda}_{t_0}$ is a l.r. solution by the open embedding result, Proposition 2D, (for $t_0 = a$ we can take $\Omega^{\Lambda}_a = \{\omega : x_a(\omega) \in U_0\}$). Restricting our attention to ω in $\Omega^{\Lambda}_{t_0}$, it follows from Lemma 2B, and Remark 2B, that there exists α in A with $\alpha(\omega) \geq \tau^{\Lambda}_{t_0}(\omega)$ a.s. But x_α is affirmed by Λ and so $\tau^{\Lambda}_{t_0}(\omega) = \tau(U_0, t_0)(\omega)$ a.s., whence $\alpha(\omega) \geq \tau(U_0, t_0)(\omega)$ almost surely. From this we have

$$x | [t_0, \xi \wedge \tau(U_0, t_0)) \times \Omega^{\Lambda}_{t_0} \sim x_\alpha | [t_0, \alpha \wedge \tau(U_0, t_0)) \times \Omega^{\Lambda}_{t_0} .$$

Therefore Λ affirms x as well.

To show $\xi > a$ a.s. take a countable cover of M by regular localizations $\{\Lambda^i\}_{i=1}^{\infty}$ with $\Lambda^i = ((U^i, \phi^i), U^i_o, \lambda^i)$. Taking $\Omega^i_a = \{\omega : x_a(\omega) \in U^i_o\}$ we have $\Omega = \bigcup_i \Omega^i_a$. Using the open embedding result as above we see that for each i there is an α_i in A with $\alpha_i(\omega) \geq \tau^{\Lambda^i}_a(\omega) > a$ almost surely on Ω^i_a. The result follows. //

A solution $x : [a,\xi) \times \Omega \to M$ as in the statement of Theorem 2E will be called a *maximal solution*, and the corresponding stopping time ξ will be called the *explosion time* of the S.D.S. (X,z) on M. The S.D.S. is *complete* or *conservative* or *non-explosive* if we can take $\xi(\omega) = b$ all ω. From the proof of theorem we have:

<u>Remark 2E</u> <u>If</u> $x : [a,\xi) \times \Omega \to M$ <u>is a maximal solution and</u> Λ <u>a regular localization then for each</u> $t_0 \in [a,b)$ <u>we have</u>

$$\xi | \Omega^{\Lambda}_t \geq \tau^{\Lambda}_t \qquad \text{a.s.} \qquad //$$

(F) As mentioned in §5 of Chapter V it will often be convenient to choose the filtration $\{F_t\}$ to take account of a given initial function x_a. In particular set

$$G^a = \sigma\{z(t) - z(a) : t \geq a\}$$

and let G_t^a be the σ-algebra generated by $z(s) - z(a) : a \leq s \leq t$ together with all sets of measure zero in G^a. Then, if $x_a : \Omega \to M$ is independent of G^a we can take

$$F_t = G_t^a \vee \sigma\{u\}$$

and replace $z(t)$ by $z'(t) = z(t) - z(a)$. Solutions of (X,z'), for $t \geq a$, are then solutions of (X,z) and because of the independence, by Lemma 5A of Chapter V, z' satisfies our standing conditions for $\{F_t : t \geq a\}$ provided z did for $\{\sigma\{z(s) : a \leq s \leq t\} : t \geq a\}$.

§3 SUBMANIFOLDS

The following theorem will be used frequently. It will often enable us to replace a problem about an S.D.S. on a manifold by one about a stochastic integral equation on a linear space.

Theorem 3 Let $i : N \to M$ be a C^3 embedding of a manifold N into M. Suppose the S.D.S. (X,z) on M has $X|i(N)$ tangent to $i(N)$, thereby inducing an S.D.S. (Y,z) on N with $i_*(Y) = X|i(N)$.

(i) Let y be an l.r. solution to $dy = Ydz$, then $x = i \circ y$ is an l.r. solution to $dx = Xdz$.

(ii) Conversely if i is a closed embedding any l.r. solution $x : [a,\xi) \times \Omega \to M$ of $dx = Xdz$ with $x(a)(\Omega) \subset i(N)$ is equivalent to one of the form $i \circ y$ where $y : [a,\xi) \times \Omega \to N$ is an l.r. solution of $dy = Ydz$.

Proof. Assume first that i is a closed embedding. Suppose $y : [a,\eta) \times \Omega \to N$, and $x : [a,\xi) \times \Omega \to M$ are maximal solutions to $dy = Ydz$ and $dx = Xdz$ with $x(a) = i \circ y(a)$. It suffices, for both (i) and (ii) to prove that $x \sim i \circ y$.

Cover $i(N)$ by a countable family A of regular localizations of X of the form

$$\Lambda = ((U,\phi), U_0, \lambda)$$

where

$$\phi(U) = W = W' \times W'' \subset H' \times H'' = H$$

$$\phi(U_0) = W_0 = W_0' \times W_0'' \subset H' \times H''$$

$$\phi(iN \cap U) = W' \times \{0\}$$

with H' × H" a splitting of H.

Then Λ restricts to a regular localization $\Lambda|N$ of $X|N$. Moreover

$$M(X_\Lambda)|H' = M((X|N)_{\Lambda|N})$$

since

$$M(X_\Lambda) = X_\Lambda + DX_\Lambda \circ X_\Lambda$$

and the last term in the expression is a differentiation in the H' direction at each point of H'.

It follows that every regular solution to the stochastic integral equation

$$y(t) = y(t_0) + \int_{t_0}^{t} M((X|N)_{\Lambda|N})\, dz$$

on H' is a regular solution to

$$x(t) = y(t_0) + \int_{t_0}^{t} M(X_\Lambda) dz.$$

From this we see $i\circ y$ is an l.r. solution of $dx = Xdz$ (using the fact that $i(N)$ is closed to get affirmations covering $M - i(N)$), so by the maximality of x:

$$\eta \prec \xi \text{ and } x|[a,\eta) \times \Omega \sim y.$$

To show that $\eta = \xi$ almost surely let $B = \{\omega \in \Omega : \eta(\omega) < \xi(\omega)\}$. Almost surely $x(-,\omega)$ is continuous on $[a,\xi(\omega))$ and so by the equivalence with y, if $\omega \in B$ then, a.s., $x(\eta(\omega),\omega) \in N$; in which case for some $\Lambda \in A$, $x(\eta(\omega),\omega) \in U_0 \cap N$. Then for a fixed countable dense $T' \subset T$ with $a \in T'$, there exists $t_0 \in T'$ with $x(t_0,\omega) \in U_0 \cap N$ and $t_0 < \eta(\omega) < \tau(U_0,t_0)(\omega)$ But except possibly for a measure zero set depending only on Λ and t_0 we have, by Remark 2E and the comments above, that if $\omega \in \Omega_{t_0}^{\Lambda|N}$

$$\tau(U_0,t_0)(\omega) = \tau_{t_0}^{\Lambda}(\omega) = \tau_{t_0}^{\Lambda|N}(\omega) \prec \eta(\omega).$$

Thus B has measure zero and $\eta = \xi$ a.e. as required.

It remains to treat the case where $i(N)$ is not closed in M. The problem here is again that of regular localizations about points in $\overline{i(N)} - i(N)$. However since embeddings are locally closed maps we can write $N = \bigcup_i N_i$ where the N_i are open in N, $\bar{N}_i \subset N_{i+1}$, and $i(\bar{N}_i)$ is closed in M. If we let τ_i be the first exit time of y from N_i and set $y_i = y|[a,\xi \wedge \tau_i) \times \Omega \to N$, as in the proof of Proposition 2D we can see that each $i\circ y_i$ satisfies dx = Xdz and then deduce that $i\circ y$ does also. //

Examples 3 (i) c.f. (Itô 1950). Define $X:\mathbb{R}^n - \{0\} \to \mathbb{L}(\mathbb{R}^n;\mathbb{R}^n)$ by

$$X(x)e = e - \langle x,e\rangle\, |x|^{-2}x$$

where $\langle , \rangle$ is the inner product of $\mathbb{R}^n$. Then $\langle x,X(x)e\rangle = 0$ for all $x \neq 0$ and $e \in \mathbb{R}^n$, so that X(x)e is always tangent to the sphere $S^{n-1}(0,|x|)$ about 0 radius $|x|$. Thus any solution to

$$dx_t = X(x_t)dz_t$$

where z is $\mathbb{R}^n$-valued, if it starts on such a sphere stays on it.

Note that the integral equation here is

$$x_t = x_0 + \int_0^t \{dz_s - \langle x_s,dz_s\rangle |x_s|^{-2}\, x_s\}$$

$$+ \tfrac{1}{2} \int_0^t \{2|x_s|^{-4}x_s\langle x_s,dz_s\rangle^2 - |x_s|^{-2}x_s\langle dz_s,dz_2\rangle -$$

$$- \langle x_s,dz_s\rangle |x|^{-2}\, dz_s\}.$$

When z is a Brownian motion this reduces to

$$x_t = x_0 + \int_0^t \{dz_s - \langle x_s,dz_s\rangle |x_s|^{-2}x_s\} - \tfrac{1}{2}(n-1) \int_0^t |x_s|^{-2}x_s\, ds,$$

or in Itô differential form:

$$(\text{Itô}) \quad dx_t = dz_t - \langle x_t,dz_t\rangle |x_t|^{-2}\, x_t - \tfrac{1}{2}(n-1)|x_t|^{-2}\, x_t\, dt.$$

Note that application of the Ito formula to $|x_t|^2$ when x_t is given by the integral equation also yields $|x_t|^2 = |x_0|^2$ almost surely.

(ii) Let z have values in o(n), the space of skew-symmetric n × n-matrices. Let $g\ell(n)$ denote the linear space of n × n-matrices and consider the equation

$$dA_t = A_t\, dz_t$$
$$A_0 = I$$

for A with values in $g\ell(n)$, where the product is matrix multiplication. This corresponds to the S.D.S. (X,z) with

$$X: g\ell(n) \to \mathbb{L}(o(n); g\ell(n))$$

given by

$$X(A)e = Ae.$$

But then X(A)e is in the tangent space $T_A O(n)$ to the space of orthogonal matrices O(n) whenever $A \in O(n)$, and so our solution A_t will have $A_t(\omega)$ in O(n) almost surely. The S.D.S. induced on O(n) is an example of a left invariant S.D.S. as described in §1A(v).

§4 MAIN THEOREM FOR COMPACT MANIFOLDS

Theorem 4 Let M be compact. Then for every F_a-measurable map $x_a: \Omega \to M$ there is a locally regular solution $x:[a,b) \times \Omega \to M$ of $dx = Xdz$ with $x(a) = x_a$. The solution is unique up to equivalence.

Moreover given $b_0 \in (a,b)$, for each partition $\Pi = (t_1,\ldots,t_n)$ of $[a,b_0]$ let

$$z_\pi:[a,b_0] \times \Omega \to E$$

be the piecewise linear approximation

$$z_\pi(s,\omega) = (\Delta_j t)^{-1}[(t_{j+1}-s)z(t_j,\omega) + (s-t_j)z(t_{j+1},\omega)]$$

for $t_j < s < t_{j+1}$ and let

$$x_\pi:[a,b_0] \times \Omega \to M$$

denote the solution of the family of ordinary differential equations

$$\frac{dx_\pi}{dt} = X(x_\pi)\frac{dz_\pi}{dt}, \quad x_\pi(a)(\omega) = x_a(\omega).$$

Then $\{x_\pi\}_\Pi$ converges to x uniformly on $[a,b_0]$ in measure as mesh $\Pi \to 0$.

Proof. Take a C^3 embedding $i:M \to \mathbb{R}^n$ for some n. For each submanifold chart (U_α,ϕ_α) for i(M) there is a C^2 map $X_\alpha:U_\alpha \to \mathbb{L}(E;\mathbb{R}^n)$ which agrees with $i_*(X)$ on $U_\alpha \cap i(M)$. Using a partition of unity these can be combined to give a C^2 map

$$\bar{X}:\mathbb{R}^n \to \mathbb{L}(E;\mathbb{R}^n)$$

which agrees with $i_*(X)$ on $i(M)$ and has compact support. In particular $M(\bar{X})$ will be C^1 with compact support, and hence globally Lipschitz.

By Theorem 3 the regular solution $\bar{x}:[a,b) \times \Omega \to R^n$ of $d\bar{x} = \bar{X}dz$ with $\bar{x}(a) = ix_a$ is equivalent to the inclusion under i of a maximal solution to $dx = Xdz$, $x(a) = x_a$. Thus we have the first part of the theorem, and applying Addendum 4.1 of Chapter VI to $\bar{x}$ we obtain the second part. //

§5 A LEMMA ON EXIT TIMES

Lemma 5 Let $M = H$, with X and DX bounded and Lipschitz. Suppose $W \subset H$ is open and $A \subset W$ is a closed subset with $d(A,H-W) > 0$. Let τ_t^W be the first exit time from W after time t of a regular solution x to $dx = Xdz$. Then for each $n = 1,2,\ldots$ there is a constant C_n depending only on the geometry of (W,A) up to rigid motions in H, the bounds for X and M(X), and the constants in Condition A(2) for the interval $[t,t+1] \cap [a,b)$ on z, such that if $\tilde{\Omega} \in F_t$ has $x(t)(\tilde{\Omega}) \subset A$ then

$$\mu\{\omega \in \tilde{\Omega}: \tau_t^W(\omega) \leq t + s\} < C_n\, \mu(\tilde{\Omega})s^n \qquad 0 \leq s \leq 1.$$

Proof. Assuming $\mu(\tilde{\Omega}) \neq 0$ we can take $\tilde{\Omega} = \Omega$ by normalizing $\mu|\tilde{\Omega}$. This makes no difference to the constants of Condition A(2) for times after time t. We can also take $t = a$.

Let $\delta = \frac{1}{2}\, d(A,H-W) > 0$. There are points $\{x_i\}_{i=1}^\infty$ in A such that the open balls $\{B(x_i;\frac{1}{2}\delta)\}_{i=1}^\infty$ cover A, while each $B(x_i;\delta) \subset W$. It therefore suffices to prove the theorem for $A = \bar{B}(0;\frac{1}{2}\delta)$, $W = B(0;\delta)$.

To do this choose a C^∞ map $f:H \to [0,1]$ with

(i) $f(y) = 0 \qquad |y| \leq \frac{2}{3}\delta$

(ii) $f(y) = 1 \qquad |y| \geq \frac{3}{4}\delta$

(iii) $D^r f$ uniformly bounded in H, $r = 1,2,3$.

Then

$$f(x_t) = f(x_a) + \int_a^t Df(x_s)X(x_s)dz_s$$
$$+ \frac{1}{2}\int_a^t D^2f(x_s)\Big(X(x_s)dz_s, X(x_s)dz_s\Big)$$
$$+ \frac{1}{2}\int_a^t Df(x_s)DX(x_s)(X(x_s)dz_s, dz_s)$$

whence by Estimates 5C of Chapter IV there is a constant γ such that if $a \leq t \leq a + 1$

$$\mu\{\omega \sup_{a\leq u\leq t} |f(x(u,\omega))-f(x(s,\omega))| > \tfrac{1}{2}\} \leq 4\gamma \int_a^t \|B(x_s)\|^2 ds$$

where

$$B:H \to \mathbb{P}_2(E;H)$$

is given by

$$B(y)e = Df(y)M(X)(y)(e) + \tfrac{1}{2} D^2f(y)(X(y)e,X(y)e).$$

Since $B(y) = 0$ for $|y| \geq \delta$, if K_1 is a bound for X and $M(X)$ on H and $|y_0| \leq \frac{1}{2}\delta$:

$$\begin{aligned}|B(y)| &\leq K_1|Df(y)| + \tfrac{1}{2} K_1^2 |D^2f(y)| \\ &= K_1|Df(y)-Df(y_0)| + \tfrac{1}{2} K_1^2 |D^2f(y) - D^2f(y_0)| \\ &\leq K(n) |y-y_0|^n \qquad n = 1,2,\ldots\end{aligned}$$

for some constant $K(n)$, since f has bounded derivatives and is identically zero near y_0.

Thus, using Corollary 3 of Chapter III: for $a \leq t \leq a + 1$, there is a constant β with

$$\begin{aligned}\int_a^t \|B(x_s)\|^2 ds &\leq \int_a^t K(n)^2 \|x_s - x_a\|^{2n} ds \\ &\leq \beta^{2n}K(n)^2 \int_a^t \left\{\int_a^s \|M(X)(x_u)\|^2 du\right\}^n ds \\ &\leq K_1^{2n} \beta^{2n}K(n)^2 \int_a^t (s-a)^n ds \\ &= \tilde{K}(n) (t-a)^{n+1} \text{ say.}\end{aligned}$$

Thus

$$\mu\{\omega: \sup_{a\leq u\leq t} |f(x(u,\omega))-f(x(a,\omega))| > \tfrac{1}{2}\} \leq 4\gamma \tilde{K}(n)(t-a)^{n+1}$$

for constants $\gamma, \tilde{K}(n)$; and the result follows. //

Going back to a manifold M with S.D.S. (X,z); for a neighbourhood W of a subset A of M, a family of regular localizations $\{\Lambda^\alpha\}_{\alpha\in A}$,

$\Lambda^\alpha = ((U^\alpha, \phi_\alpha), U_o^\alpha, \lambda^\alpha)$, of X will be called a *uniform cover* of A in W (with radii r and bound K) if for all α

$$B(0;2r) \subset \phi_\alpha(U_o^\alpha)$$

$$|M(X_{\Lambda^\alpha})(y)| \leq K \qquad y \in H$$

$$U_o^\alpha \subset W$$

and $\{\phi_\alpha^{-1}[B(0;r)]\}_{\alpha \in A}$ covers A. Note that if A is compact it admits a uniform cover in every neighbourhood of it.

Corollary 5 For a manifold M with S.D.S. (X,z); suppose W is an open neighbourhood of $A \subset M$ such that A admits a uniform cover $\{\Lambda^\alpha\}_{\alpha \in A}$ in W. Let

$$x:[a,\xi) \times \Omega \to M$$

be a maximal l.r. solution to dx = Xdz. Then for each n = 1,2,... there is a constant K_n such that for $t \geq a$ if $\tilde{\Omega} \in F_t$ with $x(t,\omega) \in A$ for $\omega \in \tilde{\Omega}$ then

$$\mu\{\omega \in \tilde{\Omega}: \tau_t^W(\omega) \wedge \xi(\omega) \leq t + s\} \leq \mu(\tilde{\Omega})s^n K_n \qquad 0 \leq s \leq 1.$$

If z only satisfies A(2) on compact subsets of [a,b) then K_n may depend on t.

Proof. We can assume A is countable. With the notation above let $U_1^\alpha = \phi_\alpha^{-1}[B(0;r)]$ each α. Split $\tilde{\Omega}$ into a disjoint union $\tilde{\Omega} = \bigcup_\alpha \tilde{\Omega}^\alpha$, $\tilde{\Omega}^\alpha \in F_t$, with $x(t)(\tilde{\Omega}^\alpha) \subset U_1^\alpha$. By Remark 2E, $\tau_t^{\Lambda^\alpha}(\omega) \leq \tau_t^W(\omega) \wedge \xi(\omega)$ for $\omega \in \tilde{\Omega}^\alpha$. But by the lemma there are constants C_n, n = 1,2,..., with

$$\mu\{\omega \in \tilde{\Omega}^\alpha: \tau_t^{\Lambda^\alpha}(\omega) \leq t + s\} \leq C_n\, \mu(\tilde{\Omega}^\alpha)s^n \qquad 0 \leq s \leq 1.$$

The result follows by summing over A. //

§6 COMPLETENESS, AND BEHAVIOUR NEAR THE EXPLOSION TIME

Theorem 6 Let $x:[a,\xi) \times \Omega \to M$ be a maximal l.r. solution to dx = Xdz. Suppose that the subset A of M admits a uniform cover in M.

Then if $D = \{\omega: \xi(\omega) < b\}$, for almost all ω in D there exists $\gamma(\omega) > 0$ with

$$x(s,\omega) \notin A \text{ for } \xi(\omega) - \gamma(\omega) < s < \xi(\omega).$$

Proof. Let R denote the set of $\omega \in D$ for which no such $\gamma(\omega)$ exists. By enlarging A if necessary, we may assume that A is open and that $\bar{A}$ admits a uniform cover in M. We can also assume without loss of generality that z satisfies A(2) on [a,b).

Let $\{\Lambda^\alpha: \alpha \in A\}$ be such a cover, $\Lambda^\alpha = ((U^\alpha,\phi_\alpha), U^0_\alpha,\lambda^\alpha)$.

Set $U^\alpha_1 = \phi_\alpha^{-1}[B(0;r)]$, where r is a radius for the cover. For $a \leq t < t + \delta < b$, and $\delta < 1$, set

$$\Omega(t,t + \delta) = \{\omega \in \Omega: t < \xi(\omega) \leq t + \delta\}$$

and let $\{N^i\}_{i=1}^\infty$ be a family of finite subsets of $(t, t + \delta)$ with $N^i \subset N^{i+1}$, $i = 1,2,\ldots$ and $\bigcup_i N^i$ dense in $(t, t + \delta)$. Set

$$\Omega^i = \{\omega \in \Omega: x(s,\omega) \in A \text{ some } s \in N^i\}.$$

Then, if $N^i = \{s_1,\ldots,s_p\}$ we can write Ω^i as a disjoint union

$$\Omega^i = \bigcup_{j=1}^p \Omega^{i,j}$$

with $\Omega^{i,j} \in F_{s_j}$ and

$$x(s_j)(\Omega^{i,j}) \subset A.$$

By Corollary 5 there is a constant K, $(K \equiv K_2)$, such that

$$\mu(\Omega^{i,j} \cap \Omega(t,t+\delta)) \leq K\, \mu(\Omega^{i,j})\delta^2 \qquad j = 1 \text{ to } p$$

giving

$$\mu\Big(\Omega^i \cap \Omega(t,t + \delta)\Big) < K\mu\,(\Omega^i)\delta^2 \leq K\delta^2 \qquad i = 1,2,\ldots .$$

Consequently, if $\bar{\mu}$ denotes outer measure, since A is open:

$$\bar{\mu}(R \cap \Omega(t,t+\delta)) \leq \mu\Big(\bigcup_i \Omega^i \cap \Omega(t,t+\delta)\Big)$$

$$= \sup_i \mu\Big(\Omega^i \cap \Omega(t,t+\delta)\Big) < K\delta^2.$$

Now, for any $\varepsilon \in (0,1)$:

$$R = \bigcup_{n=0}^{\infty} R \cap \Omega(a + \varepsilon S_n, a + \varepsilon S_{n+1})$$

where $S_0 = 0$ and $S_n = 1 + \frac{1}{2} + \ldots + \frac{1}{n}$.

$$\text{Thus } \bar{\mu}(R) \leq \sum_{n=0}^{\infty} \bar{\mu}\left(R \cap \Omega\left(a + \varepsilon S_n, a + \varepsilon S_{n+1}\right)\right)$$

$$\leq \sum_{n=0}^{\infty} C \frac{\varepsilon^2}{(n+1)^2} = K\varepsilon^2 \sum_{n=1}^{\infty} \frac{1}{n^2}.$$

Since $\varepsilon \in (0,1)$ was arbitrary, $\bar{\mu}(R) = 0$. //

The proofs of the following corollaries are immediate from the theorem: the first is essentially due to Itô (1950) see also (Clark 1973).

Corollary 6.1 Suppose M admits a uniform cover. Then (X,z) is complete (i.e. all maximal solutions go on for all time). //

It will often be convenient to introduce a *coffin state* Δ. This is an 'ideal' point disjoint from M to which processes are sent after they have exploded, i.e. we consider $x:[a,\xi) \times \Omega \to M$ as the same as a map $x:[a,b) \times \Omega \to M \cup \{\Delta\}$ such that $x(t,\omega) = \Delta$ if and only if $t \geq \xi(\omega)$.

Corollary 6.2 Let M^+ be the 1-point compactification of M, with the coffin state Δ identified with the point at infinity, i.e. we set $x(t,\omega) = \Delta$ for $\xi(\omega) < t < b$, and Δ the point at infinity of M^+. Then

$$x:[a,b) \times \Omega \to M^+$$

has almost all its sample paths continuous. //

Remark 6 In 6.2 we could alternatively give $M \cup \{\Delta\}$ the topology such that $U \subset M \cup \{\Delta\}$ is open if either $U \subset M$ is open or $\Delta \in U$ and $M-U$ is a closed set admitting a uniform cover in M. This has the advantage of making $M \cup \{\Delta\}$ Hausdorff even when M is infinite dimensional.

Examples (i) Let M be a Hilbert Lie group and z a process on $E = T_eM$ satisfying our standard conditions. By Corollary 6.1 the left (or right) invariant S.D.S. (X,z) determined by z, §1A(v), is complete. Similarly right invariant stochastic dynamical systems on suitable diffeomorphism groups of compact manifolds will be complete (Elworthy 1975, 1978): a result we shall discuss and apply in the next chapter.

(ii) Suppose $M = H$. Assume that X is C^2 (or C^{1+}) and satisfies the *linear growth* condition

$$|M(X)(x)| \leq K(1 + |x|) \qquad x \in H$$

for some constant K. As observed in (Clark 1973) the S.D.S. (X,z) has a uniform cover, radius 1, obtained by taking each

$$U^0_\alpha = \{x : |x - p_\alpha| < 2\, |p_\alpha|\}$$

and

$$\phi_\alpha(x) = |p_\alpha|^{-1}\,(x - p_\alpha)$$

for points $p_\alpha \in H - \{0\}$.

Thus linear growth is enough to ensure non-explosion. See also (Gikhman & Skorohod 1972; Friedman 1975); linear growth is often used as a basic hypothesis.

§7 A LOCALIZATION TECHNIQUE: SOME MEASURE THEORETIC LEMMAS

(A) Many of our proofs will depend on a reduction to the globally Lipschitz case with $M = H$. As a step towards this we shall often consider M as a closed submanifold of a separable Hilbert space $\bar{H}$ by choosing an embedding $i: M \to \bar{H}$. As in the proof of Theorem 4 we will then take an extension of (X,z) to an S.D.S. $(\bar{X},z)$ on $\bar{H}$ such that $\bar{X}|M = X$. The corresponding flow map for $(\bar{X},z)$ will be denoted by $\bar{F}$. By Theorem 3 if $u \in M$ then $F(u) \sim \bar{F}(u)$ and consequently we shall not usually distinguish between the two processes.

(B) The extension $\bar{X}$ will not in general be globally Lipschitz. For $\dim M < \infty$ we could approximate $\bar{X}$ by multiplying it with bump functions with larger and larger compact supports. However we shall use a slightly different method based on the following lemma from measure theory:

Lemma 7B For a complete separable metric space M, and a finite measure space (Ω,F,μ), suppose $y \in \mathcal{L}^0(\Omega,F;C([a,b];M))$. Then, for each $\varepsilon > 0$, there is a compact subset K_ε of M, and an $\Omega_\varepsilon \in F$ such that

(i) $\mu(\Omega_\varepsilon) > \mu(\Omega) - \varepsilon$

(ii) for all $\omega \in \Omega_\varepsilon$ and $t \in [a,b], y(\omega)(t) \in K_\varepsilon$.

Proof. Since $C([a,b];M)$ is a complete separable metric space the measure $\nu = y(\mu)$ induced on it by y is tight. Thus, for $\varepsilon > 0$, there is a compact subset C_ε of $C([a,b];M)$ with $\nu(C_\varepsilon) > \mu(\Omega) - \varepsilon$. By Ascoli's theorem $\{f(t) \in M: f \in C_\varepsilon, t \in [a,b]\}$ is compact in M. Let K_ε denote this subset and set $\Omega_\varepsilon = y^{-1}(C_\varepsilon)$. //

Given a solution $x:[a,\xi) \times \Omega \to M$ of $dx = Xdz$ the lemma shows that for $\varepsilon > 0$ and $t \in (a,b)$ there is a compact $K_\varepsilon \subset M$ and a subset $\Omega_{t,\varepsilon}$ of Ω_t with (i) $\Omega_{t,\varepsilon} \in F_t$

(ii) $\mu(\Omega_{t,\varepsilon}) > \mu(\Omega_t) - \varepsilon$ and

(iii) $x(s,\omega) \in K_\varepsilon$ for $(s,\omega) \in [a,t] \times \Omega_{t,\varepsilon}$.

If we have $M = H$, for example using §A, we can then take a suitable function λ with support in a neighbourhood of K_ε to get an S.D.S. (X_Λ, z) which agrees with (X,z) on a neighbourhood of K_ε but has $M(X_\Lambda)$ globally Lipschitz: in other words we take a regular localization $\Lambda = ((U,id),U_0,\lambda)$ with K_ε in U_0.

(C) The definition of convergence in measure was given in terms of a metric on the target space: the topological property we give now is well known, but is included for completeness.

Lemma 7C Let M,N be metric spaces and K a compact subset of M. Suppose $\theta:M \to N$ is continuous and $\varepsilon > 0$. Then there is a neighbourhood U of K in M and $\delta > 0$ such that

$$d(x,y) < \delta \Rightarrow d(\theta(x),\theta(\eta)) < \varepsilon \quad \text{all } x,\ y \in U.$$

Proof. Suppose no such U,δ exist. Then for each $i = 1,2,\dots$ there is a pair (x_i,y_i) with

$$d(x_i,y_i) < 2^{-i},\ d(\theta(x_i),\theta(y_i)) \geq \varepsilon,$$

$$d(x_i,K) < 2^{-i} \text{ and } d(y_i,K) < 2^{-i}$$

Choose $\bar{x}_i$, $\bar{y}_i$ in K with $d(x_i,\bar{x}_i) < 2^{-i}$, $d(y_i,\bar{y}_i) < 2^{-i}$.

By compactness, taking subsequences if necessary, we see there is a point k in K with $x_i \to k$ and $y_i \to k$, making $d(\theta(x_i), \theta(y_i)) \geqslant \varepsilon$ impossible. //

Theorem 7C Let (Ω, F, μ) be a finite measure space and $\theta: M \to N$ a continuous map of metric spaces. Assume that M admits a complete metric and is separable. Then the map

$$\theta_*: \mathcal{L}^0(\Omega, F, \mu; M) \to \mathcal{L}^0(\Omega, F, \mu; N)$$

$$\theta_*(f) = \theta \circ f$$

is continuous.

Proof. Set $A = \mathcal{L}^0(\Omega, F, \mu; M)$, $B = \mathcal{L}^0(\Omega, F, \mu; N)$. Suppose $f \in A$ and $f_i \to f$ in A for a sequence $\{f_i\}_{i=1}^{\infty}$. Given $\varepsilon > 0$, since $f(\mu)$ is a tight measure, there is a compact set K in M and $\Omega_1 \in F$ with

$$\mu(\Omega_1) < \varepsilon/2$$

and

$$f(\omega) \in K \text{ if } \omega \notin \Omega_1.$$

By the lemma, given $\alpha > 0$, there exists $\delta > 0$ such that if $m \in M$ and $\omega \notin \Omega_1$

$$d(m, f(\omega)) < \delta \Rightarrow d(\theta(m), \theta_*(f)(\omega)) < \alpha$$

But then there exists I such that

$$i > I \Rightarrow \mu\{\omega \in \Omega : d(f_i(\omega), f(\omega)) \geqslant \delta\} < \varepsilon/2.$$

Consequently, if $i > I$,

$$\mu\{\omega \in \Omega : d(\theta_*(f_i)(\omega), \theta_*(f)(\omega)) \geqslant \alpha\}$$

$$\leqslant \mu(\Omega_1) + \mu\{\omega \in \Omega : d(f_i(\omega), f(\omega)) \geqslant \delta\}$$

$$< \varepsilon.$$

Thus $\theta_*(f_i) \to \theta_*(f)$, as required. //

Corollary 7C For (Ω, F, μ) and M as in the theorem, the topology of $\mathcal{L}^0(\Omega, F, \mu; M)$ depends only on the topology of M, not on the choice of metric. //

§8 DEPENDENCE ON THE INITIAL DISTRIBUTION

(A) This section can be seen as a step towards regularity results for solutions to parabolic and semi-parabolic equations in both finite and infinite dimensions. This is one reason for giving results which include the Hilbert space case. There are some stronger theorems in finite dimensions: some of these are discussed in §§1-3 of the next chapter where related questions are dealt with in more detail. The notation δX introduced in §D will be used frequently in future.

Complications arise particularly from allowing the possibility of finite explosion times. The case of compact M, or similarly the globally Lipschitz case with M = H, is especially much more straightforward.

We have restricted our attention to stochastic differential equations, but the methods can be applied equally well to stochastic integral equations on H.

(B) We continue with the standing assumptions and notation. For a fixed stochastic dynamical system (X,z) on the manifold M let F be a "flow map": for each $u \in \mathcal{L}^0(\Omega, \mathcal{F}_a; M)$

$$F(u): [a, \xi^u) \times \Omega \to M$$

denotes a maximal locally regular solution to

$$dx = Xdz$$

with

$$x(a) = u.$$

In general we shall be mainly interested in initial points $u \in M$, rather than initial distributions $u \in \mathcal{L}^0(\Omega, \mathcal{F}_a; M)$.

For $t \in T$ set

$$\Omega_t^u = \{\omega: t < \xi^u(\omega)\}.$$

We will adjoin to every state space M our coffin state Δ to obtain

$$M^+ = M \cup \{\Delta\},$$

(no topology is given to M^+ now). We can then extend the definition of F(u) to obtain

$$F(u): [a,b) \times \Omega \to M^+$$

by

$$F(u)(t,\omega) = \Delta \qquad \text{if } t \geq \xi^u(\omega).$$

Also if d is a metric on M we shall consider it extended over M^+ by

$$d(m,\Delta) = d(\Delta,m) = \infty \qquad m \in M$$

and

$$d(\Delta,\Delta) = 0.$$

(C) Theorem 8C Let $u_0 \in M$ and $t \in [a,b)$. Then the flow map $u \to F(u)|[a,t] \times \Omega_t^{u_0}$ is continuous on M at u_0 in the sense that, for any metric d on M, given $\varepsilon > 0$ and $\delta > 0$ there is a neighbourhood U of u_0 in M such that if $u \in U$ then

$$\mu\{\omega \in \Omega_t^{u_0}: \sup_{a \leq s \leq t} d\Big(F(u)(s,\omega), F(u_0)(s,\omega)\Big) > \delta\} < \varepsilon.$$

Proof. Suppose $\varepsilon > 0$ and $\delta > 0$. Using Lemma 7B take a compact set K of M and a measurable subset Ω_1 of $\Omega_t^{u_0}$ such that $\mu(\Omega_1) < \varepsilon/2$ and

$$F(u_0)(s,\omega) \in K \text{ all } (s,\omega) \in [a,b] \times (\Omega_t^{u_0} - \Omega_1).$$

Consider M as a submanifold of $\bar{H}$ as in §7A. Then since K is compact there is a regular localization $\Lambda = ((W,\text{id}),W_0,\lambda)$ for $\bar{X}$ with $K \subset W_0$. If $u \in \bar{H}$ let

$$F^\Lambda(u):[a,b) \times \Omega \to \bar{H}$$

be a regular solution to the stochastic integral equation

$$y(s) = u + \int_a^s M(\bar{X}_\Lambda)(y)dz$$

and let $\tau^\Lambda(u): \Omega \to [a,b]$ be its exit time from W_0.

By Corollary 2B of Chapter VI the map

$$M \to \mathcal{L}^0(\Omega,F,\mu;C([a,t],\bar{H}))$$

defined by $u \mapsto F^\Lambda(u)|[a,t] \times \Omega$ is continuous. Thus if $\delta_1 > 0$ and

$$A(u) = \{\omega \in \Omega: \sup_{a \leq s \leq t} |F^\Lambda(u)(s,\omega) - F^\Lambda(u_0)(s,\omega)| > \delta_1\}$$

there is a neighbourhood U of u_0 in M such that

$$\mu(A(u)) < \varepsilon/2 \quad \text{all } u \in U.$$

For a given metric d on M choose δ_1 so that

(a) $\{x \in \bar{H} : d_{\bar{H}}(x,K) \leq \delta_1\} \subset W_0$

where $d_{\bar{H}}$ is the usual metric on $\bar{H}$ and

(b) if $(k,m) \in K \times M$ then

$$d_{\bar{H}}(k,m) \leq \delta_1 \Rightarrow d(k,m) \leq \delta.$$

This is possible by the compactness of K, using Lemma 7C.

Then, since Λ affirms $F(u_0)$ by Proposition 2C,

$$F^\Lambda(u_0)(\omega)([a,t]) \subset K \text{ for almost all } \omega \in \Omega_t^{u_0} - \Omega_1$$

whence if $u \in U$, $t < \tau^\Lambda(u)$ a.s. on $\Omega_t^{u_0} - (\Omega_1 \cup A(u))$. Thus if $u \in U$,

$$F^\Lambda(u)|[a,t] \times (\Omega_t^{u_0} - (A(u) \cup \Omega_1))$$

is equivalent to

$$F(u)|[a,t] \times (\Omega_t^{u_0} - (A(u) \cup \Omega_1)).$$

Consequently, using (b) above, if $u \in U$

$$\mu\{\omega \in \Omega_t^{u_0}: \sup_{a \leq s \leq t} d(F(u)(s,\omega), F(u_0)(s,\omega)) > \delta\}$$

$$\leq \mu(A(u) \cup \Omega_1) < \varepsilon$$

as required. //

Remark 8C The theorem still holds for initial distributions i.e. if we allow u to vary in $\mathcal{L}^0(\Omega,\mathcal{F}_a,\mu;M)$ instead of just in M. One way to see this is to take the embedding of M in $\bar{H}$ to be bounded as well as closed; which is possible even when M is non-compact if $\bar{H}$ is taken to be infinite dimensional (for example take a diffeomorphism of $\bar{H}$ with its unit sphere using Bessaga's theorem (Bessaga 1966)). Then $\mathcal{L}^0(\Omega,\mathcal{F}_a,\mu;M)$ is naturally continuously included into $\mathcal{L}^2(\Omega,\mathcal{F}_a,\mu;\bar{H})$ and we can apply Corollary 2B of Chapter VI just as in the proof of the theorem.

Although the following 'lower semi-continuity' result for explosion times appears rather weak, it nevertheless contains the standard result for ordinary differential equations; **see also §7 of Chapter VIII:**

Corollary 8C Let $t \in [a,b)$ and $u_o \in M$. For $u \in M$ let χ_t^u denote the characteristic function $\chi_t^u = \chi_{\Omega_t^u}$. Then:

(i) The map $M \to L^1(\Omega,F,\mu;\mathbb{R})$ given by

$$u \mapsto \chi_t^{u_o}(1 - \chi_t^u)$$

is continuous at u_o.

(ii) Every sequence $\{u_i\}_{i=1}^{\infty}$ in M converging to u_o contains a subsequence $\{v_j\}_{j=1}^{\infty}$ with the property that for each $\varepsilon > 0$ there is an $\Omega_\varepsilon \in F_t$ and an integer J such that

$$\mu(\Omega_\varepsilon) < \varepsilon$$

and

$$\Omega_t^{u_o} - \Omega_\varepsilon \subset \Omega_t^{v_j} \text{ all } j > J.$$

Proof.

(i) In the statement of the theorem take $\delta = 1$ and $\varepsilon > 0$. Then since $\{\omega \in \Omega_t^{u_o} : \xi^u(\omega) < t\} \subset \{\omega \in \Omega_t^{u_o} : d(F(u)(t,\omega),F(u_o)(t,\omega)) > 1\}$ there is a neighbourhood U of u_o in M such that if $u \in U$ then

$$\mu\{\omega \in \Omega_t^{u_o} : \xi^u(\omega) < t\} < \varepsilon.$$

But
$$\{\omega \in \Omega_t^{u_o} : \xi^u(\omega) < t\} = \Omega_t^{u_o} \cap (\Omega - \Omega_t^u)$$

$$= \{\omega: \chi_t^{u_o}(\omega)(1 - \chi_t^u(\omega)) = 1\},$$

giving (i).

(ii) If $\lim_{i\to\infty} u_i = u_o$ we have shown by (i) that $\chi_t^{u_o}(1 - \chi_t^{u_i}) \to 0$ in L^1 as $i \to \infty$. Consequently there is a subsequence $\{v_j\}_{j=1}^{\infty}$ of $\{u_i\}_{i=1}^{\infty}$ such that $\chi_t^{u_o}(1 - \chi_t^{v_j}) \to 0$ almost everywhere on Ω. But Egoroff's theorem, e.g. (Kingman & Taylor 1966), states that such almost everywhere convergence implies almost uniform convergence: in other words for each $\varepsilon > 0$ there exists $\Omega_\varepsilon \in F_t$ with $\mu(\Omega_\varepsilon) < \varepsilon$ such that $\chi_t^{u_0}(1 - \chi_t^{v_i}) \to 0$ uniformly on $\Omega - \Omega_\varepsilon$. Since $\chi_t^{u_0}(1 - \chi_t^{v_j})$ takes only the values 0,1 the result follows. //

(D) Gâteaux differentiability does not make sense for functions defined on manifolds. We introduce a slightly stronger notion which does:

Definition Let $f:M \to N$ be a map between differentiable manifolds modelled on Hausdorff topological vector spaces. For $m \in M$ we say f is *strongly Gâteaux differentiable* at m if there exists a map $GT_m f:T_m M \to T_{f(m)}M$ such that for every differentiable curve $\sigma:(-\delta,\delta) \to M$ with $\sigma(0) = m$ and $\sigma'(0) = v \in T_m M$ the curve $f\circ\sigma$ is right differentiable at 0 with right derivative $d/dt^+ \, (f\circ\sigma)(0) = GT_m f(v)$.

Lemma 8D Let E and F be real Hausdorff topological vector spaces with U on open subset of E. Suppose $f:U \to F$ is Gâteaux differentiable at the point x_0 of U and is also somewhat Lipschitz at x_0 (see Chapter VI, §2D). Then f is strongly Gâteaux differentiable at x_0.

Proof. For $v \in E$ and $\sigma:(-\delta,\delta) \to U$ with $\sigma(0) = x_0$ and $\sigma'(0) = v$ we have, if $t > 0$,

$$\sigma(t) = x_0 + tv + o(t^+)$$

whence

$$\begin{aligned} f(\sigma(t)) &= f(x_0 + tv + o(t^+)) \\ &= f(x_0 + tv) + o(t^+) \\ &= f(x_0) + tGDf(x_0)v + o(t^+) \end{aligned}$$

as required. //

(E) For a differentiable section X of $\mathbb{L}(\underline{E},TM)$ define a section δX of $\mathbb{L}_{TM}(\underline{E},TM)$, (where now $\underline{E}$ refers to $\underline{E} \times TM$), or equivalently a map

$$\begin{array}{ccc} E\times TM & \xrightarrow{\delta X} & TTM \\ & \searrow \quad \swarrow & \\ & TM & \end{array}$$

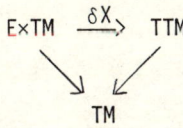

by $\delta X(e,v) = \alpha \circ T(X_e)(v)$ where $X_e: M \to TM$ is the vector field $X_e(m) = X(m)e$, and $\alpha: T^2M \to T^2M$ is given in local charts as

$$\alpha: U \times H \times H \times H \to U \times H \times H \times H \text{ by } \alpha(x,u,v,w) = (x,v,u,w).$$

Thus a sufficiently smooth stochastic dynamical system (X,z) on M determines a stochastic dynamical system $(\delta X, z)$, *the derivative S.D.S.*, on TM.

Lemma 8E When $M = H$ and X is C^2

$$M(\delta X): H \times H \to \mathbb{P}_2(E, H \times H)$$

is given by $M(\delta X)(x,v)e = (M(X)(x)e,\ D[M(X)](x)(v)e)$.

Proof. As a map $H \times H \to \mathbb{L}(E; H \times H)$, $\delta(X)$ is given by

$$\delta X(x,v)e = (X(x)e, DX(x)(v)e)$$

whence

$$M(\delta X)(x,v)(e) = \delta X(x,v)e + \tfrac{1}{2}\Big(DX(x)(X(x)e)e,\ D^2X(x)(X(x)e,v)e + DX(x)(DX(x)(v)e)e\Big)$$

which is as required by direct computation. //

Theorem 8E Suppose X is C^3, (or C^{2+}). For $v \in T_uM$ let

$$\delta F(v) : [a, \xi^v) \times \Omega \to TM$$

be a maximal locally regular solution to

$$dy = (\delta X)dz$$

with

$$y(a) = v.$$

Then (i) if $\pi: TM \to M$ is the projection

$$\pi \circ \delta F(v) \sim F(u).$$

In particular $\xi^v = \xi^u$ a.e., and

(ii) for each $t \in [a,b)$ and $u_o \in M$ the map

$$\delta F_t|\Omega_t^{u_o} : T_{u_o}M \to L^o(\Omega_t^{u_o}, F_t, \mu; (TM)^+)$$

defined by

$$v \mapsto \delta F(v)(t)|\Omega_t^{u_o}$$

is the strong Gâteaux derivative at u_o of the map

$$F_t|\Omega_t^{u_o} : M \to L^o(\Omega_t^{u_o}, F_t;\mu;M^+)$$

defined by

$$u \mapsto F(u)(t)|\Omega_t^{u_o}$$

in the sense that for any C^1 map $f:M \to G$ of M into a Banach space G the composed map

$$f_*(F_t|\Omega_t^{u_o}) : M \to L^o(\Omega_t^{u_o}, F_t, \mu; G^+)$$

is strongly Gâteaux differentiable at u_o with

$$GT_{u_o} f_*(F_t|\Omega_t^{u_o}) = (df)_*(\delta F_t|\Omega_t^{u_o})|T_{u_o}M : T_{u_o}M \to L^o(\Omega_t^{u_o}, F_t, \mu; G^+)$$

i.e. for each smooth $\sigma:(-\alpha,\alpha),0 \to M, u_o$ with $\sigma'(0) = v$, and for each $\delta > 0$

$$\mu\{\omega \in \Omega_t^{u_o} : |f o F_t(\sigma(r))\omega - f o F_t(u_o)\omega - r df o \delta F_t(v)\omega| > r\,\delta\} \to 0$$

as $r \to 0+$.

[By convention $f(\Delta) = df(\Delta) = \Delta$.]

Proof. Suppose $\varepsilon > 0$, $\delta > 0$. As in the proof of Theorem 8C take a compact set K in M and a measurable subset Ω_1 of $\Omega_t^{u_o}$ such that

$$\mu(\Omega_1) < \varepsilon/3$$

and

$$F(u_o)(s,\omega) \in K \qquad (s,\omega) \in [a,t] \times (\Omega_t^{u_o} - \Omega_1).$$

Also continue to take M to be a C^3 submanifold of $\bar{H}$ with X a restriction of $\bar{X}:\bar{H} \to \mathbb{L}(E;\bar{H})$, and let $\Lambda = ((W,id),W_o,\lambda)$ be a regular localization for $\bar{X}$

with $K \subset W_0$ and λ of class C^3. For $u \in \bar{H}$ let

$$F^\Lambda(u) : [a,b] \times \Omega \to [a,b]$$

be as before, with

$$\tau^\Lambda(u) : \Omega \to [a,b]$$

its exit time from W_0.

Furthermore, as in the proof of Theorem 8C let U be a neighbourhood of u_0 in M such that for each u in U there is a subset A(u) in F_t with

$$\mu(A(u)) < \varepsilon/3$$

and

$$t < \tau^\Lambda(u)(\omega) \qquad \text{if } \omega \in \Omega_t^{u_0} - (\Omega_1 \cup A(u)).$$

Now $M(\bar{X}_\Lambda):\bar{H} \to \mathbb{P}_2(E;\bar{H})$ is C^2 and so in particular is $\tilde{C}^1$ in the sense of Theorem 2D of Chapter VI. That theorem therefore implies that

$$F_t^\Lambda:\bar{H} \to L^2(\Omega,F_t,\mu;\bar{H})$$

$$u \mapsto F^\Lambda(u)(t)$$

is Gâteaux differentiable on $\bar{H}$, and for $(u,v) \in \bar{H} \times \bar{H}$

$$(F_t^\Lambda(u),\ GDF_t^\Lambda(u)(v)) = \delta F_t^\Lambda(u,v)$$

where $\delta F^\Lambda(u,v)$ is the regular solution of the stochastic integral equation on $\bar{H} \times \bar{H}$:

$$(x(s),y(s)) = (u,v) + \left(\int_a^s M(X_\Lambda)(x)dz,\ \int_a^s DM(X_\Lambda)(x)(y)dz\right).$$

Alternatively, by Lemma 8E, $\delta F_t(u,v)$ is an l.r. solution of the S.D.S. $(\delta X_\Lambda,z)$ on $\bar{H} \times \bar{H}$. In particular Theorem 2D of Chapter VI (or more precisely Lemma 2C, and Corollary 1C, of Chapter VI) ensures that the trajectories of $(\delta X_\Lambda,z)$ have infinite explosion times.

The open embedding theorem, Proposition 2D now shows that for $u \in M$ and $v \in T_uM$

$$\delta F^\Lambda(u,v)|[a,t] \times \{\omega\colon t < \tau^\Lambda(u)\}$$

is a l.r. solution to the S.D.S. $(\delta X, z)$ on TM.

Therefore, by maximality, up to a set of measure zero

$$[a,t] \times \{\omega: t < \tau^{\Lambda}(u)(\omega)\} \subset [a,\xi^{v}) \times \Omega$$

i.e. $$\{\omega: t < \tau^{\Lambda}(u)(\omega)\} \subset \{\omega: t < \xi^{v}(\omega)\}.$$

But, taking, $u = u_0$,

$$\mu\{\omega \in \Omega_t^{u_0} : t \geq \tau^{\Lambda}(u)(\omega)\} \leq \mu(\Omega_1) < \varepsilon/3$$

and so, since $\varepsilon > 0$ was arbitrary $\xi^{v}|\Omega_t^{u_0} \geq t$ a.e. each $t \in [a,b)$. Thus $\xi^{v} \geq \xi^{u}$ a.e. on Ω.

Since clearly $\xi^{u} \geq \xi^{v}$ a.e. and

$$\pi\circ\delta F(v)|[a,\xi^{u}) \times \Omega \sim F(u)$$

part (i) is true.

Now $F_t^{\Lambda}:\bar{H} \to L^2(\Omega, F_t, \mu; \bar{H})$ is Lipschitz by Theorem 2B of Chapter VI, as well as being Gâteaux differentiable. Consequently by Lemma 8D it is strongly Gâteaux differentiable, as is its restriction to M and as is also

$$F_t^{\Lambda}|M \times \Omega_t^{u_0} : M \to L^2(\Omega_t^{u_0}, F_t, \mu; \bar{H}).$$

Let $\sigma:(-\alpha,\alpha) \to M$ be C^1 with

$$\sigma(0) = u_0 \text{ and } \sigma'(0) = v \in T_{u_0}M;$$

we may assume also that $\sigma[(-\alpha,\alpha)] \subset U$.

Then for $r \in (-\alpha,\alpha)$, $F(\sigma(r))$ and $F^{\Lambda}(\sigma(r))$ are equivalent when restricted to

$$[a,t] \times \left(\Omega - (\Omega_1 \cup A(\sigma(r)))\right)$$

as are $\delta F(v)$ and $\delta F^{\Lambda}(u_0,v)$ when restricted to

$$[a,t] \times (\Omega_t^{u_0} - \Omega_1).$$

Choose $\alpha_1 \in (0,\alpha)$ with

$$\|\frac{1}{r}\left(F_t^\Lambda(\sigma(r)) - F_t^\Lambda(u_0)\right) - \delta F_t(u_0,v)\| < \delta\sqrt{\frac{\varepsilon}{3}} \text{ if } 0 < r < \alpha_1.$$

Then, if $0 < r < \alpha_1$,

$$\mu\{\omega \in \Omega : |F_t^\Lambda(\sigma(r))\omega - F_t^\Lambda(u_0)\omega - r\,\delta F_t^\Lambda(u_0,v)\omega| \geq r\delta\} < \varepsilon/3.$$

Whence

$$\mu\{\omega \in \Omega_t^{u_0} : |F_t(\sigma(r))\omega - F_t(u_0)\omega - r\delta F_t(v)\omega| \geq r\delta\}$$

$$< \mu(\Omega_1) + \mu(A(\sigma(r))) + \varepsilon/3 < \varepsilon.$$

Thus (ii) is proved for the special case $f = i$ of a closed C^3 embedding into a Hilbert space.

In general, if $f:M \to G$ is C^1, we can certainly extend f to a C^1 map $\bar{f}:\bar{H} \to G$, and we can insist that W_0 is a sufficiently small neighbourhood of the compact set K so that $D\bar{f}|W_0:W_0 \to \mathbb{L}(\bar{H};G)$ is bounded: suppose

$$|D\bar{f}(x)| \leq \beta \quad \text{all } x \in W_0.$$

If necessary we can also take our neighbourhood U of u_0 in M to be smaller to ensure that for $u \in U$ and $\omega \in \Omega_t^{u_0} - \Omega_1 \cup A(u)$ the line segment from $F_t(u)(\omega)$ to $F_t(u_0)(\omega)$ lies in W_0 almost surely. But then, using the mean value theorem, if $\omega \in \Omega_t^{u_0} - \Omega_1 \cup A(\sigma(r))$

$$|f\circ F_t(\sigma(r))\omega - f\circ F_t(u_0)\omega - r\,df\circ\delta F_t(u_0)(v)\omega|$$

$$\leq |\bar{f}\circ F_t(\sigma(r))\omega - \bar{f}\circ F_t(u_0)\omega - D\bar{f}\left(F_t(u_0)\omega\right)\left(F_t(\sigma(r))\omega - F_t(u_0)\omega\right)|$$

$$+ |D\bar{f}\left(F_t(u_0)\omega\right)\left(F_t(\sigma(r))\omega - F_t(u_0)\omega - r\,\delta F_t(u_0)(v)\omega\right)|$$

$$\leq 2\beta|F_t(\sigma(r))\omega - F_t(u_0)\omega| + \beta|F_t(\sigma(r))\omega - F_t(u_0)\omega - r\,\delta F_t(u_0)(v)\omega|.$$

The result now follows from the special case proved above. //

Remark 8E We have insisted that X be at least C^{2+} since we require δX to be C^{1+} in order to apply our previous existence and uniqueness results to the S.D.S. $(\delta X, z)$ on TM. However in the proof we represented δX locally by δX_Λ and our existence theorems for the stochastic integral equations determined by $(\delta X_\Lambda, z)$ only require $M(X_\Lambda)$ to be $\tilde{C}^1$: which follows if X is C^2. Presumably the theorem is also true if X is C^2 (even if some of the terms in its statement have not then been precisely defined).

§9 GLOBAL FORMULATIONS OF THE ITO FORMULA

(A) Given our stochastic dynamical system (X,z) on M, for each e in E let

$$(t,m) \mapsto S(t,m)e$$

denote the flow of the vector field X_e on M, $X_e(m) = X(m)e$. It is defined on some neighbourhood of $\{0\} \times M$ in $\mathbb{R} \times M$.

For a C^2 map $f:M \to G$, where G is a separable Hilbert space,

$$\frac{d}{dt} f \circ S(t,m)e = df \circ X_e(S(t,m)e)$$

and

$$\frac{d^2}{dt^2} f \circ S(t,m)e = d^2 f \circ TX_e\Big(X_e(S(t,m)e)\Big)$$

$$= d^2 f \circ \delta X_e\Big(X_e(S(t,m)e)\Big)$$

where $d^2 f = d(df):TTM \to G$ (non-standard notation!).
Thus at $t = 0$, for each $m \in M$ we obtain linear and bilinear maps of E into G, which we shall write

$$\frac{d}{dt} f \circ S(t,m)\Big|_{t=0} \in \mathbb{L}(E;G)$$

$$\frac{d^2}{dt^2} f \circ S(t,m)\Big|_{t=0} \in \mathbb{L}(E,E;G).$$

(B) Lemma 9D *Let* (X,z) *be a* C^2 *stochastic dynamical system on* M *and* $f:M \to G$ *a* C^2 *map. Let*

$$x:[a,\xi) \times \Omega \to M$$

be a maximal solution of $dx = Xdz$ *then* ξ *is predictable and almost surely on* Ω_t, *in the sense of Lemma 7D of Chapter III*:

(i) $$f(x_t) = f(x_a) + \int_a^{t\wedge\xi} df\circ X(x_s)dz_s + \tfrac{1}{2}\int_a^{t\wedge\xi} d^2f\circ\delta X(X(x_t)dz_s)dz_s$$

(ii) $$f(x_t) = f(x_a) + \int_a^{t\wedge\xi} \frac{d}{dt} f\circ S(t,x_s)\Big|_{t=0} dz_s + \tfrac{1}{2}\int_a^{t\wedge\xi} \frac{d^2}{dt^2} f\circ S(t,x_s)\Big|_{t=0}(dz_s,dz_s)$$

(iii) Given a linear connection on M with covariant derivative operator ∇

$$f(x_t) = f(x_a) + \int_a^{t\wedge\xi} df\circ X(x_s)dz_s + \tfrac{1}{2}\int_a^{t\wedge\xi} \nabla df\Big(X(x_s)dz_s, X(x_s)dz_s\Big) + \tfrac{1}{2}\int_a^{t\wedge\xi} df\circ\nabla X\Big(X(x_s)dz_s, dz_s\Big)$$

a.s. on Ω_t.

Here ∇df is a section of

$$\mathbb{L}(TM;\mathbb{L}(TM;\underline{G})) \simeq \mathbb{L}(TM,TM;\underline{G})$$

and ∇X is a section of

$$\mathbb{L}(TM,\underline{E};TM).$$

Proof. (i) By Chapter III, §D it is enough to show that there is a sequence $\{\eta^i\}_{i=1}^{\infty}$ of stopping times, $a \leq \eta^i < \xi$, with $\xi = \sup \eta^i$, such that (i) holds on $\{\omega : t < \eta^i(\omega)\}$ when ξ is replaced by η^i. By Chapter III, Proposition 7A the integrals with ξ replaced by η^i do exist.

To do this we can use the technique of §7B taking a family $\{\xi^i\}$ of exit times of x from suitable neighbourhoods of compact subsets of M, and setting $\eta^i = \xi^i \wedge t_i$ where $t_i \to b$ as $i \to \infty$. As in §8C we can then assume that M is the Hilbert space H and the explosion time is infinite. In this case, since

$$df: H \times H \to \mathbb{R} \text{ is given by } df(x,u) = Df(x)u$$

and

$$d^2f: H \times H \times H \times H \to \mathbb{R} \text{ by } d^2f(x,u,v,w) = D^2f(x)(v,u) + Df(x)w$$

while

$$\delta X(x,u)e = (x,u,X(x)e,DX(x)(u)e)$$

the result (i) follows immediately from the Itô formula, Theorem 3 of Chapter V.

Part (ii) follows immediately from (i) by the formulae in §9A above, and (iii) follows from (ii) since

$$\frac{d}{dt}\, df\circ X(S(m,t)e)e = \frac{D}{\partial t}(df)\circ X(S(m,t)e)e + df\circ\frac{D}{\partial t}\, X(S(m,t)e)e$$

$$= \nabla df\Big(X_e(S(m,t)e), X_e(S(m,t)e)\Big) + df\circ\nabla X_e\Big(X_e(S(m,t)e)\Big). \quad //$$

Corollary 9B Let η be a stopping time with $a \leq \eta < \xi$. Then

$$f(x(\eta(\omega),\omega)) = f(x(a,\omega)) + \int_a^{\eta(\omega)} df\circ X(x(s,\omega))dz(s,\omega)$$

$$+ \tfrac{1}{2}\int_a^{\eta(\omega)} d^2f\circ\delta X\Big(X(x(s,\omega))dz(s,\omega)\Big)dz(s,\omega) \qquad \text{a.s.}$$

with corresponding formulae for (ii) and (iii) of the lemma.

Proof. Apply Remark 7D of Chapter III to the conclusions of the lemma. //

(C) In order to do the computations we shall need later, with F_t replaced by δF_t, we must consider the flow $(t,v) \mapsto \delta S(t,v)e$ on TM, determined by the vector fields $(\delta X)_e$ on TM. The following formulae will be useful:

Lemma 9C For a torsion free linear connection on M, for all $y \in M$, $v \in T_yM$ and $e \in E$, when $t = 0$:

$$\frac{D}{\partial t}\, \delta S(t,v)e = \nabla X_e(v)$$

and

$$\frac{D^2}{\partial t^2}\, \delta S(t,v)e = \nabla^2 X_e(X_e(y),v) + \nabla X_e(\nabla X_e(v))$$

$$= \nabla\Big(\nabla X_e(X_e(-)\Big)v + R(X_e(y),v)X_e(y)$$

<u>where</u> $R:TM \times TM \times TM \to TM$ <u>is the curvature tensor, with the sign convention of</u> (Kobayashi & Nomizu 1963).

<u>NOTE</u>: <u>We use</u> $\nabla^2 X(u,v)$ <u>for</u> $\nabla_u \nabla_v X$, <u>when</u> u,v <u>are constant</u>.

<u>Proof</u>. Let $\exp_y$ denote the exponential map of the connection, at the point y. Then

$$\delta S(t,v)e = \frac{d}{ds} S(t,\exp_y sv)e\Big|_{s=0}.$$

Therefore at $s = 0$:

$$\frac{D}{\partial t}\delta S(t,v)e = \frac{D}{\partial t}\frac{d}{ds} S(t,\exp_y sv)e$$

$$= \frac{D}{ds}\frac{d}{dt} S(t,\exp_y sv)e$$

$$= \frac{D}{\partial s} X_e(S(t,\exp_y sv))$$

$$= \nabla X_e(\delta S(t,v)e)$$

giving the first equation.

Also we have

$$\frac{D^2}{\partial t^2}\delta S(t,v)e = \frac{D}{\partial t}\nabla X_e(\delta S(t,v)e)$$

$$= \nabla^2 X_e\Big(X_e(S(t,y)e), \delta S(t,v)e\Big)$$

$$+ \nabla X_e\Big(\frac{D}{\partial t}\delta S(t,v)e\Big)$$

giving the second equation.

Finally there is the identity, for $u, v \in T_y M$

$$\nabla^2 X_e(u,v) - \nabla^2 X_e(v,u) = R(u,v)X_e(y),$$

(see Appendix B), since

$$\nabla\Big(\nabla X_e(X_e(-))\Big)v = \nabla^2 X_e(v,X_e(y)) + \nabla X_e(\nabla X_e(v))$$

setting $u = X_e(y)$ we have the last equality. //

Proposition 9C Let $\phi:TM \to L$ be a C^2 linear form on TM with values in a Hilbert space L. Then, for a torsion free linear connection on M, if $v(\omega) \in T_{x(a,\omega)}M$ with v F_a-measurable,

$$\phi\delta F_t(v) = \phi(v) + \int_a^{t\wedge\xi} \nabla\phi\Big(X(F_s(x))dz_s\Big)\delta F_s(v)$$

$$+ \int_a^{t\wedge\xi} \phi\nabla X(\delta F_s(v))dz_s$$

$$+ \tfrac{1}{2}\int_a^{t\wedge\xi} \nabla^2\phi\Big(X(F_s(x))dz_s, X(F_s(x))dz_s\Big)\delta F_s(v)$$

$$+ \int_a^{t\wedge\xi} \nabla\phi\Big(X(F_s(x))dz_s\Big)\nabla X(\delta F_s(x))dz_s$$

$$+ \tfrac{1}{2}\int_a^{t\wedge\xi} \phi\nabla\Big(\nabla X(X(-)dz_s)dz_s\Big)o\delta F_s(v)$$

$$+ \tfrac{1}{2}\int_a^{t\wedge\xi} \phi R\Big(X(F_s(x))dz_s, \delta F_s(v)\Big)X(F_s(x))dz_s$$

a.s. on Ω_t.

Proof. The formula follows straightforwardly from Lemmas 9B and 9C. //

(D) We can now go back to the problem raised in §1C concerning the relationship between stochastic integral equations and stochastic differential equations. In fact a direct application of Lemma 9B(i) when M = G = H and f is the identity yields:

Proposition 9D Let $x:[a,\xi) \times \Omega \to H$ be an l.r. solution to a stochastic dynamical system (X,z) on H. Then x satisfies the stochastic integral equation (a.s. on Ω_t):

$$x_t = x_a + \int_a^t M(X)(x_s)dz_s \ . \qquad //$$

Now is also a good place to show the equivalence of our definition of a solution to (X,z) on M with that used by some other authors e.g. Ikeda & Watanabe (1981).

<u>Theorem 9D</u> <u>Let</u> ξ_1 <u>be a stopping time and</u>

$y:[a,\xi_1) \times \Omega \to M$

<u>an admissible process such that for all</u> C^2 <u>functions</u> $\theta:M \to G$ <u>into some separable Hilbert space</u> G <u>we have</u>

$$\theta(y_t) = \theta(y_a) + \int_a^{t\wedge\xi} d\theta\circ X(y_s)dz_s$$

$$+ \tfrac{1}{2}\int_a^{t\wedge\xi} d^2\theta\circ\delta X(X(y_s)dz_s)dz_s$$

a.s. on Ω_t

<u>for each</u> $t \in [a,b)$.

<u>Then</u> y <u>is a locally regular solution to</u> (X,z).

<u>Proof</u>. Let $\Lambda = ((U,\phi),U_0,\lambda)$ be a regular localization for X. Take a C^2 function $\mu:M \to [0,1]$ with support in U and which is identically 1 on a neighbourhood of $\bar{U}_0$ (for example by transporting a suitable function defined on H to one on M via ϕ). Define $\theta:M \to H$ by

$$\theta(m) = \mu(m)\phi(m) \qquad m \in U$$
$$= 0 \qquad m \notin U.$$

Since the image of θ is bounded $\theta(y_t)$ lies in $\mathcal{L}^2$, and writing out in full the expression for $d^2\theta\circ\delta X(X(y_s)e)e$, as in the proof of Lemma 9B, our hypothesis immediately implies that Λ affirms y as a locally regular solution. //

Of course when dim $M < \infty$ we need only take $G = \mathbb{R}$.

(E) Next we give an invariance property which will come in useful later. For this let $f:B \to M$ be a C^2 map of Hilbert manifolds. Then an S.D.S. $(\tilde{X},z)$ on B will be said to be *f-related* to an S.D.S. (X,z) on M if

$$Tf(\tilde{X}(b)e) = X(f(b))e$$

for all $b \in B$ and $e \in E$, i.e. if the diagram

$$\begin{array}{ccc} TB & \xrightarrow{\;Tf\;} & TM \\ \tilde{X}_e \big\downarrow & & \big\downarrow X_e \\ B & \xrightarrow{\;\theta\;} & M \end{array}$$

commutes, for all $e \in E$. Note that the space E and process z are the same for both $(\tilde{X},z)$ and (X,z). An example is the projection $f:TM \to M$ with $\tilde{X} = \delta X$. More interesting examples can be obtained when $f:B \to M$ is the projection of a Lie Group onto a coset space (homogeneous space).

Proposition 9E *Let* $(\tilde{X},z)$ *be f-related to* (X,z) *as described. Suppose* $\tilde{x}:[a,\tilde{\xi}) \times \Omega \to B$ *and* $x:[a,\xi) \times \Omega \to M$ *are maximal locally regular solutions of* $d\tilde{x} = \tilde{X}dz$ *and* $dx = Xdz$ *with* $f(\tilde{x}_a) = x_a$ a.s.. *Then*

$$x|[a,\tilde{\xi}) \times \Omega \sim f(\tilde{x})$$

and in particular $\tilde{\xi} \prec \xi$ a.s.

Proof. Let $\tilde{S}(t,b)e$ and $S(t,m)e$ give the flow of $\tilde{X}_e$ and X_e as described in 9A. By differentiating $f(\tilde{S}(t,b)e)$ with respect to t we see

$$f(\tilde{S}(t,b)e) = S(t,f(b)e \qquad b \in B, \quad e \in E.$$

Let $\theta:M \to G$ be a C^2 map into a separable Hilbert space G. Applying Lemma 9B(ii) we see that if we set $y_t = f(\tilde{x}_t)$ then a.s. on $\{\omega : t < \tilde{\xi}(\omega)\}$

$$\theta(y_t) = \theta(x_a) + \int_a^{t\wedge\tilde{\xi}} \frac{d}{dt}\theta\circ S(t,y_s)\Big|_{t=0} dz_s$$

$$+ \tfrac{1}{2}\int_a^t \frac{d^2}{dt^2}\theta\circ S(t,y_s)\Big|_{t=0} dz_s\, dz_s.$$

By the computation in the proof of Lemma 9B the hypothesis of Proposition 9D is satisfied. The result follows from that proposition by the uniqueness of solutions. //

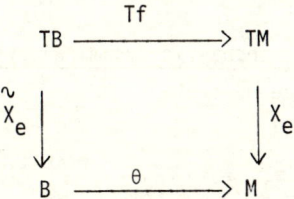

(F) The following version of Itô's formula will be particularly useful when M is finite dimensional and f has compact support.

Lemma 9F Let $(\dot{X},z)$ be a C^2 stochastic dynamical system on M and $f:M \to G$ a C^2 map into a Hilbert space G. Let

$$x:[a,\xi) \times \Omega \to M$$

be a maximal locally regular solution of dx = Xdz. Suppose that the support of f admits a uniform cover in M and that both $df \circ X(m) \in \mathbb{L}(E;G)$ and $d^2 f \circ \delta X(X(m) -)(-) \in \mathbb{L}(E,E;G)$ are bounded uniformly in $m \in M$. Then, with the convention that $f(\Delta) = 0$ we have almost surely on Ω:

$$f(x(t\wedge\xi)) = f(x_a) + \int_a^{t\wedge\xi} df \circ X(x_s) dz_s$$
$$+ \tfrac{1}{2} \int_a^{t\wedge\xi} d^2 f \circ \delta X\Big(X(x_s) dz_s\Big) dz_s \qquad (*)$$

with corresponding formulae to (ii), (iii) of Lemma 9B.

Proof. Because of the separability of M we can take a sequence $\{U_i\}_{i=1}^{\infty}$ of open subsets of M such that

(i) $\operatorname{supp} f \subset U_1$

(ii) $\bar{U}_i \subset U_{i+1}$ each i

(iii) $\bigcup_{i=1}^{\infty} U_i = M$

(iv) each $\bar{U}_i$ admits a uniform cover in M.

Let ξ^i be the first exit time of x from U_i. Then $\xi^i < \xi$ almost surely by (iv) and Theorem 6. Also $\xi^i < \xi^{i+1}$ almost surely for each i. Set $\bar{\xi} = \sup_i \xi^i$. Every compact subset of M lies in some U_i by (iii); consequently if $\bar{\xi}(\omega)$ were less than $\xi(\omega)$ then almost surely the image under x of $[a,\bar{\xi}(\omega)]$ would lie in some U_i, giving the almost sure contradiction $\bar{\xi}(\omega) < \xi^i(\omega)$. Thus $\bar{\xi} = \xi$ almost surely. This is, in itself, a useful point.

From (i) we now see:

(v) $\lim_{i\to\infty} f(x(\xi^i)) = f(\Delta)$ almost surely on $\{\omega : \xi(\omega) < \infty\}$.

The rest of the proof will use only properties (ii) - (v).

By Corollary 9B equation (*) holds with ξ replaced by ξ_i, for each i because $t \wedge \xi_i < \xi$ almost surely by Theorem 6. Since $t \wedge \xi_i$ converges to $t \wedge \xi$ almost surely (*) itself is seen to hold by combining Theorem 6 and the continuity of f, for the left hand side, with Theorem 7A of Chapter III for the right hand side. //

Remark 9F The proof works equally well to show the validity of the formulae given (X,z) and $f:M \to G$ both C^2 with a maximal solution x assuming only that $df \circ X(x_s)$ and $d^2 f \circ \delta X(X(x_s)(-))(-)$ are bounded on $[0,\xi)$ a.s. and that there exists a sequence $\{U_i\}_{i=1}^{\infty}$ of open subsets of M together with a number $f(\Delta)$ satisfying (ii) - (v).

Corollary 9F Under the conditions of the lemma the conclusion (*) holds with ξ replaced by η whenever ξ is a stopping time with $a \leqslant \eta \leqslant \xi$.

Proof. Apply Remark 7B of Chapter III to the lemma. //

§10 PIECEWISE LINEAR APPROXIMATION: GENERAL CASE

There is the following extension of Theorem 4, Chapter VI and the approximation part of Theorem 4:

Theorem 10 For infinite partitions $\Pi = (t_1, t_2, \ldots)$ of $[a,b)$ with $a = t_1 < t_2 < \ldots < b$ and $t_j \to b$ as $j \to \infty$ let z_π be the piecewise linear approximation to z as in Theorem 4. Suppose X is C^2 and $x_a : \Omega \to M$ is F_a-measurable.

Let $x:[a,\xi) \times \Omega \to M$ be a maximal locally regular solution to $dx = X\,dz$, and let

$$x\ :[a,\xi_\pi) \times \Omega \to M$$

be maximal l.r. solutions to the families of ordinary dynamical systems $dx\ = X dz_\pi$, $x_\pi(a) = x_a$.

Then $\{x_\pi\}_\Pi$ converges in measure to x as mesh $\Pi \to 0$ in the sense that for each $t \in [a,b)$, and any metric d on M:

$$\mu\{\omega \in \Omega_t : \sup_{a \leqslant s \leqslant t} d(x_\pi(s,\omega), x(s,\omega)) > \delta\} \to 0$$

as mesh $\Pi \to 0$, for each $\delta > 0$.

In particular $t \wedge \xi_\pi \to t$ in measure on Ω_t as mesh $\Pi \to 0$.

Proof. As in §§7, 8 we may take $M = H$ and for $\varepsilon > 0$, find a compact set K of H with a measurable subset $\tilde{\Omega}$ of Ω_t such that

$$\mu(\tilde{\Omega}) < \varepsilon/2$$

and

$$x(s,\omega) \in K \quad \text{all} \quad (s,\omega) \in [a,t] \times (\Omega_t - \tilde{\Omega}).$$

For a regular localization $\Lambda = ((W,\mathrm{id}), W_0,\lambda)$ for X, with $K \subset W_0$, and X_Λ bounded, we let $\hat{x}$, $\hat{x}_\pi$ correspond to x, x_π with X replaced by X_Λ. Also let

$$\tau_\pi^\Lambda : \Omega \to [a,b]$$

be the first exit time of x_π from W_0.

Suppose that $0 < \delta < d(K, H-W_0)$. By the Addendum to Theorem 4 of Chapter VI there is $\gamma > 0$ such that if

$$\Omega(\Pi,\delta) = \{\omega \in \Omega: \sup_{a \leq s \leq t} d(\hat{x}_\pi(s), \hat{x}(s)) > \delta\}$$

then

$$\text{mesh } \Pi < \gamma \Rightarrow \mu(\Omega(\Pi,\delta)) < \varepsilon/2.$$

Therefore, for mesh $\Pi < \gamma$, if $\omega \in \Omega_t - \Omega(\Pi,\delta) - \tilde{\Omega}$, then

$$\tau_\pi^\Lambda(\omega) > t$$

whence

$$\hat{x}_\pi(s)(\omega) = x_\pi(s)(\omega) \qquad a \leq s \leq t,$$

almost surely. Consequently if mesh $\Pi < \gamma$

$$\mu\{\omega \in \Omega_t: \sup_{a \leq s \leq t} d(x_\pi(s), x(s)) > \delta\} \leq \mu(\Omega(\Pi,\delta) \cup \tilde{\Omega}) < \varepsilon$$

proving the theorem. //

§11 THE STOCHASTIC DEVELOPMENT: BROWNIAN MOTION ON A FINITE DIMENSIONAL MANIFOLD

(A) We will first consider a slightly more general version of Example 1A(iii). Let N^n be a finite dimensional manifold with a reduction of its tangent bundle to a subgroup G of GL(n); e.g. G = O(n), in which case N is just a Riemannian manifold. Let $\pi:G(N) \to N$ be the principle G-bundle of N, and suppose we have a connection on G(N). As in Example 1A(iii), and Appendix B, this determines an element $X \in \tilde{\mathbb{L}}(G(N) \times \mathbb{R}^n; TG(N))$.

Let $x_0 \in N$ and suppose $z:[a,b) \times \Omega \to T_{x_0}N$ is a process satisfying all our standing conditions. If $u_0 \in \pi^{-1}(x_0)$, we can consider u_0 as a linear isomorphism $u_0:R^n \to T_{x_0}N$ and define $z_{u_0}:[a,b) \times \Omega \to R^n$ by $z_{u_0} = u_0^{-1} o z$. In this way we get an S.D.S. (X, z_{u_0}) on G(N). Take a maximal solution $u:[a,\xi) \times \Omega \to G(N)$ of $du = Xdz_{u_0}$ with $u(a) = u_0$. Define $\mathcal{R}(z)_{u_0}:[a,\xi) \times \Omega \to N$ by $\mathcal{R}(z)_{u_o} = \pi o u$.

<u>Lemma 11A</u> $\mathcal{R}(z)_{u_0}$ <u>is independent of the choice of</u> u_0 <u>in</u> $\pi^{-1}(x_0)$

<u>Proof</u>. Suppose $v_0 \in \pi^{-1}(x_0)$ also. Then there exists $g \in G$ with $R_g(u_0)=v_0$ where R_g denotes the right action of G on G(N):

$$R_g(\alpha) = \alpha.g \qquad \alpha \in G(N).$$

By the invariance of connections, for each $\alpha \in G(N)$,

$$X(\alpha.g) = T_\alpha R_g o X(\alpha) o g: R^n \to T_{\alpha.g} G(N).$$

Thus

$$X(\alpha.g)z_{v_0}(t) = T_\alpha R_g o X(\ )(g.z_{v_0}(t))$$

$$= T_\alpha R_g o X(\alpha)(g(u_0.g)^{-1}z(t))$$

$$= T_\alpha R_g o X(\alpha)(z_{u_0}(t)).$$

It follows from the diffeomorphism invariance theorem, Theorem 1F, that $v = u.g:[0,\xi) \times \Omega \to G(N)$ is a maximal solution to $dv - Xdz_{v_0}$ with $v(a)=v_0$. Since $\pi o v = \pi o u$ we have $\mathcal{R}(z)_{u_0} = \mathcal{R}(z)_{v_0}$ as required. //

According to Lemma 11A, from our original process z on $T_{x_0}N$ we obtain a process on N, which we may denote by $R(z)$. In the case $z(a)(\omega) = 0$ with probability one, $R(z)$ will be called the *stochastic development* of z. When z has sufficiently smooth sample paths for the stochastic differential equation to become a family of ordinary differential equations $R(z)$ becomes the composition of z with the *Cartan development* R mapping smooth paths in $T_{x_0}N$ with initial value 0 to smooth paths on N starting at x_0, see (Kobayashi & Nomizu 1963) Chapter III §4. In fact when G(N) is compact and $\Omega = C([0,\infty), 0;T_{x_0}N,0)$, the space of continuous paths σ in $T_{x_0}N$ with $\sigma(0) = 0$, with the compact open topology, the map $z(\omega) \mapsto R(z)(\omega)$ determines, by Theorem 4, a Borel map

$$\tilde{R}:C([0,\infty),0;T_{x_0}N,0) \to C([0,\infty),0;N,x_0)$$

uniquely determined up to sets of μ-measure zero. Moreover, by that theorem, $\tilde{R}$ is the limit in measure of the maps obtained by piecewise linear approximations composed with the classical development as described by Eells & Elworthy: see Elworthy (1974, 1975) and (Eells & Elworthy 1976).

The compactness of N is not enough alone to assure that $R(z)$ has infinite explosion time: for example there exist compact Lorentz manifolds which are not complete and for which therefore the classical development is not defined for all time. We could define the connection to be *stochastically complete* if the explosion time is infinite when z is a Brownian motion.

(B) In case z is Brownian motion on $T_{x_0}N$ and $G = O(n)$ i.e. N is a Riemannian manifold, we shall call $R(z)$ *Brownian motion on N starting at* x_0. If we take the standard model of Brownian motion this then has a concrete realization as

$$R(z): [0,\xi) \times C([0,\infty),0;T_{x_0}N,0) \to N.$$

When $\xi = \infty$, e.g. if N is compact, for each $a \in (0,\infty)$ we can take the restrictions of R to obtain

$$\tilde{R}:C([0,a],0;T_{x_0}N,0) \to C([0,a],0;N,x_0).$$

The probability measure induced on $C([0,a],0;N,x_0)$ by $\tilde{R}$ could be called *Wiener measure* on $C([0,a],0;N,x_0)$, following (Eells & Elworthy 1970). The use of the Cartan development, or rolling of the manifold along its tangent plane (see §C below), to construct Brownian motion on a Riemannian manifold was inspired by work of Gangolli (1964), see (Eells & Elworthy 1971), but goes back much further in special cases: see (Gorman 1960) and (McKean 1960a). A detailed examination of the role of the development map was given by a student of E. Nelson, in his doctoral disseration (Ferebee 1972). It has been used systematically by Malliavin and his students, e.g. see (Malliavin 1978a). For closely related work and generalizations see (Itô 1975; Jørgensen 1978; Meyer 1981; Pinsky 1981; Rogerson 1981).

When N has a given isometric embedding in some Euclidean space we could use Example 1A(iv) to obtain a process on N which is isonomous to such a Brownian motion. When N is a Lie group with invariant metric we could use Example 1A(v) to do the same. This is discussed more in Chapter IX when the reasons for calling $R(z)$ the "Brownian motion" should become clear. The reasons for using the stochastic development are: it does not require any additional structure on N; it is very nice geometrically; and, most important for us here, it gives not only a "Brownian motion" on N but it also gives us the 'horizontal lift' process u on $O(N)$. The latter can be used to define *stochastic parallel translation* along the sample paths of the Brownian motion in exactly the same way as the horizontal lift of a smooth path is used (Appendix B). It is used in this way in §13 below. The article (Itô 1963) on 'stochastic parallel translation' was a major stimulus for this theory, see also (Dynkin 1968).

The stochastic development also turns out to be a mathematical and global analogue of an approach by Pauli, Van-Vleck, DeWitt, and others to Feynman path integration on curved spaces in non-relativistic quantum mechanics, see (Elworthy & Truman 1981) §4E, or (DeWitt-Morette et al. 1980).

(C) When $G = O(n)$ the Cartan development has a concrete interpretation: Consider a C^1 curve σ in $T_{x_0}N$ with $\sigma(0) = 0$, and think of the manifold N, isometrically embedded in some Euclidean space, and "sitting on" its tangent space $T_{x_0}N$ (so an affine translation is used to identify $0 \in T_{x_0}N$ with $x_0 \in N$). Now "roll N without slipping, on $T_{x_0}N$ and along σ". As N rolls, its point of contact with $T_{x_0}N$, and therefore with the curve, will determine a curve $\tilde{\sigma}$ on N with $\tilde{\sigma}(0) = x_o$. It is an instructive exercise in basic differential geometry to verify that $\tilde{\sigma} = R\sigma$.

§12 SOME FORMULAE FOR BROWNIAN MOTION ON N.

(A) Let N^n be a finite dimensional Riemannian manifold with orthonormal frame bundle O(N) as in §§1,11. We seek formulae corresponding to the Itô formulae Lemma 9B for Brownian motion on N and its "derivative process" on TN. It will be more illuminating to have these in terms of operations on N rather than in terms of lifts up to O(N). Theorem 12B is especially important. It relates Brownian motion on N to the Laplace Beltrami operator, see §3 of Appendix B.

We use Kobayashi and Nomizu's curvature conventions.

(B) Let u_o be a fixed frame at $x_o \in N$. We consider the canonical S.D.S. (X,z) on O(N), as described in §1A. In particular z is a Brownian motion on $\mathbb{R}^n$. Let

$$u:[0,\xi) \times \Omega \to O(N),$$

be a maximal l.r. solution to

$$du = Xdz$$

with $u(o) = u_o$. Set $x = \pi \circ u:[0,\xi) \times \Omega \to N$. By Lemma 11A this process x depends only on the Riemannian structure and the Brownian motion $u_o \circ z$ on $T_{x_o}N$.

We will use the notation $S, \delta S, F, \delta F$ from §9. For example $F_t(u_o) = u(t)$; also $e_1,\ldots,e_n$ denotes an orthonormal base in $\mathbb{R}^n$ and for $\alpha \in O(N)$, $S^i(t,\alpha) = S(t,\alpha)e_i$ while

$$\frac{d}{dt} S(t,\alpha)e = X_e(S(t,\alpha))$$

and

$$S(0,\alpha)e = \alpha.$$

Set $$\gamma^i(t,\alpha) = \pi \circ S^i(t,\alpha).$$

The following lemma is immediate from the definitions: (See §2, Appendix B).

<u>Lemma 12B</u> <u>The curves</u> $\gamma^i(-,\alpha)$ <u>are geodesics in</u> N <u>with</u>

$$\frac{d\gamma^i}{dt}(0,\alpha) = \alpha e_i. \quad //$$

<u>Theorem 12B</u> <u>Let</u> $f:N \to \mathbb{R}$ <u>be</u> C^2. <u>Then, almost surely on</u> Ω_t

$$f(x_t) = f(x_0) + \int_0^{t\wedge\xi} df(u_s dz_s) + \tfrac{1}{2}\int_0^{t\wedge\xi} \Delta f(x_s)ds$$

<u>where</u> Δ <u>denotes the Laplace-Beltrami operator of</u> N:

$$\delta f = \text{trace } \nabla df.$$

<u>Proof</u>. We compare with the formulae of Lemma 9B, replacing $f(x_t)$ by $(f\circ\pi)(u_t)$. Since, for $e \in R^n$,

$$d(f\circ\pi)\ (X(u_s)e = df\circ T\pi\circ X(u_s)e$$
$$= df(u_s(e))$$

our first integral agrees with that in 9B(i). Also, at $t = 0$,

$$\frac{d^2}{dt^2}[f\circ\pi\circ S^i(t,u_s)] = \frac{d^2}{dt^2}[f\circ\gamma^i(t,u_s)]$$
$$= \frac{d}{dt}[df\circ\frac{d\gamma^i}{dt}(t,u_s)]$$
$$= \nabla df(u_s e_i)(u_s e_i)$$

by Lemma 12B since the second time derivative of a geodesic is zero.

Thus, since z is a Brownian motion: almost surely on Ω_t

$$\int_0^{t\wedge\xi} \frac{d^2}{dt^2}\Big(f\circ\pi\circ S(t,u_s)\Big)\Big|_{t=0} (dz_s,d\dot{z}_s)$$
$$= \int_0^{t\wedge\xi} \sum_i \frac{d^2}{dt^2}\Big(f\circ\pi\circ S^i(t,u_s)\Big)\Big|_{t=0} ds$$
$$= \int_0^{t\wedge\xi} \Delta f(x_s)ds$$

giving the second integral as required, by 9B(ii). //

(C) Recall that a vector field J along a geodesic γ in N is a *Jacobi field* if it satisfies the differential equation

$$\frac{D^2 J}{\partial t^2} + R(J, \frac{d\gamma}{dt}) \frac{d\gamma}{dt} = 0$$

where R denotes the curvature tensor of N. We will show that as x is related to the geodesics of N so the derivative process is related to the Jacobi fields.

For $\alpha \in O(N)$ and $V \in T_\alpha O(N)$ we have the field $\delta S^i(-,V)$ along $S^i(-,\alpha)$, $i = 1,\ldots,n$. Set

$$J^i(t,V) = Tx \circ \delta S^i(t,V).$$

Thus $J^i(-,V)$ is a field along $\gamma^i(-,\alpha)$.

Let ϖ denote the connection form of N, so ϖ is a smooth 1-form on O(N) with values in o(n), the Lie algebra of O(n), considered as the space of skew symmetric linear endomorphisms of $\mathbb{R}^n$.

Lemma 12C $J^i(-,V)$ is a Jacobi field along $\gamma^i(-,\alpha)$ with

$$J^i(0,V) = T\pi(V)$$

and

$$\left.\frac{D}{\partial t} J^i(t,V)\right|_{t=0} = \alpha\varpi(V)e_i.$$

Proof. Let hV denote the horizontal component of V and consider

$$g:(-1,1) \to O(n)$$

$$g(s) = \exp s\varpi(V).$$

Let $\sigma:(-1,1) \to O(N)$

be a horizontal path with $\sigma(0) = \alpha$ and $\alpha'(0) = hV$. Then

$$\left.\frac{d}{ds}\Big(\sigma(s)\cdot g(s)\Big)\right|_{s=0} = V.$$

Therefore

$$J^i(t,V) = Tx \circ \delta S^i(t,V)$$
$$= \left.\frac{\partial}{\partial s}\Big(\pi\, S^i(t,\sigma(s)g(s))\Big)\right|_{s=0} = \left.\frac{\partial}{\partial s}\gamma^i(t,\sigma(s)g(s))\right|_{s=0}.$$

This expresses $J^i(-,V)$ as an infinitesimal variation of $\gamma^i(-,\alpha)$ through geodesics and therefore it is a Jacobi field; see for example (Milnor 1963) Lemma 14.3 (but his curvature tensor has a different sign from ours) or (Eliasson 1967) Theorem 3.3.

Certainly $J^i(0,V) = T\pi(V)$ while for $t = 0$, $s = 0$

$$\begin{aligned}\frac{D}{\partial t} J^i(t,V) &= \frac{D}{\partial t}\ \frac{\partial}{\partial s}\ \gamma^i(t,\sigma(s)g(s)) \\ &= \frac{D}{\partial s}\ \frac{\partial}{\partial t}\ \gamma^i(t,\sigma(s)g(s)) \\ &= \frac{D}{\partial s}\left(\sigma(s)g(s)e_i\right) \\ &= \sigma(s)\ \frac{d}{ds}\ g(s)e_i \\ &= \alpha\widetilde{\omega}(V)e_i\end{aligned}$$

as required. //

(D) We next relate the derivative process to the de Rham-Hodge Laplacian Δ on 1-forms, $\Delta = -(d\delta + \delta d)$. [*This is the negative of the operator usually considered in harmonic form theory.*] For this we need to consider the *Ricci curvature*

$$K: TN \oplus TN \to \mathbb{R}$$

$$K(v_1,v_2) = \text{trace}\ [v \mapsto R(v,v_1)v_2].$$

We also need:

<u>Weitzenböck Formula</u> <u>(for 1-forms)</u>. <u>Let</u> ϕ <u>be a</u> C^2 <u>1-form on</u> N. <u>Then</u>

$$\Delta\phi = \text{trace}\ \nabla^2\phi - K(-,\phi^\#)$$

<u>where</u> $\phi^\#$ <u>is the vector field corresponding to</u> ϕ.

A proof is in (Goldberg 1962) §2.12.

<u>Lemma 12D</u> <u>Let</u> $y \in N$ <u>and</u> $B \in \mathbb{L}(T_yN;T_yN)$. <u>Suppose</u> $g_1,\dots,g_n$ <u>is an orthonormal base for</u> T_yN. <u>Then if</u> B <u>is skew adjoint, for any smooth</u> <u>1-form</u> ϕ <u>on</u> N:

$$\sum_i \nabla\phi(g_i)(Bg_i) = \sum_i d\phi(g_i,Bg_i).$$

Proof. Let $(\nabla\phi)^{\#}:TN \to TN$ correspond to $\nabla\phi:TN \to T^*N$. Then

$$\sum_i \nabla\phi(g_i)(Bg_i) = \sum_i \langle(\nabla\phi)^{\#} g_i, Bg_i\rangle_y$$
$$= -\sum_i \langle B(\nabla\phi)^{\#} g_i, g_i\rangle$$
$$= -\text{ trace } B(\nabla\phi)^{\#}$$
$$= -\text{ trace } (\nabla\phi)^{\#}B$$
$$= -\sum_i \langle(\nabla\phi)^{\#} Bg_i, g_i\rangle_y$$
$$= -\sum_i \nabla\phi(Bg_i)(g_i).$$

But $\nabla\phi(Bg_i)(g_i) = -2d\phi(g_i, Bg_i) + \nabla\phi(g_i)(Bg_i)$, see (Kobayashi & Nomizu 1963), Corollary 8.6, Chapter III. //

Let $v_0 \in T_{x_0}N$ with $\tilde{v}_0$ its lift to $HT_{u_0}O(N)$. By the *derivative process* from v_0 we mean the process

$$v:[0,\xi) \times \Omega \to TN$$
$$v(t,\omega) = T\pi \circ \delta F_t(\tilde{v}_0)(\omega) \qquad \omega \in \Omega_t.$$

This process can be obtained from a connection on TN associated to an indefinate metric on its tangent bundle, just as the Brownian motion on N came from a connection on N, (Caldwell 1976). Our discussion is closely related to (Malliavin 1977b).

Theorem 12D For a C^2 1-form ϕ on N, and for almost all $\omega \in \Omega_t$

$$\phi(v_t) = \phi(v_0) + \int_0^t \nabla\phi(u_s dz_s)(v_s)$$
$$+ \int_0^t \phi(u_s A_s dz_s)$$
$$+ \tfrac{1}{2}\int_0^t \Delta\phi(v_s)ds$$
$$+ \int_0^t \sum_{i=1}^n d\phi(u_s e_i, u_s A_s e_i)ds$$

where

$$A:(0,\xi) \times \Omega \to O(n)$$

is given by

$$A(s) = \widetilde{\omega}(\delta F_s(\tilde{v}_0)).$$

Proof. By Lemma 12C, for $\alpha \in O(N)$, $V \in T_\alpha O(N)$, and $\underline{V} = T\pi(V)$:

$$\frac{d}{dt}[\phi o T\pi o \delta S^i(t,V)] = \frac{d}{dt}[\phi(J^i(t,V))]$$

$$= \nabla\phi(\frac{d\gamma^i}{dt}(t,\alpha))(J^i(t,V)) + \phi(\frac{DJ^i}{\partial t}(t,V))$$

$$= \nabla\phi(\alpha e_i)(\underline{V}) + \phi(\alpha\widetilde{\omega}(V)e_i)$$

at $t = 0$.

From this

$$\frac{d^2}{dt^2}[\phi o T\pi o \delta S^i(t,V)]_{t=0} = \nabla^2\phi(\alpha e_i,\alpha e_i)(\underline{V}) + 2\nabla\phi(\alpha e_i)(\alpha\,\widetilde{\omega}(V)e_i) + \phi(\frac{D^2J^i}{\partial t^2}(0,V)).$$

Now, for $g_i = \alpha e_i$,

$$\sum_i \phi(\frac{D^2J^i}{\partial t^2}(0,V)) = -\sum_i \phi(R(\underline{V},g_i)g_i)$$

$$= -\sum_i \langle R(\underline{V},g_i)g_i,\phi^\#\rangle$$

$$= \sum_i \langle R(\underline{V},g_i)\phi^\#,g_i\rangle$$

$$= -K(\underline{V},\phi^\#)$$

while by Lemma 12D, since $\alpha\widetilde{\omega}(V)\alpha^{-1}$ is skew adjoint:

$$\sum_i \nabla\phi(\alpha e_i)(\alpha\widetilde{\omega}(V)e_i) = \sum_i d\phi(\alpha e_i,\alpha\widetilde{\omega}(V)e_i).$$

Thus, using the Weitzenböck formula above,

$$\sum_i \frac{d^2}{dt^2}[\phi o T\pi o \delta S^i(t,V)]\Big|_{t=0} = \Delta\phi(\underline{V}) + 2\sum_i d\phi(\alpha e_i, \alpha\varpi(V)e_i).$$

The result follows from Lemma 9B(ii) using the fact that z is a Brownian motion, (take $\alpha = u(s)$, $V = \delta F_s(\tilde{v}_0)$, so $\underline{V} = v(s)$). //

(E) To examine the behaviour of the processes v and A define

$$\psi : TO(N) \to \mathbb{R}^n \times o(n)$$

by

$$\psi(W) = (\theta(W), \varpi(W)) \qquad W \in TO(N)$$

where θ is the canonical 1-form

$$\theta(W) = \alpha^{-1}\, T\pi(W) \qquad W \in T_\alpha O(N).$$

Set $F = \mathbb{R}^n \times o(n)$. We have the parallelization of $TO(N)$:

$$\tilde{\psi} : TO(N) \to O(N) \times F$$
$$W \mapsto (\tau(W), \psi(W))$$

where $\tau : TO(N) \to O(N)$ is the projection. Let

$$[\delta X]^\wedge : (O(N) \times F) \times \mathbb{R}^n \to TO(N) \times F$$

be the 'principal part' of the S.D.S. induced from δX by this parallelizing diffeomorphism:

$$[\delta X]^\wedge((\alpha,\xi,A),e) = (T\tau o \delta X_e o \tilde{\psi}^{-1}(\alpha,\xi,A), \mathrm{proj}_F o T\psi o \delta X_e o \tilde{\psi}^{-1}(\alpha,\xi,A))$$

for $(\alpha,\xi,A) \in O(N) \times \mathbb{R}^n \times o(n)$ and $e \in \mathbb{R}^n$.

We use $\underline{\Omega}$ for the curvature form of the connection: so $\underline{\Omega}$ is the $o(n)$-valued 2-form on $O(N)$ defined by

$$\underline{\Omega}(V_1, V_2) = d\varpi(hV_1, hV_2)$$

where h denotes horizontal projection.

<u>Proposition 12E</u> <u>For</u> (α,ξ,A) <u>in</u> $O(N) \times \mathbb{R}^n \times o(n)$ <u>and</u> e <u>in</u> $\mathbb{R}^n$

$$[\delta X]^\wedge(\alpha,\xi,A)e = (X(\alpha)e, Ae, 2\underline{\Omega}(X(\alpha)e, X(\alpha)\xi))$$

Proof. With the notation above set

$$V = \psi^{-1}(\alpha,\xi,A) \in T_\alpha O(N).$$

We will work over a chart (U,ϕ) for $O(N)$ about α, considering U as an open subset of F. In this representation ψ can be considered as a map

$$\psi: U \to \mathbb{L}(F;\mathbb{R}^n \times o(n))$$

and

$$\delta X: U \times F \times \mathbb{R}^n \to (U \times F) \times F \times F$$

is given by $\delta X(\alpha,V)e = ((\alpha,V),\ X_0(\alpha)e,\ DX_0(\alpha)(V)e)$

where $X_0: U \to \mathbb{L}(\mathbb{R}^n;F)$ is the principal part of X.

Whence

$$[\delta X]^\wedge(\alpha,\xi,A)e = \Big(X(\alpha)e, D\psi(\alpha)(X_0(\alpha)e)V + \psi(\alpha)DX_0(\alpha)(V)e\Big).$$

Now

$$D\psi(\alpha)(X_0(\alpha)e)V = 2d\psi(X(\alpha)e,V) + D\psi(\alpha)(V)(X_0(\alpha)e)$$

and

$$D\psi(\alpha)(V)(X_0(\alpha)e) = -\psi(\alpha)DX_0(\alpha)(V)e$$

since

$$\psi(\alpha)X_0(\alpha)e = (e,0).$$

Thus

$$[\delta X]^\wedge(\alpha,\xi,A)e = (X_0(\alpha)e,\ 2d\psi(X(\alpha)e,V)).$$

There are the structure equations, for V_1, V_2 tangent vectors at a point of $O(N)$, see (Kobayashi & Nomizu 1963) Theorem 2.4, Chapter III:

$$d\varpi(V_1,V_2) = -\tfrac{1}{2}[\varpi(V_1),\ (V_2)] + \underline{\Omega}(V_1,V_2)$$

$$d\theta(V_1,V_2) = \tfrac{1}{2}\ \{\varpi(V_2).\theta(V_1) - \varpi(V_1).\theta(V_2)\}.$$

These give

$$2d\varpi(X(\alpha)e,V) = 2\underline{\Omega}(X(\alpha)e,V) = 2\underline{\Omega}(X(\alpha)e,X(\alpha)\xi)$$

(since $hV = X(\alpha)\xi$), and

$$2d\theta(X(\alpha)e,V) = Ae,$$

which completes the proof. //

Corollary 12E Let $\tilde{v}_0 \in T_{u_0}O(N)$ be the horizontal lift of $v_0 \in T_{x_0}N$. Set

$$\xi_t = \theta\circ\delta F_t(\tilde{v}_0):\Omega_t \to \mathbb{R}^n$$

and

$$A_t = \varpi\circ\delta F_t(\tilde{v}_0):\Omega_t \to \mathbb{R}^n \qquad 0 \leq t < \infty.$$

Then the processes ξ and A satisfy the equations:

$$\xi_k = \theta(\tilde{v}_0) + \int_0^t A_s dz_s - \tfrac{1}{2}\int_0^t u_s^{-1}K(v_s,-)^{\#}ds$$

$$A_t = \int_0^t u_s^{-1}R(u_s dz_s, v_s)u_s$$

$$+ \tfrac{1}{2}\int_0^t [u_s^{-1}\sum_i R(u_s e_i, u_s A_s e_i)u_s]ds$$

$$+ \tfrac{1}{2}\int_0^t [u_s^{-1}\sum_i \nabla R(u_s e_i)(u_s e_i, v_s)u_s]ds$$

where

$$v_t = u_t\xi_t$$

and

$K(v_s,-)^{\#} \in T_{x(s)}N$ corresponds to $K(v_s,-) \in T^*_{x(s)}N$.

Proof. We use Lemma 9B again. For $V \in T_\alpha O(N)$, by the proposition

$$\frac{d}{dt}[\theta\circ\delta S^i(t,V)] = \varpi(\delta S^i(t,V))e_i$$

and

$$\frac{d}{dt}(\varpi\circ\delta S^i(t,V)) = 2\underline{\Omega}(X(S^i(t,\alpha))e_i, h\delta S^i(t,V))$$
$$= S^i(t,\alpha)^{-1}R(S^i(t,\alpha)e_i, J^i(t,V))\circ S^i(t,\alpha),$$

(recall that R corresponds to $2\underline{\Omega}$), using Lemma 12C. Thus, at t = 0,

$$\frac{d^2}{dt^2}[\theta\circ\delta S^i(t,V)] = \alpha^{-1}R(\alpha e_i, T\pi(V))(\alpha e_i)$$

and

$$\frac{d^2}{dt^2}[\varpi\circ\delta S^i(t,V)] = \alpha^{-1}\frac{D}{\partial t}[R\left(S^i(t,\alpha)e_i, J^i(t,V)\right)\circ S^i(t,\alpha)]\Big|_{t=0}$$

$$= \alpha^{-1}\ \nabla R(\alpha e_i)(\alpha e_i, T\pi(V))\circ\alpha$$

$$+ \alpha^{-1}R(\alpha e_i, \alpha\varpi(V)e_i)\circ\alpha,$$

using Lemma 12C.

Since $\sum_i \langle R(\alpha e_i, T\pi(V))\alpha e_i, w\rangle_y = -\sum_i \langle R\left(\alpha e_i, T\pi(V)\right)W, \alpha e_i\rangle_y$

$$= -K(T\pi(V), w)$$

all $w \in T_yN$, $y = \pi(\alpha)$, the result follows by Lemma 9B(ii), (take $\alpha = u(s)$, $V = F_s(v_0)$). //

(F) Since we have the basic formulae here we will go on to give some estimates for $|v(t)|$ and $|A(t)|$, even though the final results belong more properly to the next chapter.

<u>Theorem 12F</u> <u>With the notation of Corollary 12E for</u> r = 1,2,... <u>and almost all</u> $\omega \in \Omega_t$:

(i)
$$|v_t|^{2r}_{x(t)} = |v_0|^{2r} + 2r\int_0^t |v_s|^{2r-2}_{x(s)} \langle v_s, u_s A_s dz_s\rangle_{x(s)}$$

$$+ 2r(r-1)\int_0^t |v_s|^{2r-4}_{x(s)} |A_s u_s^{-1} v_s|^2 ds$$

$$+ r\int_0^t |v_s|^{2r-2}_{x(s)} |A_s|^2\, ds$$

$$- r\int_0^t |v_s|^{2r-2}_{x(s)} K(v_s, v_s)ds.$$

(ii) $$|A_t|^{2r} = -2r\int_0^t |A_s|^{2r-2}\,\mathrm{trace}\,[A_s\tilde{R}(u_s dz_s, v_s)]$$

$$-r\int_0^t |A_s|^{2r-2}\,\mathrm{trace}\,[A_s \sum_i \tilde{R}(u_s e_i, u_s A_s e_i)]ds$$

$$-r\int_0^t |A_s|^{2r-2}\,\mathrm{trace}\,[A_s u_s^{-1} \sum_i \nabla R(u_s e_i)(u_s e_i, v_s)u_s]ds$$

$$+ 2r(r-1)\int_0^t |A_s|^{2r-4} \sum_i \{\mathrm{trace}\,[A_s\tilde{R}(u_s e_i, v_s)]\}^2 ds$$

$$+ r\int_0^t |A_s|^{2r-2} \sum_i |R(u_s e_i, v_s)|^2 ds$$

where

$$|A_t|^2 = \mathrm{trace}\, A_t^* A_t, \text{ etc.}$$

and

$$\tilde{R}(v_1, v_2) = u_s^{-1} R(v_1, v_2) u_s \text{ for } v_1, v_2 \text{ in } T_{x(s)}N.$$

Proof. This follows by the Itô formula from Corollary 12E using the orthogonality of the frames u(s): so that $|v(t)|_{x(t)} = |\xi(t)|$ etc.. For the second integral of (i) we also use

$$\sum_i \langle \xi_t, A_t e_i\rangle^2 = \sum_i \langle e_i, A_t \xi_t\rangle^2$$
$$= |A_t \xi_t|^2 .$$

Alternatively (i) could be proved using Lemma 12C, instead of Corollary 12E. //

Remark 12F For $y \in N$ and $w \in T_yN$ define $\Theta(w): T_yN \to T_yN$ by

$$\langle \Theta(w)v_1, v_2\rangle = \nabla K(v_2)(v_1, w) - \nabla K(v_1)(v_2, w) \qquad v_1, v_2 \in T_yN.$$

Then, if $f_1, \ldots, f_n$ is an orthonormal base for T_yN

$$\sum_{i=1}^{n} \nabla R'(f_i)(f_i, w) = \Theta(w).$$

Proof By the second Bianchi identity if $v_1, v_2 \in T_yN$

$$\nabla K(v_1)(v_2,w) = \sum_i \langle \nabla R(v_1)(f_i,v_2)\, w, f_i\rangle$$

$$= -\sum_i \langle \nabla R(v_2)(v_1,f_i)\, w, f_i\rangle$$

$$-\sum_i \langle \nabla R(f_i)(v_2,v_1)\, w, f_i\rangle$$

$$= \nabla K\,(v_2)(v_1,w) + \sum_i \langle \nabla R(f_i)(f_i,w)v_2, v_1\rangle. \quad //$$

Consequently we can replace the terms involving R in Corollary 12E and Theorem 12F by terms involving ∇K.

Corollary 12F. If R and ∇K are uniformly bounded on N then for each $t \geq 0$,

$$|v(t)|_{x(t)} \in \mathcal{L}^p(\Omega_t, \mathcal{F}_t, \mu; R) \qquad 1 \leq p < \infty$$

and

$$|A(t)| \in \mathcal{L}^p(\Omega_t, \mathcal{F}_t, \mu; R) \qquad 1 \leq p < \infty$$

Proof The proof of Theorem 3A, Chapter VI can be applied to $(v(t),A(t)) \in T_{x(t)}N \times o(n)$, with essentially no change, this yields the corollary after using the remark above. //

§13 COVARIANT LINEAR EQUATIONS ON VECTOR BUNDLES

This section is rather more technical. The theory is relevant to the study of gauge invariant Schrödinger operators (DeWitt-Morette et al 1980) and to gauge field theory (Gaveau & Trauber 1981, Asorey & Mitter 1981). Special (and simple) cases will be used in the last chapter when the Feynman-Kac and Girsanov-Cameron-Martin formulae are briefly discussed. They are also used in the next chapter.

(A) Consider a C^r vector bundle $p:B \to M$, $r \geq 0$, with Hilbert space F as fibre and structure group G. Let $\pi:G(B) \to M$ be the principal G-bundle of p. Suppose we have admissible processes

$$u:[a,\xi) \times \Omega \to G(B) \qquad u(a) = u_0$$

$$x = \pi \circ u : [a,\xi) \times \Omega \to M \qquad x(a) = x_0.$$

In particular we have maps $u(t,\omega) : F \to \pi^{-1}(x(t,\omega)) \quad (t,\omega) \in [a,\xi) \times \Omega$.

Let $J:\underline{E} \to \mathbb{L}(B;B)$ be a continuous vector bundle map over the identity map of M into the vector bundle associated to p with fibre $\mathbb{L}(F;F)$. Thus for $m \in M$ by the usual abuse of notation we have

$$J(m) \in \mathbb{L}(E;\mathbb{L}(p^{-1}(m), p^{-1}(m))).$$

We shall consider equations for admissible processes with $\eta \prec \xi$

$$v:[a,\eta) \times \Omega \to B, \qquad p \circ v = x$$

which will be written as

$$(\text{It}\hat{\text{o}}) \quad Dv_t = J(x_t)\,(dz_t)\,(v_t) \tag{1}$$

The significance of '(Itô)' is that no second order terms are to be used in the integral equations representing (1). In fact we shall say that *v is a solution of* (i) with $v(a) = v_0 \in \mathcal{L}^0(\Omega,F,p^{-1}(x_0))$ if

$$\tilde{v}:[a,\eta) \times \Omega \to F$$

$$\tilde{v}(t,\omega) = u(t,\omega)^{-1}v(t,\omega)$$

satisfies $\tilde{v}_t = v_0 + \int_a^{t\wedge\eta} u_s^{-1}\, J(x_s)\,(dz_s)(u_s\tilde{v}_s)$

almost surely on $\{\omega : t < \eta(\omega)\}$, $t \geq a$, (2)

in the sense of §7D of Chapter III.

Theorem 13A Suppose ξ is predictable. Then for each $v_0 \in \mathcal{L}^0(\Omega,F_0;p^{-1}(x_0))$ there is an admissible process $v:[a,\xi) \times \Omega \to B$ with $v(a) = v_0$ and $p \circ v = x$, satisfying (1). This solution is unique up to equivalence. When $p:B \to M$ has a Riemannian metric with $G(B)$ consisting of isometries then if J is uniformly bounded on M and $v_0 \in \mathcal{L}^2$

$$v(t) \in \mathcal{L}^2(\Omega_t,F_t,\mu;\, x(t)_*(B)) \qquad a \leq t < b$$

<u>where</u> $x(t)_*(B) \to \Omega_t$ <u>is the pull back over</u> Ω_t of $p:B \to M$ <u>by</u> $x(t):\Omega_t \to M$.

<u>Proof</u>. Define $\tilde{J}:T \times \Omega \to \mathbb{L}(E;\mathbb{L}(F;F))$ by

$$\tilde{J}_t e = \begin{cases} u_t^{-1} J(x_t)(e)u_t & t < \xi(\omega) \\ 0 & \text{otherwise.} \end{cases}$$

Then equation (2) when $\eta = \xi$ and $\tilde{J}$ is bounded is just

$$\tilde{v}_t = v_0 + \int_a^t \tilde{J}_s\,(dz_s)\tilde{v}_s \qquad t \geq a \tag{3}$$

When $\tilde{J}$ is bounded and v_0 is in $\mathcal{L}^2$ we see that (3) satisfies all the conditions of Theorem 1C, Chapter VI, except possibly Condition 1B(b). However the use of Theorem 7A of Chapter III in the proof of it shows that the conclusions of that Theorem 1C still hold with the possible exception of the continuity of the sample paths. In particular there is a solution which lies in $\mathcal{L}^2$ for each t.

For general $\tilde{J}$ let η_1^r be the first exit time of $|\tilde{J}|$ from $[-r,r]$, $r = 1,2,\ldots$, and let $\{\eta^r\}_{r=1}^{\infty}$ be a sequence of stopping times with $a \leq \eta^r < \xi$ such that $\lim_{r\to\infty} \eta^r = \xi$ almost surely. Set $\xi^r = \eta_1^r \wedge \eta^r$, $r = 1,2,\ldots$, and define $\tilde{J}^r$ by $\tilde{J}^r(t,\omega) = \tilde{J}(t\wedge\xi^r(\omega),\omega)$.

This time all the conditions of Theorem 1C, Chapter VI are satisfied and so there is an essentially unique admissible process

$$\tilde{v}^r : [a,\xi^r) \times \Omega \to F$$

satisfying

$$\tilde{v}^r_t = v_0 + \int_a^t \tilde{J}^r_s\,(dz_s)(\tilde{v}^r_s).$$

By Lemma 6B of Chapter III we can amalgamate these processes to obtain an admissible process $\tilde{v} : [a,\xi) \times \Omega \to F$. The uniqueness part of that lemma allows us to conclude that $\tilde{v}$ satisfies (2), and that the solution is unique up to equivalence. This, of course, also shows that the $\mathcal{L}^2$ solution obtained when $\tilde{J}$ is bounded can be chosen to have continuous sample paths. //

(B) The most interesting case is when $r \geq 3$ and u is a maximal solution of a stochastic differential equation

$$du = \tilde{X}\, dz \qquad u(a) = u_0,\ \pi(u_0) = x_0$$

on G(B) where $\tilde{X}(-)e$ is horizontal vector field on G(B) for each $e \in E$. We will work out an Itô formula for $\phi(v(s))$ in this case, when $\phi : B \to L$ is a C^2 linear form on B with values in a Hilbert space L. Suppose also that M is equipped with a connection and let $\tilde{S}(t,g)e$ give the flow on G(B) determined by $\tilde{X}(g)e$, $g \in G(B)$, $e \in E$.

<u>Proposition 13B</u>. <u>For</u> u <u>and</u> $\tilde{J}$ <u>as above and</u> $v_0 \in \mathcal{L}^0(\Omega,F_0;p^{-1}(x_0))$, <u>almost surely on</u> Ω_t, $a \leqslant t < b$,

$$\begin{aligned}\phi(v_t) = \phi(v_0) &+ \int_a^{t\wedge\xi} \nabla\phi\Big(T\pi o\tilde{X}(u_s)dz_s\Big)v_s \\ &+ \int_a^{t\wedge\xi} \phi\Big(J(x_s)\ (dz_s)v_s\Big) \\ &+ \frac{1}{2}\int_a^{t\wedge\xi} \nabla^2\phi\Big(T\pi o\tilde{X}(u_s)dz_s,\ T\pi o\tilde{X}(u_s)dz_s\Big)v_s \\ &+ \frac{1}{2}\int_a^{t\wedge\xi} \nabla\phi\Big(\frac{D}{t}\ T\pi o\tilde{X}(\tilde{S}(t,u_s)dz_s)dz_s\Big|_{t=0}\Big)v_s \\ &+ \int_a^{t\wedge\xi} \nabla\phi\Big(T\pi o\tilde{X}(u_s)dz_s\Big)J(x_s)(dz_s)v_s.\end{aligned}$$

<u>Proof</u> We shall take $L = \mathbb{R}$, since the general case will then follow on taking coordinates. Define

$$\tilde{\phi}:G(B) \to F^*$$

by

$$\tilde{\phi}(g)(f) = \phi(gf).$$

Then

$$\phi(v\ ) = \tilde{\phi}(u\ )\tilde{v} \qquad\qquad a \leqslant t < \xi(\omega),$$

where we are using the notation of the proof of Theorem 13A. Now by Lemma 9B almost surely on Ω_t, $a \leqslant t < b$

$$\tilde{\phi}(u_t) = \tilde{\phi}(u_0) + \int_a^{t\wedge\xi} \frac{d}{dt}(\tilde{\phi}o\tilde{S}(t,u_s)|_{t=0}\ dz_s$$

$$+ \int_a^{t\wedge\xi} \frac{d^2}{dt^2}(\tilde{\phi}o\tilde{S}(t,u_s))|_{t=0}\ (dz_s,dz_s).$$

Consequently, using the basic form of Itô's lemma for Hilbert space valued processes:

$$\tilde{\phi}(u_t)\tilde{v}_t = \tilde{\phi}(u_0)(\tilde{v}_0) + \int_a^{t\wedge\xi} \frac{d}{dt}(\tilde{\phi}o\tilde{S}(t,u_s))|_{t=0}\ (dz_s)\tilde{v}_s$$

$$+ \int_a^{t\wedge\xi} \phi(u_s)\tilde{J}_s(dz_s)\tilde{v}_s$$

$$+ \frac{1}{2}\int_a^{t\wedge\xi} \frac{d^2}{dt^2}(\tilde{\phi}o\tilde{S}(t,u_s))|_{t=0}\ (dz_s,dz_s)\tilde{v}_s$$

$$+ \int_a^{t\wedge\xi} \frac{d}{dt}(\tilde{\phi}o\tilde{S}(t,u_s))|_{t=0}\ (dz_s)\tilde{J}_s\ (dz_s)\tilde{v}_s, \quad \text{a.s. on } \Omega_t.$$

However for $g \in G(B)$, $e \in E$ and $f \in F$, if $\check{S}(t) = \tilde{S}(t,g)e(f)$ then $\check{S}(t) \in p^{-1}(\pi\tilde{S}(t,g)e)$ and since $\tilde{S}(t,g)e$ is horizontal for the connection of $G(B)$:

$$\frac{D}{\partial t}\check{S}(t) = 0.$$

Therefore

$$\frac{d}{dt}[\tilde{\phi}o\tilde{S}(t,g)e]f = \frac{d}{dt}\phi(\check{S}(t))$$

$$= \nabla\phi(T\pi o\ \frac{d}{dt}\tilde{S}(t,g)e)\check{S}(t)$$

and so at $t = 0$

$$\frac{d^2}{dt^2}[\tilde{\phi}o\tilde{S}(t,g)e]f = \nabla^2\phi(T\pi o\tilde{X}(g)e,\ T\pi o\tilde{X}(g)e)\ gf$$

$$+ \nabla\phi\Big(\frac{D}{\partial t}\ T\pi o\tilde{X}_e(S(t,g)e)|_{t=0}\Big)gf$$

which gives the required formula. //

(C) Given a stochastic dynamical system (X,z) on M we can always construct an S.D.S. $(\tilde{X},z)$ on G(B) when B has a connection by taking $\tilde{X}(g)e$ to be the unique horizontal tangent vector to G(B) at g such that

$$T\pi(\tilde{X}(g)e) = X(\pi(g))e.$$

The systems $(\tilde{X},z)$ and (X,z) are then π-related in the sense of Proposition 9E so that if

$$u:[a,\tilde{\xi}) \times \Omega \to G(B)$$

is a maximal solution for $(\tilde{X},z)$ it follows that $x = \pi \circ u$ is a solution for (X,z). We will show in a moment that it is even a maximal solution.

The process u is called the *horizontal lift* of x. To apply our existence theorems G will have to be modelled on a Hilbert space. This is no problem if F is finite dimensional. Infinite dimensional examples include the case where G is modelled on some Hilbert space of Hilbert-Schmidt operators on F, or when G is a suitable Sobolev space of maps of some finite dimensional manifold into a Lie group.

Lemma 13C Let $\pi:G(B) \to M$ be a principal bundle with group G, of class C^2. Suppose that $(\tilde{X},z)$ and (X,z) on G(B) and M respectively are π-related and that $\tilde{X}$ is G-invariant i.e.

$$\tilde{X}(R_g u)e = TR_g\tilde{X}(u)e \qquad u \in G(B),\ g \in G,\ e \in E$$

where R_g denotes the right action of G on G(B). Then, for subsets K of M, $\pi^{-1}(K)$ admits a uniform cover for $(\tilde{X},z)$ whenever K does for (X,z).

Proof Let $\{\Lambda^\alpha : \alpha \in A\}$ be a uniform cover of K with $\Lambda^\alpha = ((U^\alpha,\phi_\alpha),U^0_\alpha,\lambda^\alpha)$. Shrinking U^α if necessary we can assume it is homeomorphic to a ball in H, and in particular that $\pi^{-1}(U^\alpha)$ is trivializable. We will therefore identify $\pi^{-1}(U_\alpha)$ with $U^\alpha \times G$.

Take a chart (W,θ) for G about the identity element. For each $g \in G$ we have a new chart $(R_g(W),\theta_g)$ about g where $\theta_g = \theta \circ R_{g^{-1}}$. From these we obtain charts $\{(U^\alpha \times R_g(W),\ \phi_\alpha \times \theta_g): g \in G\}$ covering $U^\alpha \times G$.

Since $(\tilde{X},z)$ is π-related to (X,z), $\tilde{X}$ is represented in $U^\alpha \times G$ in the form

$$\tilde{X}(x,g)e = (X(x)e,\ Y(x,g)e) \in TU^\alpha \times TG$$

for some Y. By right invariance we must have

$$Y(x,g) = TR_g Y(x,1)e$$

i.e.

$$(R_g)_*(Y) = Y.$$

Therefore for

$$(y,h) \in \phi_\alpha(U) \times \theta_g(R_g(W)) = \phi_\alpha(U) \times \theta(W)$$

if $x = \phi_\alpha^{-1}(y)$ we have

$$(\phi_\alpha \times \theta_g)_*(\tilde{X})(y,h)_e = \left(\phi_{\alpha*}(X)(y)e,\ \theta_{g*}(Y(x\ -)(h)e\right)$$

However

$$\theta_{g*}(Y) = \theta_*((R_{g^{-1}})_*(Y)) = \theta_*(Y),$$

and so $M(\theta_{g*}(Y))$ has bounds which are independent of $g \in G$, at least on any compact subset of $\theta(W)$. A uniform cover of $\pi^{-1}(K)$ can therefore be obtained from

$$\bigcup_\alpha \{(U^\alpha \times R_g(W), \phi_\alpha \times \theta_g) : g \in G\}. \ //$$

We can now prove the existence of horizontal lifts for all time. For a more general discussion see (Bismut 1981b) and (Meyer 1981).

<u>Theorem 13C</u> <u>Let</u> $\pi : G(B) \to M$ <u>be a principal G-bundle with connection of class</u> C^2, <u>where</u> G <u>is modelled on a separable Hilbert space</u>. <u>Let</u> $x:[a,\xi) \times \Omega \to M$ <u>be a maximal solution for the system</u> (X,z) <u>on</u> M. <u>Then for any</u> $u_a : \Omega \to G(B)$ <u>which is</u> F_a<u>-measurable and has</u> $x_a = \pi \circ u_a$ <u>there is a horizontal lift</u> $u:[a,\xi) \times \Omega \to G(B)$ <u>of</u> x.

<u>Proof</u> Let $u:[a,\eta) \times \Omega \to G(B)$ be a maximal solution of $(\tilde{X},z)$ with u_a as given. We must show that $\eta = \xi$ a.s. Take a nested sequence $\{U_i\}_{i=1}^\infty$ of open sets in M, covering M, and such that each $\bar{U}_i$ has a uniform cover. Then, as shown in the proof of Lemma 9F, $\xi = \sup_i \xi^i$ where ξ^i is the first exit time of x from U_i. By Lemma 13C we also have $\eta = \sup_i \eta^i$ where η^i is

the first exit time of u from $\pi^{-1}(U_i)$. Since $x = \pi o u$ we have $\eta^i = \xi^i$ a.s., and so $\eta = \xi$ a.s. as required. //

(D) We need to mention another possible definition of covariant stochastic differential equations. As before we assume that B has been given some linear connection, though here we shall treat it as a connection map $K : TB \to B$, see (Eliasson 1967). In particular we shall use the vector bundle isomorphism

$$TB \xrightarrow{\alpha} p_*(B) \oplus p_*(TM)$$

determined by K.

Given J as before, but now assumed to be C^2, define

$$\check{J} : B \times E \to p_*(B)$$

by

$$\check{J}(b)e = (b, J(p(b))(e)b) \qquad\qquad b \in B$$

considering $p_*(B)$ as a subset of $B \times B$, i.e. $p_*(B) = \{(b,b'):p(b) = p(b')\}$.

Suppose that x is a solution of a stochastic differential equation $dx = Xdz$ on M. Define

$$\check{X} : B \times E \to p_*(TM)$$

by

$$\check{X}(b)e = (b, X(p(b))e)$$

and consider the stochastic dynamical system (Y,z) on B where

$$Y : B \times E \to TB$$

is given by

$$Y(b,e) = \alpha^{-1}(\check{J}(b)e, \check{X}(b)e).$$

The equation $dv = Y\,dz$ is then a good candidate to be the definition of the covariant equation

$$Dv_t = J(x_t)(dz_t)(v_t)$$

since it reduces to the equation $\frac{Dv}{\partial t} = J(x_t)(\frac{dz}{dt})(v_t)$ when z is differentiable.

By Theorem 10 the solution v can also be expressed as a limit of solutions obtained from piecewise C^1 approximations of z.

Proposition 13D With the above notation suppose $v:[a,\xi)\times\Omega\to B$ is a maximal solution to $Dv_t = J(x_t)(dz_t)v_t$. Let $\phi:B\to L$ be a C^2 linear form on B with values in a Hilbert space L. Then almost surely on Ω_t, $a \leq t < b$, given a connection on M,

$$\begin{aligned}\phi(v_t) = \phi(v_a) &+ \int_a^{t\wedge\xi} \nabla\phi\Big(X(x_s)dz_s\Big)v_s + \int_a^{t\wedge\xi} \phi\Big(J(x_s)(dz_s)v_s\Big)\\ &+ \frac{1}{2}\int_a^{t\wedge\xi} \nabla^2\phi\Big(X(x_s)dz_s, X(x_s)dz_s\Big)v_s\\ &+ \frac{1}{2}\int_a^{t\wedge\xi} \nabla\phi\Big(\nabla X(X(x_s)dz_s)dz_s\Big)v_s\\ &+ \int_a^{t\wedge\xi} \nabla\phi\Big(X(x_s)dz_s\Big)J(x_s)(dz_s)v_s\\ &+ \frac{1}{2}\int_a^{t\wedge\xi} \phi\Big(\nabla J(X(x_s)dz_s)(dz_s)v_s\Big)\\ &+ \frac{1}{2}\int_a^{t\wedge\xi} \phi\Big(J(x_s)(dz_s)(J(x_s)(dz_s)v_s)\Big).\end{aligned}$$

Proof Let S(t,m)e give the flow of the vector field $X_e(-)$ on M and $\check{S}(t,b)e$ that of Y(-)e on B. Then if m = p(b)

$$P(\check{S}(t,b)e) = S(t,m)e$$

and

$$\frac{D}{\partial t}\ \check{S}(t,b)e = J(S(t,m)e)(e)\check{S}(t,b)e.$$

Consequently

$$\frac{D^2}{\partial t^2}\check{S}(t,b)e\Big|_{t=0} = \nabla J(X_e(m))(e)b + J(m)(e)(J(m)(e)b).$$

Since

$$\frac{d}{dt}\,\phi\circ\check{S}(t,b)e = \nabla\phi(X_e(S(t,m)e))\check{S}(t,b)e$$

$$+\ \phi\left(\frac{D}{\partial t}\,\check{S}(t,b)e\right)$$

and

$$\frac{d^2}{dt^2}\,\phi\circ\check{S}(t,b)e\Big|_{t=0} = \nabla^2\phi\Big(X_e(m),X_e(m)\Big)b + \nabla\phi\Big(\nabla X_e(X_e(m))\Big)b$$

$$+\ \nabla\phi\Big(X_e(m)\Big)\Big(J(m)(e)b\Big) + \nabla\phi\Big(X_e(m)\Big)\Big(J(m)(e)b\Big)$$

$$+\ \phi\left(\frac{D^2}{\partial t^2}\,S(t,b)e\Big|_{t=0}\right)$$

the result follows by Lemma 9B. //

<u>Example 13D</u> (i) Suppose X is C^3. With the notation of the proposition take $B = TM$ and $J(m)e(v) = \nabla X(v)e$ for $v \in T_mM$. The S.D.S. (Y,z) on B then becomes the derivative S.D.S. $(\delta X,z)$ on TM, as can be seen for example by taking normal coordinates about each point and using the formula for δX which appears in the proof of Lemma 9B(i). The solution $v(t)$ to

$$Dv_t = X(x_t)(dz_t)v_t$$

with $v(a) = v_0$ will therefore be the process $\delta F_t(v_0)$. Alternatively this could be seen by comparing the formula of Proposition 13D with that of Proposition 9C.

(ii) Take $J \equiv 0$. Let v be a maximal solution of $Dv_t = 0$. Set $x_t = \pi(v_t)$ as usual. Then v_t is the *parallel translate of* v_a *along* x *to* x_t. When x has a horizontal lift u to G(B) as in §C, i.e. when G is modelled on a Hilbert space, Proposition 13E below shows that for any maximal solution x of $dx = Xdz$:

(a) if $v_a : \Omega \to B$ is $\mathcal{F}_a$-measurable with $\pi(v_a) = x_a$ then v_a has a parallel translation along x up to the time x explodes;

(b) almost surely on Ω_t

$$v_t = u_t\,u_a^{-1}\,v_a$$

and v satisfies

$$(\text{Itô})\quad Dv_t = 0.$$

In case we do not know that u exists we can nevertheless prove (a) by the method used in Theorem 13C but using Example (ii) of §6 and the linearity of Y in the fibres of B to get the analogue of Lemma 13C.

When x is as in §A and not necessarily given by some (X,z) on M we can *define* parallel translation v_t by $v_t = u_t u_a^{-1} v_a$, for Brownian motion on M given by the stochastic development we have to use a slight modification: see §F below. The advantages of a general theory become apparent: for this see (Bismut 1981b) and (Meyer 1981). Early discussions of stochastic parallel translation used different approaches: (Itô 1963, 1975; Dynkin 1968; Eells & Malliavin 1973). It has proved a very useful tool, especially in the hands of Malliavin and his students. A discussion of diffusion processes in infinite dimensional vector bundles was given in (Belopol'skaja 1975).

(E) In §A our map J could equally well be taken to be polynomial, rather than linear, in E. This makes no difference to Propositions 13A and 13B. Using this we can see that when both are defined the difference between our two definitions of covariant equations is essentially the difference between Itô and Stratonovich equations:

Proposition 13E *With the notation of the previous paragraphs let* $\tilde{X}$ *be the horizontal lift of X to G(B), giving rise to a horizontal lift u of x to G(B). Let v be a maximal solution to*

$$Dv_t = J(x_t)(dz_t)v_t \tag{1}$$

Then v satisfies

$$(\text{Itô})\ Dv_t = J(x_t)(dz_t)v_t + \tfrac{1}{2}\nabla J(X(x_t)dz_t)(dz_t)v_t$$
$$+\tfrac{1}{2}J(x_t)(dz_t)J(x_t)(dz_t)v_t \tag{2}$$

and conversely. In particular if $\tilde{v}_t = u_t^{-1}v_t$ *we have*

$$\tilde{v}_t = v_a + \int_a^{t\wedge\xi} u_s^{-1} J(x_s)(dz_s)u_s\tilde{v}_s$$

$$+ \tfrac{1}{2}\int_a^{t\wedge\xi} u_s^{-1}\ \nabla J(X(x_s)dz_s)(dz_s)u_s\tilde{v}_s$$

$$+ \tfrac{1}{2}\int_a^{t\wedge\xi} u_s^{-1}\ J(x_s)(dz_s)J(x_s)(dz_s)u_s\tilde{v}_s$$

$$\text{a.s. on } \Omega_t. \qquad (3)$$

Proof Let v' be a maximal solution to (2). Let ϕ be as in Proposition 13C and 13B. The formulae for $\phi(v_t)$ and $\phi(v_t')$ given in those propositions reduce to being the same. Since ϕ could be an extension over B of part of a local trivialization of B, and such local trivializations can be used to get a cover of B by regular localizations of Y, we see (as in Theorem 9D) that v' satisfies (1). The result follows by the uniqueness of solutions and the definition of (2). //

(F) Let B,G(B), and J be as before, but now suppose that M is a finite dimensional Riemannian manifold and that x is a Brownian motion on M defined via the stochastic development. To distinguish it from $\pi:G(B) \to M$ we shall temporarily let $\tilde{\pi}:O(M) \to M$ be the projection of the orthonormal frame bundle. Then $x_t = \tilde{\pi}(u_t)$ where u_t is the O(M)-valued horizontal lift process.

Let us first assume that G is modelled on a separable Hilbert space and construct a horizontal lift of x to G(B). For this we can take the pull back of π over $\tilde{\pi}$ to obtain

$$\begin{array}{ccc} \tilde{\pi}^*G(B) & \xrightarrow{\pi^*(\tilde{\pi})} & G(B) \\ {\scriptstyle \tilde{\pi}^*(\pi)}\downarrow & & \downarrow{\scriptstyle \pi} \\ O(M) & \xrightarrow{\tilde{\pi}} & M \end{array}$$

where $\tilde{\pi}^*G(B) \to O(M)$ is a principal G-bundle and will be furnished with the pull-back connection from that of G(B): its connection form is just the connection form of G(B) pulled-back by $\pi^*(\tilde{\pi})$. Given $g_a \in \pi^{-1}(x_a)$ we can apply Theorem 13C to obtain a horizontal lift U_t of u_t to $\tilde{\pi}^*G(B)$ with $U_a = (u_a, g_a)$. Set

$$g_t = \pi^*(\tilde{\pi})(U_t).$$

We can then *define* g to be the *horizontal lift of* x through g_a. It will have the same explosion time as x.

We are now in the situation of §A and the equation

$$(\text{Itô})\ Dv_t = J(x_t)(dz_t)(v_t)$$

is defined by equation (2) of §A (with g_t replacing u_t). Also the equation

$$Dv_t = J(x_t)(dz_t)v_t$$

can be defined to mean the Itô equation (2) of Proposition 13E, when J is C^2. Alternatively it can be defined by pulling everything back over $\tilde{\pi}$, including J. Then solving the corresponding equation for sections of $\tilde{\pi}^*(B)$ over O(M), and projecting back down to sections of $p:B \to M$. This works even if G is not modelled on a separable Hilbert space.

When the horizontal lift g_t exists we have parallel translation in B as before. Namely, for $\omega \in \Omega_t$

$$\Pi_t(\omega):p^{-1}(x_a(\omega)) \to p^{-1}(x_t(\omega))$$

with

$$\Pi_t(\omega)(v) = g_t(\omega)g_a(\omega)^{-1}(v).$$

In this case it is clear that it is an isomorphism of the fibres of p. The following is a variant of Proposition 13B. In it Δ denotes the Laplace-Beltrami, or gauge invariant Schrödinger operator, on sections of p, i.e. $\Delta = \frac{1}{2}\,\text{trace}\,\nabla^2$.

Proposition 13F Let ψ be a C^2 section of $p:B \to M$, with Π_t parallel translation of the fibres of p along the Brownian motion x of M. Then, almost surely on Ω_t, $a < t < b$

$$\Pi_t^{-1}\ \psi(x_t) = \psi(x_a) + \int_a^{t\wedge\xi} \Pi_s^{-1}\ \nabla\psi(u_s dz_s) + \frac{1}{2}\int_a^{t\wedge\xi} \Pi_s^{-1}\ \Delta\psi(x_s)ds$$

where u is the O(M)-valued horizontal lift process of x and z is the $\mathbb{R}^n$-valued Brownian motion driving u.

Proof This follows as usual by Lemma 9B(ii), using Lemma 12B as in the proof of Proposition 13B, and the standard reduction of second order integrals with respect to Brownian motion. //

Example 13F (DeWitt-Morette et al. 1980). In the gauge invariant model of a quantum particle, charge e, with configuration space M in a classical magnetic field, the field is represented by a connection on a principal bundle G(B) over M with group G = U(1), and the wave function ψ is a section of an associated complex line bundle $p:B \to M$.

Suppose that these bundles are trivializable. Choosing a trivialization we can write B(G) as $M \times U(1)$ and B as $M \times \mathbb{C}$. The equation for the horizontal lift (x_t, g_t) of the Brownian motion $x_t = \tilde{\pi}(u_t)$ on M is then

$$dg_t = -i\,\Gamma(x_t)(u_t dz_t)g_t \tag{1}$$

$(g_t(\omega) \in U(1))$ where Γ is the connection coefficient, $\Gamma = \frac{eA}{\hbar c}$ for A the vector potential, a real valued 1-form on M. The covariant differentiation operator on sections of p, represented by their principal parts $\psi:M \to \mathbb{C}$, is then

$$\psi \mapsto d\psi + i\,\frac{eA}{\hbar c} : TM \to \mathbb{C}.$$

We can solve the linear equation (1): almost surely on Ω_t

$$g_t = \exp\{-i\int_a^t \Gamma(x_s)(u_s \circ dz_s)\}g_a,$$

where the integral is Stratonovich (see page 75). In particular for our wave function ψ

$$\begin{aligned}\Pi_t^{-1}\,\psi(x_t) &= \exp\{\frac{ie}{\hbar c}\int_a^t A(x_s)(u_s \circ dz_s)\}\psi(x_t)\\ &= \exp\{\frac{ie}{\hbar c}\int_a^t A(x_s)(u_s dz_s) + \tfrac{1}{2}\frac{ie}{\hbar c}\int_a^t \operatorname{div} A(x_s)ds\}\psi(x_t)\end{aligned} \tag{2}$$

by definition of the Fisk-Stratonovich integral and using Lemma 9B(ii) to compute '$d(A(x_s)u_s)$'. By $\operatorname{div}A(x_s)$ we mean $\sum_i \nabla A(e_i)(e_i)$ for an orthonormal base $\{e_i\}_i$ of $T_{x_s}M$.

(G) Let P and Q be C^2 manifolds with P separable Hilbert. Let $y:[a,\zeta) \times \Omega \to Q$ be a process of the form

$$y_t = f(x_t)$$

where x_t is a maximal solution of a stochastic differential equation $dx = X dz$ on P, and $f:P \to Q$ is C^2. If ϕ is a C^1 form on Q with values in a Hilbert space G and $\eta:\Omega \to [a,b)$ is a stopping time which is less than ζ we can define *the integral of* ϕ *along* y *from* a *to* η by

$$\int_{y|[a,\eta]} \phi = \int_a^\eta \phi(Tf(X(x_s) \circ dz_s)).$$

In terms of Itô integrals, if S(m,t)e denotes the flow of X_e on P with S(m,a)e = m, then

$$\int_{y|[a,\eta]} \phi = \int_a^\eta \phi(Tf(X(x_s)dz_s))$$
$$+ \tfrac{1}{2} \int_a^\eta \frac{d}{dt} \phi\Big(Tf(X(S(t,x_s)(-)dz_s)\Big)\Big|_{t=0} dz_s$$

using Lemma 9B(ii) to compute '$d[\phi(Tf(X(x_s)-))]$'. There is a more general theory, for example (Kohn 1975; Ikeda & Manabe 1978; Michel 1979; Bismut 1981b; Meyer 1981), see (Ikeda & Watanabe 1981; Meyer 1981a) and also (Darling 1982).

Note that by comparing with Lemma 9B(ii), taking $\phi = dg$ for $g:Q \to G$ of class C^2 we have

$$\int_{y|[a,\eta]} dg = gf(x_\eta) - gf(x_a)$$
$$= g(y_\eta) - g(y_a)$$

almost surely. There is also the following, on the lines of a *transfer principle*, c.f. (Malliavin 1977a, 1978a,b). The proof is immediate from the Itô integral expression given above.

<u>Theorem 13G</u> <u>Suppose that for all</u> p ∈ P <u>and</u> e ∈ E

$$\phi(Tf(X(p)e)) = 0$$

<u>then</u>

$$\int_{y|[a,\eta]} \phi = 0 \qquad \text{a.s.}$$

all $a \prec \eta < \zeta$. //

When Q is the total space of a principle bundle with connection form ϖ, and if the group concerned is modelled on a separable Hilbert space, so that ϖ has values in such a space, it is natural to say that the process y is *horizontal* if for all $a \prec \eta < \xi$ we have

$$\int_{y|[a,\eta)} \varpi = 0.$$

By Theorem 13G it is a simple matter to check that all the 'horizontal lifts' constructed in the previous sections are in fact horizontal in this sense.

§14 ITÔ EQUATIONS ON MANIFOLDS. BAXENDALE'S APPROACH.

(A) Suppose that M is furnished with a linear connection, which we will denote by Γ. Given (X,z) as before on M we can define the *Itô equation*

$$(\text{Itô}) \quad dx_t = X(x_t)dz_t, \tag{1}$$

sometimes denoted by $(\Gamma)\ dx_t = X(x_t)dz_t$, to mean the equation

$$dx_t = X(x_t)dz_t - \tfrac{1}{2}\nabla X(X(x_t)dz_t)dz_t. \tag{2}$$

In fact we have not strictly speaking defined equations on manifolds which are bilinear in the 'noise' dz, but the definition is obvious and in any case in the examples we are interested in the second order term reduces to a first order term, e.g. to $-\tfrac{1}{2}$ trace $\nabla X(X(x_t)-)(-)$ when z is a Brownian motion.

This definition was given in (Elworthy 1978) following the approach used in (Baxendale 1976, 1980). In a regular localization Λ of (2) we see it is represented as

$$y_t = \phi(x_{t_0}) + \int_{t_0}^{t} M(X_\Lambda)(y_s)dz_s - \tfrac{1}{2}\int_{t_0}^{t} \lambda(y_s)D\tilde{X}(y_s)(\tilde{X}(y_s)dz_s)dz_s$$
$$- \tfrac{1}{2}\int_{t_0}^{t} \lambda(y_s)\Gamma_\phi(y_s)(\tilde{X}(y_s)dz_s)(\tilde{X}(y_s)dz_s) \tag{3}$$

for $\Lambda = ((U,\phi),U_0,\lambda)$ where $\tilde{X} = \phi_*(X)$ and Γ_ϕ is the connection coefficient (Christoffel symbol) in the chart (U,ϕ). For $t < \tau^\Lambda_{t_0}$ this reduces to

$$y_t = \phi(x_{t_0}) + \int_{t_0}^{t} X_\Lambda(y_s)dz_s$$

$$- \tfrac{1}{2} \int_{t_0}^{t} \Gamma_\phi(y_s)(X_\Lambda(y_s)dz_s)(X_\Lambda(y_s)dz_s) \qquad (4)$$

and so, since we are only interested in $t < \tau_{t_0}^{\Lambda}$, by the local uniqueness theorem, Theorem 5 of Chapter VI, we can replace equation (3) by (4) in the definition of affirmation. However (4) involves no derivatives of $\tilde{X}$, and therefore we see that (1) will have maximal locally regular solutions which will be unique given only that X is locally Lipschitz, for example C^1. Moreover when $M = \mathbb{R}^n$ with its usual connection this definition of (1) agrees with the usual definition, namely

$$x_t = x_a + \int_a^t X(x_s)dz_s.$$

(B) In (Baxendale 1976, 1980) there is a slightly different approach to stochastic differential equations on manifolds. Essentially the given data is a Gaussian field of random vector fields. To be precise, let M be a complete Riemannian manifold with a continuous vector field A and a mean zero Gaussian measure γ on the Fréchet space C(TM) of continuous vector fields on M. Associated to γ there is a Wiener process $\{W_t : t \geq 0\}$, $W_t : \Omega \to C(TM)$. Set $Z_t = W_t + tA$. For $x_0 \in M$ and a partition Π of an interval $[0,T]$, $0 = t_0 < \ldots < t_{m+1} = T$ define $x_j^\pi : \Omega \to M$ for $j = 0,\ldots,m$ by

$$x_{j+1}^\pi(\omega) = \exp_{x_j^\pi(\omega)}\left(\Delta_j Z(\omega)(x_j^\pi(\omega))\right).$$

The idea is that x_{m+1}^π should converge, as mesh Π goes to zero, to some $x_T : \Omega \to M$, to give an M-valued process $\{x_T : T \geq 0\}$. In fact the proof of the existence of an associated Markov process on M is obtained, under certain conditions on γ and A, by solving stochastic differential equations: it is essentially that corresponding to the equation

$$(\text{Itô}) \quad dx_t = X(x_t)dZ_t$$

where E = C(TM) and $X : C(TM) \times M \to TM$ is the evaluation

$$X(\xi,m) = \xi(m).$$

However, as such, (X,z) does not satisfy our standing conditions since X will not be locally Lipschitz (since ξ may not be).

If we are given a stochastic differential equation in our sense

$$(\text{Itô})\quad dx_t = XdB_t + A\,dt$$

where B is an n-dimensional Brownian motion, we can obtain a Gaussian measure γ on C(TM) as the image under

$$X_{-}(-) : \mathbb{R}^n \to C(TM)$$

$$e \to X_e$$

of the standard Gaussian measure of $\mathbb{R}^n$. The process W_t will then be just the image of B_t. This is closely related to the process on the diffeomorphism group constructed in the next chapter, and this aspect of Baxendale's approach is brought out in (Baxendale 1981).

In general the use of the exponential map in approximation schemes, as in the definition of x_r^{π} above, seems to lead to Itô calculus, while the solution of ordinary differential equations, as in piecewise linear approximations leads to Stratonovich calculus. This is born out in the integration of 1-forms in (Darling 1982).

CHAPTER VIII : REGULARITY

§1 THE INDUCED PROCESS ON THE DIFFEOMORPHISM GROUPS

(A) We shall use some results about the diffeomorphism groups of compact manifolds to obtain information about the flow F_t of an S.D.S. and the uniform convergence of piecewise linear approximations. This was described, for compact manifolds, in (Elworthy 1978). The method mimics that used by Ebin & Marsden (1970) for ordinary differential equations.

Instead of using diffeomorphism groups of compact manifolds some readers might prefer to consider groups of diffeomorphisms of $\mathbb{R}^n$ which are the identity outside the unit ball. Many of the properties are easier to see for these groups and they can be used in the compact manifold case by embedding the manifold in some open unit ball and then extending the S.D.S. so that it is the identity in a neighbourhood of the boundary of the ball. This is described in detail in (Carverhill & Elworthy 1982) where the method is used to give a unified treatment of many recent results, e.g. the generalized Itô formula (Bismut 1981a,b) and the criterion for the flow to be a diffeomorphism (Kunita 1981, 1982). A similar technique, using nuclear spaces is employed in (Ustunel).

(B) Assume that M is a C^∞ compact Riemannian manifold with dim M = m. For a C^∞, finite dimensional, Riemannian manifold N let $H^s(M;N)$ denote the space of maps $h:M \to N$ of Sobolev class H^s (sometimes written $L^{2,s}$ or $W^{2,s}$) : for $s > \frac{m}{2}$ it is well defined and has a natural C^∞ Hilbert structure, independent of the choice of Riemannian metrics on M and N, see (Eliasson 1967, Palais 1968, Omori 1974). In fact if $h:M \to N$ is C^∞ there is a natural chart (U,ϕ_h) where U is a sufficiently small neighbourhood of h and ϕ_h maps U into $H^s(h^*TN)$, the H^s sections of the pull-back of TN by h. This is given by

$$\phi_h(f)(x) = \exp^{-1}_{h(x)} f(x) \qquad f \in U, \quad x \in M,$$

where exp is the exponential map of N. The tangent space $T_h(H^s(M;N))$ to $H^s(M;N)$ at any point h can be identified with the Hilbertable space $H^s(h^*TN)$ or equivalently with the space of H^s maps $v:M \to TN$ satisfying $\pi \circ v = h$, where $\pi:TN \to N$ is the projection. The tangent bundle to $H^s(M;N)$ can then be identified with $H^s(M;TN)$.

For $r = 0,1,2,\ldots,\infty$ let $C^r(M;N)$ denote the corresponding spaces of C^r maps. There are the well known Sobolev embedding theorems which say that there are natural continuous inclusions:

$$C^s(M;N) \to H^s(M;N) \qquad s > \tfrac{1}{2}m$$

and

$$H^s(M;N) \to C^r(M;N) \qquad s > r + \tfrac{1}{2}m.$$

See (Palais 1968).

When $s > 1 + \frac{1}{2}m$ it therefore makes sense to talk about the set of diffeomorphisms in $H^s(M;N)$. This will be an open subset of $H^s(M;M)$ and we denote it by $\mathcal{D}^s$. It is a topological group under composition; furthermore:

(a) For fixed $h \in \mathcal{D}^s$ the right translation map

$$R_h:\mathcal{D}^s \to \mathcal{D}^s$$

$$R_h(f) = f \circ h$$

is C^∞, and

(b) the multiplication

$$\psi_r:H^{s+r}(M;N) \times \mathcal{D}^s \to H^s(M;N)$$

$$\psi_r(f,h) = f \circ h$$

is C^r, for $r = 0,1,2,\ldots$.

For these facts see (Ebin & Marsden 1970) and the references therein; (Bourguignon & Brezis 1974) may also be found helpful. From now on we take $s > 1 + \frac{1}{2}m$.

(C) Consider an S.D.S (X,z) on M such that every vector field $X(-)e$, for $e \in E$, lies in $H^{s+2}(M;TM)$. They will then all lie in the tangent space at the identity, $T_1\mathcal{D}^{s+2}$, to $\mathcal{D}^{s+2}$. Thus we have a continuous linear map

$$X(-):E \to T_1\mathcal{D}^{s+2}.$$

This can be extended by right translation to an S.D.S. $(\tilde{X},z)$ on $\mathcal{D}^s$. In fact define

$$\tilde{X}:\mathcal{D}^s \times E \to T\mathcal{D}^s$$

by

$$\tilde{X}(h,e) = T_1R_h(X_e).$$

Then

$$\tilde{X}(h,e)(x) = X_e(h(x)) = \psi_2(X_e,h)(x) \qquad x \in M$$

taking $N = TM$ in (b). Considered as a section of $\mathbb{L}(\underline{E};T\mathcal{D}^s)$ it follows from (b) that $\tilde{X}$ is C^2 when $\dim E < \infty$, and (b) implies by a standard argument that for general E it is C^1 with locally Lipschitz first derivative.

Let (U,θ) be a C^∞ chart for $\mathcal{D}^s$ about the identity, for example $\theta = \phi_1:U \to T_1\mathcal{D}^s$ as above. Then for each $h \in \mathcal{D}^s$ we have a new chart

$$\theta_h:R_h[U] \to T_1\mathcal{D}^s$$

given by

$$\theta_h = \theta \circ R_{h^{-1}}.$$

This is a C^∞ chart by (a). From $\{(R_h[U],\theta_h)\}_h$ we get a uniform cover of $\mathcal{D}^s$ for $(\tilde{X},z)$.

Theorem 1C For $s > 1 + \frac{1}{2}m$ the S.D.S. $(\tilde{X},z)$ on $\mathcal{D}^s$ is complete. In particular we have a locally regular solution $F:[a,b] \times \Omega \to \mathcal{D}^s$ to

$$dF = \tilde{X}dt$$

with

$$F(0) = 1.$$

This is a version of the flow of the S.D.S. (X,z) on M: for each $x_0 \in M$

the process $F(-)(x_0)$ is an l.r. solution of $dx = Xdz$.

Proof The completeness comes from Corollary 6.1 of Chapter VII. To see that F is a version of the flow of (X,z) set f equal to the evaluation map at x_0

$$f: \mathcal{D}^s \to M$$

$$f(h) = h(x_0).$$

Then f is C^∞ and it is easy to see that $\tilde{X}$ and X are f-related in the sense of Proposition 9E of Chapter VII. The result therefore follows by that proposition. //

Theorem 1C was given in (Elworthy 1978) and was stimulated by previous work of Baxendale (1976; 1980) and Malliavin (1976; 1977a; 1978a). The corollaries which follow are versions of those of Malliavin on the flow and uniform convergence of approximating solutions. However he regularized his process z by convolution rather than by using piecewise linear approximations. He then showed uniform convergence of the corresponding approximating flows, and from this was able to deduce the existence of a smooth version of the flow F_t. See also (Kunita 1980) (Bismut 1981a,b) and (Ikeda & Watanabe 1981), without forgetting (Blagoveščenskii & Freidlin 1961). The converse problem of when a process on a space of homeomorphisms or diffeomorphisms of M comes from an S.D.S. on M is discussed in (Baxendale 1981) and (Kunita 1982). The first corollary is improved in Corollary 1D below.

Corollary 1C.1 For a compact C^∞ manifold M if each X_e is an H^{s+2} vector field, where $s > 1 + \frac{1}{2}m$, then there is a version of the flow of (X,z) which is an H^s diffeomorphism at each time t and is continuous in t into the H^s topology.

Note that we already have the existence of a continuous version of the flow by Theorem 3B of Chapter VI, and the embedding technique used for Theorem 4 of Chapter VII.

Corollary 1C.2 For a compact C^∞ manifold M if X is C^∞ there is a version of the flow of (X,z) which is a C^∞ diffeomorphism at each time t and is continuous in t into the C^∞ topology.

Corollary 1C.3 For a compact C^∞ manifold M let $\{z_\pi\}$ be the piecewise linear approximations as in Theorem 4 of Chapter VII and let $F_{\pi,t}$ be the flow of (X,z_π) obtained by solving the ordinary differential equations $\frac{dx_\pi}{dt} = X(x_\pi)\frac{dz_\pi}{dt}$. If X is C^∞ then as mesh $\pi \to 0$ so $F_{\pi,t}$ converges in measure, in the C^∞ topology of $C^\infty(M;M)$, uniformly on each subinterval $a \leq t \leq b_0 < b$, with limit a version of the flow of (X,z).

If X is H^{s+2} for some $s > 1 + \frac{1}{2}m$ the same holds but with convergence in the H^s topology. Consequently if each $F_{\pi,t}$ lies a.s. in some closed subset A of $\mathcal{D}^s$ then F_t lies in A almost surely.

Proof of the Corollaries

The first is immediate from the theorem. The second follows from the first. The last one comes from the theorem via Theorem 10 of Chapter VII. //

As an example of a closed subset A in Corollary 1C.3 we could take A to be the set of volume preserving diffeomorphisms in $\mathcal{D}^s$. We then see that F_t is almost surely volume preserving if each vector field X_e is divergence free. In fact in this case A is a submanifold of $\mathcal{D}^s$ and $(\tilde{X},z)$ restricts to an S.D.S. on A.

(D) We will look at some equations for the derivative flow δF_t under the assumptions of Theorem 1C. A useful reference here is (Eliasson 1967) §6, but we shall not go into all the formalism. The case of the unit disc in $\mathbb{R}^n$ is very much simpler.

Let F_t be the flow of (X,z): by this we shall mean a version which is continuous, and so in our case we can assume that $F_t(\omega) \in \mathcal{D}^s$ for all ω in Ω. In particular we can, and will, take $\delta F_t(\omega) = TF_t(\omega)$ for each ω to get a continuous version of δF_t. Let U_t be an H^s version of the flow on the frame bundle O(M) obtained from a horizontal lift of X as in §13C of Chapter VII. Thus $U_t(u_0)$ is a horizontal lift of $F_t(x_0)$ for u_0 a frame at x_0. Correspondingly for each $\omega \in \Omega$ we have a parallel translation along the flow:

$$\Pi_t(\omega) \in H^s(\mathbb{L}(TM;F_t(-,\omega)^*TM))$$

which is that element of the space of H^s sections of the vector bundle $\mathbb{L}(TM;F_t(-,\omega)^*TM)$, with $F_t(-,\omega)^*TM$ the pull back of TM, given by

$$\Pi_t(\omega)(x_0)v_0 = u_0\, U_t(u_0,\omega)^{-1}\, v_0 \qquad v_0 \in T_{x_0}M$$

where u_0 is an arbitrary frame at $x_0 \in M$. In fact we will often consider $\Pi_t(\omega)$ as a map of TM into itself. There is also the inverse

$$\Pi_t(\omega)^{-1} \in H^s(\mathbb{L}(F_t(-,\omega)^*TM;TM)).$$

The next proposition follows from Proposition 13E and Example 13D of Chapter VII: in Proposition 13E we will have

$$\tilde{v}(t) = u_0^{-1}\, \tilde{V}_t(x_0)v_0.$$

Proposition 1D For $\omega \in \Omega$ define $\tilde{V}_t(\omega) \in H^{s-1}(\mathbb{L}(TM;TM))$ by

$$\tilde{V}_t(\omega) = \Pi_t(\omega)^{-1}\circ\delta F_t(-,\omega).$$

Then

$$\tilde{V}_t = I + \int_a^t \Pi_s^{-1}\, \widetilde{\nabla X}(\Pi_s\cdot\tilde{V}_s)dz_s + \int_a^t \Pi_s^{-1}\, \widetilde{\nabla^2 X}\Big(\tilde{X}(F_s)dz_s, \Pi_s\cdot\tilde{V}_s\Big)dz_s$$

$$+ \int_a^t \Pi_s^{-1}\, \widetilde{\nabla X}\Big(\widetilde{\nabla X}(\Pi_s\cdot\tilde{V}_s)dz_s\Big)dz_s \quad (*)$$

where

$$\widetilde{\nabla X}(\alpha)(e)(x_0)v_0 = \nabla X_e(\alpha(x_0)v_0)$$

and

$$\widetilde{\nabla^2 X}(\beta,\alpha)(e)(x_0)v_0 = \nabla^2 X_e(\beta(x_0),\alpha(x_0)v_0)$$

for

$$e \in E,\ x_0 \in M,\ v_0 \in T_{x_0}M,\quad \alpha = \Pi_s(\omega)\cdot\tilde{V}_s(\omega)$$

and

$$\beta = \tilde{X}(F_s(-,\omega))e. \quad //$$

We can now continue to mimic Ebin and Marsden and so improve Corollary 1C.1.

Corollary 1D. The flow F_t lies in $\mathcal{D}^{s+1}$ almost surely and is almost surely continuous in t into the H^{s+1} topology of $\mathcal{D}^{s+1}$.

Proof The equation (*) for $\tilde{V}_t$ can be considered as an equation for $\tilde{V}_t \in H^s(\mathbb{L}(TM;TM))$. It is a linear equation and satisfies all of Conditions 1B of Chapter VI except that the Lipschitz constant L may be a function of ω. It follows by similar arguments to those we have used before that an admissible solution exists in $H^s(\mathbb{L}(TM;TM))$ defined for all time, for example see (McShane 1974) Chapter V, Theorem 1.5, (alternatively it should be possible to produce a uniform cover of $\mathcal{D}^s \times H^s(\mathbb{L}(TM;TM))$ for the pair $(F_t,\tilde{V}_t)$).

Letting $\tilde{V}_t'$ denote such a solution we have $\tilde{V} = \tilde{V}'$ a.s. and so $\Pi_t\cdot\tilde{V}_t$ is equivalent to δF_t which is just TF_t. Thus TF_t is continuous into H^s, almost surely, and the result follows. //

Remark 1D The corresponding statements to Corollaries 1C and 1D can be made about the flow F_t of an S.D.S. (X,z) on a non-compact finite dimensional manifold M when X has compact support and is of class H^{s+2}, or C^∞. This can be reduced to the compact manifold case by enclosing the support of X in an open relatively compact submanifold N of M with smooth boundary. We can then take the double N' of N (i.e. glue together two copies of N along their boundary) and extend $X|N$ to give an S.D.S. (X',z) on N'.

§2 THE FLOW FOR NON-COMPACT FINITE DIMENSIONAL MANIFOLDS

(A) We shall continue to assume that M is finite dimensional. A map f defined on M will be said to be in H^s_{loc} if each point of M is contained in some relatively compact domain N of M with smooth boundary such that $f|\bar{N}$ is of class H^s. Spaces of such maps are given the smallest topology such that the map $f \mapsto f|\bar{N}$ is continuous into the H^s topology for every such domain N. See (Palais 1968). The topology on the space of C^∞ maps will be defined likewise.

(B) The C^∞ and C^0 cases of parts (i) and (ii) of the theorem we give next seem to have appeared first in (Kunita 1980).

Theorem 2B Let (X,z) be an S.D.S. on a finite dimensional manifold M. Suppose X is in H^{s+2}_{loc}, where $s > 1 + \frac{1}{2}\dim M$. Then there is a version of the explosion time map $u \mapsto \xi^u$ defined for u in M and a version of the flow $F_t(u,\omega)$, $t \geq a$ defined when $t < \xi^u(\omega)$ such that if $M(t,\omega) = \{u \in M : t < \xi^u(\omega)\}$ then for each $(t,\omega) \in [a,b) \times \Omega$:

(i) $M(t,\omega)$ <u>is open in</u> M

(ii) $F_t(-,\omega):M(t,\omega) \to M$ <u>is in</u> H^{s+1}_{loc} <u>and is a diffeomorphism onto an open subset of</u> M. <u>Moreover the map</u> $\alpha \mapsto F_\alpha(-,\omega)$ <u>of</u> [a,t] <u>into</u> $H^{s+1}_{loc}(M(t,\omega);M)$ <u>is continuous</u>.

<u>Also, for any subset</u> K <u>of</u> M <u>let</u>

$$\xi^K(\omega) = \inf\{\xi^u(\omega) : u \in K\}$$

<u>and</u>

$$\Omega^K_c = \{\omega \in \Omega : c < \xi^K(\omega)\}$$

(for the measurability of such sets see Remark 2B(c) below). <u>Then</u>:

(iii) <u>When</u> K <u>is compact we have</u>

$$\sup\{d(u_0,F_t(u,\omega)):u \in K\} \to \infty \text{ as } t \to \xi^K(\omega)$$

<u>whenever</u> $\xi^K(\omega) < b$, <u>for any complete metric</u> d <u>on</u> M <u>and any point</u> u_0 <u>of</u> M.

(iv) <u>Let</u> $\{z_\pi\}_\Pi$ <u>be the piecewise linear approximations of</u> z <u>with corresponding flow</u> $F_{\pi,t}$ <u>as in Theorem 4 of Chapter VII</u>. <u>Then</u> $F_{\pi,t}$ <u>converges to</u> F_t <u>in</u> H^s_{loc} <u>in measure, uniformly in each closed time interval</u> $a \leqslant t \leqslant c \leqslant b$, <u>in the sense that for each</u> $\delta > 0$ <u>and relatively compact domain</u> N <u>of</u> M <u>with smooth boundary then for a metric</u> d^s <u>on</u> $H^s(\bar{N};M)$ <u>we have</u>

$$\lim_{\text{mesh }\Pi\to 0} \mu\{\omega \in \Omega^{\bar{N}}_c : \sup_{a\leqslant t\leqslant c} d^s(F_{\pi,t}(-,\omega);F_t(-,\omega)) > \delta\} = 0$$

<u>where the distance</u> d^s <u>is taken to be infinite when</u> $F_{\pi,t}(-,\omega)$ <u>is not defined on</u> N.

<u>Finally, when</u> X <u>is</u> C^∞ <u>the corresponding results to</u> (i), (ii), (iv) <u>hold with the</u> C^∞ <u>topology replacing the</u> H^s <u>topology</u>.

If the explosion time map can be chosen with $\xi^u = b$ for all u we will say that (X,z) is *strongly complete*. The term *strictly conservative* is also used.

<u>Proof</u> Let N_r, r = 1,2,... be a sequence of relatively compact subdomains of N with smooth boundaries ∂N_r and

$$\bar{N}_r \subset N_{r+1} \subset N = \bigcup_r N_r.$$

Take C^∞ maps $\lambda_r: M \to [0,1]$ with compact support such that λ_r is identically one on $\bar{N}_r$. Set $X_r(x) = \lambda_r(x)X(x)$ for $x \in M$, and consider the S.D.S. (X_r, z) on M. By Remark 1D it has a solution flow $F_t^r(-,\omega): M \to M$ defined for all time and satisfying (i), (ii) and (iv), (for (iv) we are using (X_r, z_π) with corresponding flow $F_{\pi,t}^r$).

Let $\tau^r(u,\omega)$ be the first exit time of $F_t^r(u,\omega)$ from N_r, for each u in M. Then, for each u and any choice of the explosion time ξ^u of F(u), we have

$$\tau^r(u,\omega) < \xi^u(\omega) \qquad \text{a.s.}$$

and

$$F(u)|[a,\tau^r(u,-)) \times \Omega \sim F^r(u)|[a,\tau^r(u,-)) \times \Omega .$$

Therefore we get an equivalent version of ξ^u if we redefine it by

$$\xi^u(\omega) = \sup \{\tau^r(u,\omega): r = 1,2,\ldots\}. \tag{1}$$

With this definition assertion (i) is immediate by the sample continuity in (t,u) of $F_t^r(u,-)$ since

$$M(t,\omega) = \bigcup_r \{u \in M: t < \tau^r(u,\omega)\}. \tag{2}$$

To get a good version of F_t we define

$$F_t(u,\omega) = \lim_{r\to\infty} F_t^r(u,\omega) \qquad t < \xi^u(\omega). \tag{3}$$

The limit exists for all (u,ω) with $t < \xi^u(\omega)$ except for a set of measure zero independent of (t,u) because $F_t^{r+1}(u,\omega)$ agrees with $F_t^r(u,\omega)$ if $t < \tau^r(u,\omega)$ except on such a set, $\Omega(r)$ say, of measure zero: to see this take a countable dense subset $\{u_i\}_{i=1}^\infty$ in M, then

$$F^r(u_i,-)|[a,\tau^r(u_i,-)) \times \Omega \sim F^{r+1}(u_i,-)|[a,\tau^r(u_i,-)) \times \Omega$$

and we can use the sample continuity of F^r and F^{r+1} and the lower semi-continuity of $\tau^r(-,\omega)$ to go to a general point u of M. For $\omega \in \bigcup_r \Omega(r)$ we will redefine each $F_t^r(u,\omega)$ and define $F_t(u,\omega)$ so that

$$F_t(u,\omega) = u = F_t^r(u,\omega) \qquad (t,u) \in [a,b) \times M, \quad r = 1,2,\ldots$$

and

$$\xi^u(\omega) = b.$$

Now let N be a relatively compact subdomain with smooth boundary. Suppose $\bar{N} \subset M(t,\omega)$. By the compactness of $\bar{N}$ and equality (2) it follows that $t < \tau^{r_1}(u,\omega)$ for all $u \in \bar{N}$ for some r_1. This means that $F_\alpha(u,\omega) = F_\alpha^{r_1}(u,\omega)$ for $(\alpha,\omega) \in [0,t] \times \bar{N}$ and so assertion (ii) follows (if $F_t(u_1,\omega) = F_t(u_2,\omega)$ we could choose N to contain both u_1 and u_2, but then $u_1 = u_2$ since $F_t^{r_1}(-,\omega)$ is injective).

Assertion (iii) is immediate from (1) and (3) since bounded subsets of M with respect to any complete metric are relatively compact and so lie in some N_r. For (iv) first note that by the above argument we have a map $r_2:\Omega_c^{\bar{N}} \to \{1,2,\ldots\}$ with $c < \tau^{r_2(\omega)}(u,\omega)$ for all $(u,\omega) \in \bar{N} \times \Omega_c^{\bar{N}}$. Also r_2 is measurable with respect to the completion $\bar{F}$ of F: see Remark 2B(c) below. Given $\varepsilon > 0$ we can therefore choose $r_3 \in \{1,2,\ldots\}$ and $A \in F$ with $\mu(A) < \varepsilon/3$ and

$$c < \tau^{r_3}(u,\omega) \text{ for } (u,\omega) \in \bar{N} \times (\Omega_c^{\bar{N}} - A).$$

Next use Lemma 7B of Chapter VII but with $[a,c] \times \bar{N}$ replacing the interval [a,b] to obtain a compact set K in N_{r_3} and a measurable set B in $\Omega_c^{\bar{N}} - A$ with $\mu(B) < \varepsilon/3$ such that

$$F_t^r(u,\omega) \in K \text{ if } (t,u,\omega) \in [a,c] \times \bar{N} \times (\Omega_c^{\bar{N}} - A - B).$$

Given $\delta > 0$, set $\delta_1 = \min(\delta, \operatorname{dist}(K, \partial N_{r_3}))$ and choose $\lambda > 0$ so that mesh $\Pi < \lambda$ implies

$$\mu\{\omega \in \Omega: \sup_{a\leq t\leq c} d^s(F_t^r(-,\omega), F_{\pi,t}^r(-,\omega)) \geq \delta_1\} < \varepsilon/3.$$

Since $d^s(F_t^r(-,\omega), F_{\pi,t}^r(-,\omega)) < \delta_1$ for $0 \leq t \leq c$ ensures that $F_{\pi,t}^r(u,\omega) = F_{\pi,t}(u,\omega)$ for $(t,u) \in [0,c] \times \bar{N}$ when $\omega \in \Omega_c^{\bar{N}} - A - B$, the assertion follows.

The C^∞ cases can be proved in exactly the same ways, or they can be deduced by noting that $C^\infty = \bigcap_{s=1}^{\infty} H_{loc}^s$ with the direct limit topology. //

<u>Remark 2B</u> (a) For X only of class C^{1+} we could also prove the corresponding statements to (i), (ii) and (iii) for the C_{loc}^0 topology using Theorem

3B of Chapter VI.

(b) From the proof we see that $\xi^u(\omega)$ can be taken to be measurable in $(u,\omega) \in M \times \Omega$ with respect to the product σ-algebra Borel $(M)^* F$.

(c) In order to prove (iv) we could have set up an S.D.S. $(\tilde{X},z)$ on $H^s(\bar{N};M)$ as in §1A. This need no longer be complete but the maximal solution, $\tilde{F}_t$, say, starting from the inclusion of $\bar{N}$ into M will have an explosion time, η say. We will show that $\eta = \xi^{\bar{N}}$ a.s. We do not claim that $\xi^{\bar{N}}$ is measurable: in fact general results e.g. see (Dellacherie & Meyer 1978) Chapter III, Theorem No. 44, together with the previous Remark (2B(b)), imply that $\xi^{\bar{N}}$ is $\bar{F}$-measurable where $\bar{F}$ is the completion of F and for the rest of the proof we shall work with $\bar{F}$ rather than F. However we need not use the measurability in our proof and then it will come out as a consequence of the result $\eta = \xi^{\bar{N}}$ $\bar{F}$-almost surely which we are about to prove.

Clearly $\eta(\omega) \leq \xi^{\bar{N}}(\omega)$ a.s.. On the other hand, using part (ii) of the theorem, for any $\varepsilon > 0$ by Lemma 7B of Chapter VII there is a compact set K_ε of $H^{s+1}(\bar{N};M)$ and an $\Omega_\varepsilon \in F$ with $\mu(\Omega_\varepsilon) < \varepsilon$ such that if $a \leq t < \xi^{\bar{N}}(\omega)$ and $\omega \notin \Omega_\varepsilon$ then $F_\alpha(-,\omega)|\bar{N} \in K_\varepsilon$ for $a \leq \alpha \leq t$. Considering K_ε as a subset of $H^s(\bar{N};M)$, according to Theorem 6 of Chapter VII if $\eta(\omega) < b$ then $\bar{F}(-,\omega)$ leaves this subset, almost surely, at some positive time before $\eta(\omega)$. Since $F_\alpha(-,\omega) = \bar{F}_\alpha(-,\omega)$ for $0 \leq \alpha < \eta(\omega)$, a.s., it follows that $\eta(\omega) = \xi^{\bar{N}}(\omega)$ a.s. with respect to $\bar{F}$. Thus, with the obvious abuse of notation,

$$F|[0,\xi^{\bar{N}}) \times \bar{N} \times \Omega \sim \bar{F}.$$

This fact can be useful in other ways: for example it follows that the isonomy class of $\xi^{\bar{N}}$ depends only on the isonomy class of z. Consequently whether or not we can choose $\xi \equiv b$ depends only on the isonomy class of z.

(d) Assertion (iii) of the theorem is a maximality condition which ensures *uniqueness* even if it only holds almost surely: suppose we have versions F^1 and F^2 with corresponding explosion time maps ξ_1^u and ξ_2^u which almost surely satisfy (i), (ii) and (iii) for the C^0_{loc} topology, with $F_t^1(-,\omega)$ and $F_t^2(-,\omega)$ possibly only continuous and not necessarily even local homeomorphisms and allowing for the possibility of exceptional negligible sets in (iii) depending on K. Then, for all $u \in M$, $\xi_1^u = \xi_2^u$ almost surely and, for all t, $F_t^1(u,\omega) = F_t^2(u,\omega)$ almost surely on $\{\omega: t < \xi_1^u(\omega)\}$, (we are assuming M has at most countably many connected components).

To see this we can first use the separability of M to observe that the flows certainly agree up to the time $\xi_1^u \wedge \xi_2^u$ except on some negligible set independent of u and t. Next let

$$N(\omega) = \{u \in M: \xi_1^u(\omega) < \xi_2^u(\omega)\}.$$

For a compact K in M, if K intersects $N(\omega)$ our weakened version of (iii) assures us that ω lies in some negligible set Ω_K say. The result follows since M is a countable union of compact sets.

Thus the explosion time map and flow are essentially independent of the smoothness class we are considering. In particular if $M = \mathbb{R}^n$ and X is also globally Lipschitz then we can take $\xi^u \equiv b$ for all $u \in M$ by Theorem 3B of Chapter VI.

(C) *Warning* 2C:

(a) With the notation of Theorem 2B we could set $F_t(u,\omega) = \Delta$ for $t \geqslant \xi^u(\omega)$ where Δ is the point at infinity. The resulting map $F_t(-,\omega): M \to M^+$ may not be continuous a.s.

(b) Suppose $f: M \to \mathbb{R}$ is continuous with compact support. Consider $f \circ F_t$ as a map defined on $M \times \Omega$ by

$$f(F_t(u,\omega)) = 0 \text{ if } t \geqslant \xi^u(\omega).$$

Then there may be a positive probability that $f(F_t(-,\omega))$ is not continuous on M.

(c) Even if (X,z) is complete in the sense that $\xi^u(\omega) = b$ a.s. for each $u \in M$, we may not be able to take $\xi^u \equiv b$ in the theorem i.e. complete does not imply strongly complete.

To see these assertions take $M = \mathbb{R}^2 - \{0\}$ and take $X: M \times \mathbb{R}^2 \to \mathbb{R}^2$ as $X(u,e) = e$, with z a Brownian motion on $\mathbb{R}^2$ as in (Elworthy 1978). A locally regular solution with $x(0) = u$ is given by $x(t) = u + z(t)$. It is well known that there is zero probability of a Brownian path in $\mathbb{R}^2$ hitting a given point (at $t > 0$) during its lifetime, see §6B Chapter V: therefore for each $u \in \mathbb{R}^2$

$$\mu\{\omega: u + z(t,\omega) = 0 \text{ for some } 0 < t < \infty\} = 0$$

and so solutions are defined for all time, i.e. (X,z) is complete. On the other hand by continuity in u we must have $F_t(u,\omega) = u + z(t,\omega)$ if $t < \xi^u(\omega)$ for all $u \in \mathbb{R}^2 - \{0\}$, except possibly on a set of measure zero in Ω. Therefore, almost surely, $\xi^u(\omega) \leqslant t$ if $u = -z(t,\omega)$. In particular for almost all ω we have $\xi^u(\omega) < \infty$ for some u in $\mathbb{R}^2 - \{0\}$. This proves assertion (c). In fact we must have

$$\xi^u(\omega) = \inf\{t : 0 < t < \infty \text{ and } u = -z(t,\omega)\}.$$

To prove (b) let Z be a countable dense subset of $\mathbb{R}^2 - \{0\}$. Then there is a subset Ω_0 in F of full measure such that $-z(-,\omega)$ does not enter Z if $\omega \in \Omega_0$. Let $f : M \to \mathbb{R}$ be continuous and satisfy

$$\begin{aligned} f(u) &= 0 \qquad \text{if } |u| \leqslant 1 \text{ or } |u| \geqslant 4 \\ &= 1 \qquad \text{if } 2 \leqslant |u| \leqslant 3. \end{aligned}$$

For fixed $t > 0$ let Ω_1 be the set of ω in Ω_0 for which there exists u in M satisfying

$$2 < |u + z(t,\omega)| < 3$$

and

$$u = -z(s_0,\omega) \quad \text{some } s_0 \in (0,t).$$

Then Ω_1 has positive measure (it corresponds to an open subset in the space of continuous paths). For $\omega \in \Omega_1$ and u as described we have $\xi^u(\omega) < t$ so that

$$f(F_t(u,\omega)) = 0.$$

However we can choose a sequence $\{u_i\}_{i=1}^{\infty}$ in Z converging to u and with $2 < |u_i + z(t,\omega)| < 3$ for each i. But then $\xi^{u_i}(\omega) = \infty$ so that

$$f(F_t(u_i,\omega)) = 1.$$

Thus $f(F_t(-,\omega))$ is not continuous at u.

This proves (b), and (a) follows. //

(D) At the moment the following conjecture seems plausible:

<u>Conjecture 2D(i)</u> If the stochastic dynamical system (X,z) is complete (i.e. for each point of M) and if each ordinary dynamical system

(X,z_π) is complete for piecewise linear approximations $\{z_\pi\}$ then we can take $\xi^u \equiv b$ for all u in Theorem 2B.

There is a somewhat weaker conjecture:

Conjecture 2D(ii) If M has a uniform cover for (X,z) then we can take $\xi^u \equiv b$ for all u.

(E) Even when (X,z) is strongly complete the flow F_t may not consist of diffeomorphisms onto all of M. An example of this is furnished by the Bessel processes described in §6B of Chapter V, as can be seen using the criteria in (Kunita 1981) or (Elworthy 1982).

Surjectivity holds when $M = \mathbb{R}^n$ and M(X) is globally Lipschitz. This was proved by Kunita and Varadhan in (Kunita 1980). For an equation $dx = YdB + Adt$ with B a Brownian motion and Y and A suitably smooth surjectivity holds if and only if both the equation and its *adjoint* $dx{:}YdB\text{-}Adt$ are strongly complete (Kunita 1981, 1982), see also (Carverhill & Elworthy 1982). There is another criterion in (Baxendale 1980) and a necessary condition in (Elworthy 1982). See also (Bismut 1981a,b). In (Carverhill 1982) an S.D.S. is constructed on $\mathbb{R}^2 - \{0\}$ which is strongly complete and for which each $\{F_t(u):t \geq 0\}$ is isonomous to $\{u + B_t:t \geq 0\}$ where $\{B_t:t \geq 0\}$ is a Brownian motion on $\mathbb{R}^2$.

(F) A left invariant S.D.S. on a Lie group G is strongly complete. In fact let $\{g_t:a \leq t < b\}$ be a solution with g_a equal to the identity element. Such a non-explosive solution exists by Proposition 13C of Chapter VII, taking M to be a point so that G(B) = G. Then set $F_t(u) = u \cdot g_t$. This gives a version of the flow by the diffeomorphism invariance of solutions, Theorem 1F of Chapter VII. This example is related to the considerations in (Kunita 1980).

§3 THE C^r CASE, $0 < r < \infty$.

It is a simple matter to obtain the C^r versions of some of the results above from the C^0 and C^∞ versions. As before we start by considering the case of a compact manifold.

Lemma 3 Let (X,z) be a C^2 stochastic dynamical system on a compact manifold M. Then there is a version of the solution flow

$F_t(-,\omega):M \to M$, $a \leq t < b$ which is almost surely C^1 and continuous in t into $C^1(M;M)$.

Proof By embedding M in some $\mathbb{R}^n$ as usual we can reduce to the case of a C^2 S.D.S. (X,z) on $\mathbb{R}^n$ with compact support. By standard smoothing techniques there is an S.D.S. $(\tilde{X},z)$ on $\mathbb{R}^{n+1}$ such that

(a) $\tilde{X}$ has compact support and is C^2

(b) $(\tilde{X},z)$ restricts to an S.D.S (X^α,z) on $\mathbb{R}^n \times \{\alpha\}$ for each $\alpha \in \mathbb{R}$

(c) X^α is C^∞ if $\alpha \neq 0$

(d) $X^0 = X$.

For example first set $\tilde{X}^\alpha(x)e = (2\pi\alpha^2)^{-\pi/2} \int_{\mathbb{R}^n} \exp(-\frac{|x-y|^2}{2\alpha^2}) X_e(y)dy$, to obtain some $\tilde{X}$ on $\mathbb{R}^n \times \mathbb{R}$ and then multiply by a suitable C^∞ function of compact support.

Let $\tilde{F}$ and $\delta\tilde{F}$ be C^0 versions of the flows of $\tilde{X}$ and δX. These exist by Theorem 3B and Lemma 2C, both of Chapter VI. Modifying on a set of measure zero, by the uniqueness of flows and Remark 1D we can suppose that $\tilde{F}$ and $\delta\tilde{F}$ restrict to give flows F^α and δF^α for X^α and δX^α, when $\alpha = 0$ or $\frac{1}{n}$, n = 1,2,..., and that these are continuous in time into the C^∞ topology when $\alpha = \frac{1}{n}$, n = 1,2,..., with $\delta F_t^\alpha(-,\omega)$ the derivative of F_t^α. Using the strong form of Blagoveščenskii & Freidlin's theorem (which we have not proved) that $F(-,\omega)$ and $\delta F(-,\omega)$ can be taken to be jointly C^0 in space and time we see that both $F_t^{1/n}(-,\omega)$ and $\delta F_t^{1/n}(-,\omega)$ converge to $F_t^0(-,\omega)$ and $\delta F_t^0(-,\omega)$ uniformly in space and time. However this implies that $F_t^{1/n}(-,\omega)$ converges in the C^1 topology uniformly in time and the limit is $F_t^0(-,\omega)$. The latter is therefore a version satisfying our requirements. //

Using (Kunita 1980) for the C^0 case we are assured that $\tilde{F}$ and $\delta\tilde{F}$ in the proof of the lemma can be chosen to consist of homeomorphisms. From this we see that the inverses of F and δF converge uniformly as $n \to \infty$. Therefore F^0 consists of C^1 diffeomorphisms, as is also shown more directly in the same articles at least when X is C^3. The following is due to Kunita (1980), although he loses 2 degrees of differentiability there against our 1 degree.

Theorem 3 Let (X,z) be a C^r stochastic dynamical system on the finite dimensional manifold M, with $r \geq 2$. Then there is an explosion time map $\{\xi^u : u \in M\}$ and a version F of the flow which satisfy the analogues of (i), (ii) and (iii) of Theorem 2B but with H^{s+1}_{loc} replaced by C^{r-1}_{loc}.

Proof When r = 2 the result is immediate by the method of proof of Theorem 2B together with Lemma 3 and the remark above. The derivative of F_t will be the flow δF_t of $(\delta X, z)$ since that is its derivative in probability. The general case follows by induction: if X is C^r then δX is C^{r-1} and if δF_t is C^{r-1} then F_t is C^r. //

§4 EQUATIONS DEPENDING ON A PARAMETER

Consider a family (X^α, z) of stochastic dynamical systems on M depending on a parameter α which is a point on some manifold A. Given sufficient regularity the easiest way to deal with this is to consider the S.D.S. $(\tilde{X}, z)$ on $M \times A$ with

$$\tilde{X}(u,\alpha)e = (X^\alpha(u)e, 0) \in TM \times TA = T(M \times A).$$

The behaviour of solutions of (X^α, z) as α is varied then is covered by the behaviour of the flow of $(\tilde{X}, z)$ as the initial point is varied, and so we can apply the results we have already discussed.

For example consider the equation

$$dx^\mu_t = \mu Y(x^\mu_t)dB_t + A(x^\mu_t)dt$$

on M where $x^\mu_0 = x_0, \mu \in \mathbb{R}$. When M is finite dimensional and compact (to be safe), if Y and A are C^∞, we can apply Theorem 2B to obtain versions of x^μ for $\mu \in \mathbb{R}$ which are C^∞ in μ. At $\mu = 0$ we have the solution x^0 of the ordinary differential equation $dx^0/dt = A(x^0_t)$. The behaviour of x^μ for small μ can be considered as the behaviour when the ordinary differential equation is perturbed by a small noise. There is an extensive literature on this subject see for example (Friedman 1975) Volume II Chapter 14, (Doss 1980) and (Azencott 1980) and the references therein. For applications to quantum mechanical tunnelling see the article by Jona-Lasinio et al. in (DeWitt-Morette & Elworthy 1981). In (Elworthy & Truman 1981) the first two derivatives of x^μ are computed at $\mu = 0$ for an S.D.S. on an orthonormal frame bundle O(N).

The use of $\tilde{X}$ leads to unnecessary losses of differentiability and more precise results can be obtained by working from first principles.

See (Gikhman & Skorohod 1972) for L^2 and L^0 behaviour and (Baxendale 1980) for almost sure behaviour.

§5 L^p REGULARITY OF THE DIFFERENTIATED PROCESSES

(A) Throughout this section we consider M with a Riemannian metric and induced Levi-Civita connection. It need no longer be finite dimensional.

Proposition 5A Assume that $X, \nabla X, \nabla^2 X$ are all uniformly bounded on M, or more generally that ∇X is uniformly bounded on M and so is the trilinear map over M

$$\tilde{\delta}^2 X : TM \oplus \underline{E} \oplus \underline{E} \to TM$$

$$(x,(v,e,f)) \mapsto \nabla^2 X_f(X_e(x),v) + \nabla X_f(\nabla X_e(v)).$$

Then for each $x_0 \in M$ and $v_0 \in T_{x_0}M$

$$|\delta F_t(v_0)|_{F_t(x_0)} \in \mathcal{L}^p(\Omega_t, F, \mu; R) \qquad 1 \leq p < \infty.$$

In fact

$$\|\delta F_t(v_0)\|_{\mathcal{L}^{2p}} \leq 2^{1-\frac{1}{2}p^{-1}} e^{(t-a)\Lambda} |v_0|$$

where

$$\Lambda = 2^{2p-4}[(p-1)L^2 + 2L]^2 p\beta^2$$

for

$$L = |\nabla X|_{C^0} + |\tilde{\delta}^2 X|_{C^0}.$$

Proof From Example 12D and Proposition 12E of Chapter VII we can consider $\delta F_t(v_0)$ as the solution of a stochastic integral equation of the type considered in Theorem 3A of Chapter VI. The only difference is in the possibility of explosion: but the proof goes over equally well to take that into account. //

Remark 5A

$$\tilde{\delta}^2 X(v,e,f) = \nabla(\nabla X_f \cdot X_e)v + R(X_e(x),v)X_f(x) \text{ for } v \in T_x M.$$

In particular when M is a Hilbert space with its standard inner product we have $\delta^2 X(v,e,f) = D(DX(X(x)e)f)v$ and the conditions of the proposition reduce to those of the basic differentiation result: Theorem 2D of Chapter VI where we required a global Lipschitz condition on X and M(X) as well as the twice differentiability of X.

(B) To discuss higher derivatives e.g. $\delta^2 F_t = \delta\delta F_t$) it is convenient to use the notation of covariant derivatives. However note that these reduce to ordinary derivatives for M = H with its trivial Riemannian metric.

The tangent bundle TTM to TM splits, by the connection, into the sum $HT^2M \oplus VT^2M$ of horizontal and vertical subbundles both naturally isomorphic to the pull back of TM by the projection $TM \to M$. In particular for $v \in T_xM$ we can write $T_vTM = T_xM \times T_xM$ and so an element of TTM is determined by a triple (v,u,w) of tangent vectors to some common point x of M. Using this decomposition it turns out that

$$\delta^2 F_t(v,u,w) = \left(\delta F_t(v),\ \delta F_t(u),\ \nabla_1 \delta F_t(u,v) + \delta F_t(w)\right)$$

for a map $\nabla_1 \delta F_t$ with

$$\nabla_1 \delta F_t(u,v)(\omega) \in T_{F_t(x)(\omega)}M \qquad u,v \in T_xM \text{ and } \omega \in \Omega_t.$$

See for example (Eliasson 1967). However he would write $\nabla_1 \delta F_t(v,u)$, reversing the order of u and v. If $f: M \to \mathbb{R}$ is C^2 then the second derivative in measure

$$d^2(f \circ F_t) \equiv d(d(f \circ F_t)) : TTM \to \mathcal{L}^0(\Omega_t, \mathcal{F}; \mathbb{R})$$

is given by

$$d^2(f \circ F_t)(v,u,w) = \nabla df(\delta F_t(u),\ \delta F_t(v)) + df \circ \nabla_1 \delta F_t(u,v) + df \circ \delta F_t(w).$$

Compare the formula for $d^2 f$ in the proof of Lemma 9B of Chapter VII.

For fixed u_0 and v_0 in $T_{x_0}M$ we can consider $t \mapsto \nabla_1 \delta F_t(u_0,v_0)(\omega)$

as a vector field along $t \mapsto F_t(x_0)(\omega)$, for $\omega \in \Omega$. When we have almost surely smooth flows $F_t(\ )(\omega)$ it is given by

$$\nabla_1 \delta F_t(u_0,v_0)(\omega) = \frac{D}{\partial s}\,\delta F_t(v(s))(\omega)$$

where $v(s)$ is the parallel translate of v_0 along the curve $s \mapsto \exp_{x_0} su_0$. With the notation of Proposition 5A we have:

<u>Lemma 5B</u> <u>For</u> $\omega \in \Omega_t$ <u>set</u> $V(t,\omega) = \nabla_1 \delta F_t(u_0,v_0)(\omega)$. <u>Then</u> V <u>is an l.r. solution to</u>

$$DV_t = \nabla^2 X(\delta F_t(u_0),\delta F_t(v_0))dz_t + \nabla X(V_t)dz_t$$

$$+ R(X(F_t(x_0))dz_t,\delta F_t(u_0))\delta F_t(v_0) \qquad (*)$$

<u>with</u> $V_a = 0$. <u>By this we mean</u>

$$\begin{aligned}\tilde{V}_t = &\int_a^{t\wedge\xi} U_s^{-1}\,\nabla^2 X(u_s,v_s)dz_s\\ &+ \int_a^{t\wedge\xi} U_s^{-1}\,R(X(x_s)dz_s,u_s)v_s\\ &+ \int_a^{t\wedge\xi} U_s^{-1}\,\nabla X(V_s)dz_s\\ &+ \tfrac{1}{2}\int_a^{t\wedge\xi} U_s^{-1}\,\tilde{\delta}^3 X(u_s,v_s)(dz_s,dz_s)\\ &+ \tfrac{1}{2}\int_a^{t\wedge\xi} U_s^{-1}\,\tilde{\delta}^2 X(V_s)(dz_s,dz_s) \qquad \text{a.s. on } \Omega_t. \qquad (**)\end{aligned}$$

<u>where (this time) we have let</u> U_t <u>denote the horizontal lift of</u> $F_t(x_0)$ <u>and have set</u> $\tilde{V}_t = U_t^{-1}V_t$ with

$$x_t = F_t(x_0),\ v_t = \delta F_t(v_0),\ u_t = \delta F_t(u_0)$$

<u>and</u>

$$\tilde{\delta}^3 X:(TM \oplus TM) \oplus (\underline{E} \oplus \underline{E}) \to TM$$

<u>is the quadrilinear map, symmetric in the last two variables with</u>

$$\tilde{\delta}^3X(u,v)(e,e) = \nabla^3X_e(X_e(x),u,v) + \nabla^2X_e(\nabla X_e(u),v)$$
$$+ \nabla^2X_e(u,\nabla X_e(v))$$
$$+ \nabla X_e(\nabla^2X_e(u,v) + R(X_e(x),u)v)$$
$$+ \nabla R(X_e(x))(X_e(x),u)v$$
$$+ R(\nabla X_e(X_e(x)),u)v$$
$$+ R(X_e(x),\nabla X_e(u))v + R(X_e(x),u)\nabla X_e(v).$$

Proof Consider the map

$$f:O(M) \times T^2M \to H$$

given by

$$f(U,(v,u,w)) = U^{-1}w.$$

On $O(M) \times T^2M$ we have the S.D.S. given by

$$(U,(v,u,w),e) \mapsto (\tilde{X}(U)e,\delta^2X(v,u,w)e)$$

where $\tilde{X}$ is the horizontal lift of X to $O(M)$. It has $Z_t = (U_t,\delta^2F_t(v_o,u_o,0))$ as a solution. Since $\tilde{V}_t = f(Z_t)$ we can apply the Itô formula, Lemma 9B of Chapter VII, to check that $\tilde{V}$ satisfies (**).

To do this let $\delta S_t(\ )e$, $\delta^2S_t(\ )e$ and $\tilde{S}_t(\ )e$ be the deterministic flows of X_e, δ^2X_e and $\tilde{X}_e$. We shall often omit to write the 'e' in what follows. Then

$$\frac{D}{\partial t}\,\delta S_t(v) = \nabla X(\delta S_t(v)) \qquad v \in TM$$

and the third component of $\delta^2S_t(v,u,w)$ is $\nabla_1\delta S_t(u,v) + \delta S_t(w)$ with

$$\frac{D}{\partial t}\,\nabla_1\delta S_t(u,v) = \nabla^2X(\delta S_t(u_o),\delta S_t(v_o)) + \nabla X(\nabla_1\delta S_t(u,v)) +$$
$$R(X(S_t(x_o)),\delta S_t(u_o))\delta S_t(v_o).$$

See Appendix B, Proposition 2F. This is a reason for using the symbolic equation (*) to describe V.

The result follows from the Itô formula after observing that

$$\frac{d^r}{dt^r}\tilde{S}_t(U)^{-1}(\nabla_1\delta S_t(u,v) + \delta S_t(w))$$

$$= \tilde{S}_t(U)^{-1}\frac{D^r}{\partial t^r}(\nabla_1\delta S_t(u,v) + \delta S_t(w)) \qquad r = 1,2.$$

Remarks 5B

(i) The above proof was a good example of the usefulness ot the Ito formula in its form (ii): it gives a straightforward way to get (**) from the 'Stratonovich equation' (*) which we knew V had to satisfy in some sense. We have resisted the temptation to go into a general theory of equations of the type of (*).

(ii) Alternatively, omitting to write the 'e's:

$$\delta^3X(u,v) = \nabla\tilde{\delta}^2X(u)v + \nabla X\Big(R(X(x),u)v\Big)$$

$$+ \nabla R\Big(X(x)\Big)\Big(X(x),u\Big)v + R\Big(\nabla X(X(x)),u\Big)v$$

$$+ R\Big(X(x),\nabla X(u)\Big)v + 2R\Big(X(x),u\Big)\nabla X(v).$$

(C) Proposition 5C Suppose X and its first three covariant derivatives are uniformly bounded on M together with R and ∇R, or more generally suppose that ∇X, $\tilde{\delta}^2 X$, $\nabla^2 X$, $R(X,-)(-)$ and $\tilde{\delta}^3 X$ are uniformly bounded. Then for u_0 and v_0 tangent vectors at $x_0 \in M$ we have

$$|\nabla_1\delta F_t(u_0,v_0)|_{F_t(x_0)} \in \mathcal{L}^p(\Omega_t,\mathcal{F},\mu;\mathbb{R}) \qquad \begin{array}{l}1 \leqslant p < \infty\\ a \leqslant t < b.\end{array}$$

In fact

$$|\nabla_1\delta F_t(u_0,v_0)|_{F_t(x_0)}$$

is continuous and 'bilinear' in $(u_0,v_0) \in T_{x_0}M \times T_{x_0}M$ into $\mathcal{L}^p(\Omega_t,\mathcal{F},\mu;\mathbb{R})$ for each $1 \leqslant p < \infty$ and $a \leqslant t < b$.

Proof We will use the notation of Lemma 5B. First note that both u(t) and v(t) lie in $\mathcal{L}^p$ for $1 \leqslant p < \infty$ by Proposition 5A and are even continuous in t into $\mathcal{L}^p$ by the estimate in that proposition extended to the case where v_0 is random. It follows as in Corollary 3.1 of Chapter V that if

$$a_t = \int_a^{t\wedge\xi} U_s^{-1}\nabla^2 X(u_s,v_s)dz_s$$

$$+ \int_a^{t\wedge\xi} U_s^{-1} R(X(x_s)dz_s,u_s)v_s$$

$$+ \tfrac{1}{2}\int_a^{t\wedge\xi} U_s^{-1}\tilde{\delta}^3 X(u_s,v_s)(dz_s,dz_s)$$

then $|a_t|$ is in $\mathcal{L}^p$ for all $1 \leq p < \infty$.

Consequently equation (**) for $\tilde{V}$ satisfies the conditions of Theorem 3A of Chapter VI (apart from the possibility of explosion, which makes no real difference to the proof). The result follows from that theorem: for the continuity we use the $\mathcal{L}^p$-estimates for u and v in Proposition 5A to get an estimate for $\|a_t\|_{\mathcal{L}^p}$ and hence one of the form

$$\|\nabla_1\delta F_t(u_0,v_0)\|_{\mathcal{L}^p} \leq c_p(t)\;|u_0|\;|v_0|$$

for a suitable function $c_p(t)$. //

Remarks 5C

(i) Clearly there are corresponding theorems valid for higher covariant 'derivatives' $\nabla_1^r\delta F_t$, $r \geq 1$. The conditions can most easily be written as uniform boundedness of

$$\frac{D}{\partial t}\nabla_1^j\delta S_t,\ \frac{D^2}{\partial t^2}\nabla_1^j\delta S_t,$$

for $j = 0,\ldots,r$; or as boundedness of X and its first (r + 1) covariant derivatives together with R and its first (r - 1) covariant derivatives.

(ii) It seems very likely that when M is complete the conditions on X and R are enough to ensure non-explosion as well.

§6 COMPLETE SYSTEMS: REGULARITY OF THE TRANSITION PROBABILITIES AND OPERATORS

(A) Again (X,z) is a fixed S.D.S. on M, and we continue with the notation of the previous sections. Let G denote a separable Banach space, and BC(M,G) the Banach space of all bounded continuous functions $f:M \to G$. We continue to allow M to be infinite dimensional. For each Borel subset B of M and each $(u,t) \in M \times [a,b)$, [or more generally with $u \in \mathcal{L}^0(\Omega,F_a,\mu;M)$], write

$$p_t(u,B) = \mu\{\omega \in \Omega_t^u : F_t(u)(\omega) \in B\}$$

For $f:M \to G$ and $a < t < b$ define

$$P_t f:M \to G$$

by

$$P_t f(u) = \int_{\Omega_t^u} f \circ F_t(u)(\omega) d\mu(\omega)$$

whenever f is measurable and the integral exists. For non-negative real valued f allow $P_t f:M \to \mathbb{R} \cup \{\infty\}$.

Equivalently we could define $p_t(u,B)$ by letting $p_t(u,-)$ be the measure on M^+ given by

$$p_t(u,-) = F_t(u)(\mu).$$

Then $$P_t f(u) = \int_M f(m) p_t(u,dm).$$

The main importance of these concepts occurs in the Markov situation discussed later, but we are in a good position to have a first look at their regularity properties now.

(B) Let $\mathcal{M}(M)$ denote the space of finite, non-negative, Borel measures on M. Recall that the *narrow topology* on $\mathcal{M}(M)$ is the weak topology determined by the maps

$$\mu \mapsto \int f \, d\mu$$

for all $f \in BC(M;\mathbb{R})$. In particular if $\{\mu_i\}_{i=1}^{\infty}$ is a sequence in $\mathcal{M}(M)$ then

$$\mu_i \to \mu \text{ narrowly}$$

iff for all $f \in BC(M)$

$$\int f \, d\mu_i \to \int f \, d\mu.$$

The narrow topology is sometimes called the *weak* or *Bernoulli* topology. Useful discussions of this topology can be found in (Parthasarathy 1967; Bauer 1972; Schwartz 1973).

A good example to see the relevance of completeness in what follows (at least for degenerate systems) is furnished by the ordinary dynamical system consisting of the flow $F_t(x)(w) \equiv x + tv$ for fixed $v \in \mathbb{R}^2$, acting on $\mathbb{R}^2-\{0\}$ when defined. For this, if $f \equiv 1$ we see $P_t f(x) = 1$ for all $x \neq -sv$, all $0 < s < t$, but $P_t f(-tv) = 0$.

We need the following standard result:

Theorem 6B For every complete separable metrizable space M and finite measure space (Ω, F, μ):

(i) the map

$$L^0(\Omega, F, \mu; M) \to M(M)$$
$$f \mapsto f(\mu)$$

is continuous into the narrow topology.

(ii) if $\mu \in M(M)$ and $f: M \to G$ is a bounded measurable map which is continuous except possibly at a set of μ-measure zero on M, then for any sequence $\{\mu_i\}_{i=1}^{\infty}$ in $M(M)$ converging narrowly to μ we have

$$\int f \, d\mu_i \to \int f \, d\mu.$$

Proof. Special cases of the theorem are proved in the references cited above. A proof of (i) is given in Appendix C and also in Schwartz's book (1973) and a proof of (ii) is in the Appendix to that book. //

Theorem 6C Suppose that (X,z) is complete. Then

(i) the map

$$M \to M(M)$$
$$u \mapsto p_t(u,-)$$

is continuous into the narrow topology;

(ii) P_t <u>restricts to a continuous linear map</u>

$$P_t : BC(M;G) \to BC(M;G)$$

<u>with</u>

$$|P_t| = 1$$

<u>and for each</u> $f \in BC(M,G)$

$$P_t f \to f \text{ as } t \to a,$$

<u>uniformly on compacta</u>. <u>In fact, for each</u> $f \in BC(M;G)$ <u>and</u> $b_0 \in [a,b)$, <u>the map</u>

$$M \to C([a,b_0];G)$$
$$u \to P_{\cdot} f(u)$$

<u>is continuous, whence so are the corresponding maps</u>

$$M \times [a,b_0] \to G$$
$$(u,t) \mapsto P_t f(u)$$

<u>and</u>

$$[a,b_0] \to C(M;G)$$
$$t \mapsto P_t f$$

<u>provided</u> $C(M;G)$ <u>is given the compact open topology.</u>

<u>Proof</u> By Theorem 8C of Chapter VII the flow map

$$F : M \to L^0(\Omega, F, \mu; C([a,b_0];M))$$

is continuous. Consequently, by Theorem 6B(i) so is the map

$$M \to M(C([a,b_0];M))$$
$$\mu \mapsto F(u)(\mu).$$

In particular, for each $t \in [a,b)$ the map $u \mapsto F_t(u)(\mu) = p_t(u,-)$ is continuous; giving (i).

If $f \in BC(M;G)$ it induces, by composition, a continuous map

$$f_* : C([a,b_0];M) \to C([a,b_0];G),$$

whence, by Theorem 6B(ii), the map

$$M \to C([a,b_o];G)$$

given by

$$u \mapsto \int_\Omega f_* d(F(u)(\mu))$$

is continuous. However since evaluation at t is a continuous linear map $C([a,b_o];G) \to G$

$$\int_\Omega f_* d(F(u)(\mu)) = P_. f(u)$$

and so we have shown that

$$M \to C([a,b_o];G)$$
$$u \to P_. f(u)$$

is continuous. The continuity of the maps $[a,b_o] \times M \to G$ and $[a,b_o] \to C(M;G)$ follow directly from standard results about the compact open topology since $[a,b_o]$ is locally compact: e.g. see (Dugundji 1966). In particular we have the continuity of $P_t f$ and the uniform convergence on compacta of $P_t f$ to f. Finally

$$\sup_{u \in M} |P_t f(u)| = \sup_{u \in M} \left|\int f(m) p_t(u,dm)\right|$$
$$\leq \sup_m |f(m)|,$$

and so $P_t : BC(M;G) \to BC(M;G)$ exists and has $|P_t| \leq 1$. However $P_t(1) = 1$, so $|P_t| = 1$, and the proof is complete. //

<u>Corollary 6C</u> <u>Suppose</u> (X,z) <u>is complete and for some</u> $u_o \in M$ <u>and</u> $B \in$ Borel M, $p_t(u_o,\partial B) = 0$ <u>where</u> ∂B <u>denotes the topological boundary of</u> B. <u>Then the map</u>

$$M \to \mathbb{R}$$
$$u \mapsto p_t(u,B)$$

<u>is continuous at</u> u_o.

<u>Proof</u> Immediate by part (i) of the theorem and part (ii) of Theorem 6B. //

Remarks 6C. (i) By Remark 8C of Chapter VII the theorem, and its corollary, still hold if we substitute the space of initial distributions $L^0(\Omega,F_a;\mu;M)$ for the ambient space of u.

(ii) The property that $P_t f$ is continuous whenever f is bounded and continuous is often called the *Feller property*.

§7 GENERAL SYSTEMS: SEMI-CONTINUITY AND MEASURABILITY

Theorem 7 For B open in E and $t_0 \in [a,b)$ the map

$$M \to \mathbb{R}$$

$$u \mapsto p_{t_0}(u,B)$$

is lower semi-continuous.

Proof We need only consider the case $t_0 > a$. Let $\tilde{M}$ be the open submanifold of $M \times (-\infty,\infty)$ given by

$$\tilde{M} = [M \times (-\infty,\infty)] - [(M-B) \times \{t_0\}].$$

Define $\tilde{z}:[a,b) \times \Omega \to E \times \mathbb{R}$ by

$$\tilde{z}(t,\omega) = (z(t,\omega),t)$$

and define

$$\tilde{X}:\tilde{M} \times (E \times \mathbb{R}) \to T\tilde{M} \subset TM \times T\mathbb{R} \approx TM \times (\mathbb{R} \times \mathbb{R})$$

by

$$\tilde{X}(m,t),(e,s)) = (X(m)e,(t,s)).$$

Then $(\tilde{X},\tilde{z})$ is an S.D.S. on M. Let $\tilde{F}$ be its flow map. For $u \in M$ set $\tilde{u} = (u,a) \in \tilde{M}$ and let $\xi^{\tilde{u}}$ be the explosion time of $\tilde{F}(\tilde{u})$. In fact by the uniqueness of solutions we can choose

$$\tilde{F}(\tilde{u})(t,\omega) = (F(u)(t,\omega),t) \qquad t < \xi^{\tilde{u}}(\omega)$$

with

$$t_0 < \xi^{\tilde{u}}(\omega) \quad \text{iff} \quad t_0 < \xi^{u}(\omega) \text{ and } F(u)(t_0,\omega) \in B.$$

In particular

$$\{\omega \in \Omega^u_{t_0}:F(u)(t_0,\omega) \in B\} = \{\omega \in \Omega:t_0 < \xi^{\tilde{u}}(\omega)\}.$$

Now Corollary 8C(ii) of Chapter VII implies that if $\{x_j\}_{j=1}^{\infty}$ is a sequence in M converging to u then

$$\mu\{\omega:t_o < \xi^{\tilde{u}}(\omega)\} \leq \underline{\lim}_{j\to\infty}\ \mu\{\omega:t_o < \xi^{\tilde{x}_j}(\omega)\}$$

(in that corollary choose $\{u_i\}$ to be a subsequence of $\{\tilde{x}_j\}$ such that $\mu\{\omega:t_o < \xi^{u_j}(\omega)\}$ converges to the lower limit). However this is precisely the required semi-continuity condition:

$$p_{t_o}(u,B) \leq \underline{\lim}_{j\to\infty}\ p_{t_o}(x_j,B)$$

or equivalently: for each $\alpha \in \mathbb{R}$ the set of u in M with $p_{t_o}(u,B) > \alpha$ is open in M. //

<u>Corollary 7</u> (i) <u>For each Borel set</u> B <u>of</u> M <u>and each</u> $t \in [a,b)$ <u>the map</u> $u \mapsto p_t(u,B)$ <u>of</u> M <u>into</u> $\mathbb{R}$ <u>is measurable.</u>

(ii) <u>If</u> $f:M \to G$ <u>is measurable and</u> P_tf <u>exists then</u> P_tf <u>is measurable.</u>

<u>Proof</u> Part (ii) follows immediately from (i) by approximating f by simple functions. For (i) observe that $p_t(u,-)$ is necessarily outer regular:

$$p_t(u,B) = \inf \{p_t(u,U):B \subset U \& U \text{ open}\}.$$

However by the theorem $p_t(u,U)$ is measurable in u for each open set U. //

§8 DIFFERENTIABILITY OF P_tf

(A) Unfortunately differentiability in measure is rather far from differentiability in $\mathcal{L}^1$, and from being able to differentiate under the integral sign, as the following example shows.

<u>Example 8A</u> For $\Omega = [0,1]$ with Lebesgue measure μ define $f:[0,1] \to \mathcal{L}^0(\Omega;\mathbb{R})$ by $f(t)(\omega) = \chi_{[0,t]}(\omega)$. Then f is differentiable in measure on [0,1] with $f' \equiv 0$. However,

$$\frac{d}{dt}\int_{\Omega} f(t,\omega)d\mu(\omega) = 1$$

Proof For $\delta > 0$, $s \neq 0$, $t \in [0,1]$, $t + s \in [0,1]$:

$$\mu\{\omega: \left|\frac{f(t+s)(\omega) - f(t)(\omega)}{s}\right| > \delta\} \leq \mu\{\omega: |t-\omega| \leq |s|\}$$

$$\leq 2|s| \to 0 \text{ as } s \to 0.$$

Thus f is differentiable in measure and $f' \equiv 0$. On the other hand

$$\int_\Omega f(t,\omega)d\mu(\omega) = \int_0^1 \chi_{[0,t]}(\omega)d\omega = t. \quad //$$

Note that f in the above example was Lipschitz into $\mathcal{L}^1$ but not into $\mathcal{L}^2$. In fact

Lemma 8A Suppose $f:(a,b) \to \mathcal{L}^2(\Omega,F,\mu;G)$ is Lipschitz and is differentiable in measure with derivative a bounded map $f':(a,b) \to \mathcal{L}^2(\Omega,F,\mu;G)$. Then f is differentiable into $\mathcal{L}^1$.

Proof For $a < t < b$ and $\varepsilon > 0$ there exists $\delta > 0$ so that for $|s| < \delta$, there is $\Omega_\varepsilon^s \in F$ with $\mu(\Omega_\varepsilon^s) < \varepsilon$ and

$$\left|\frac{f(t+s)(\omega) - f(t)(\omega)}{s} - f'(t)(\omega)\right| < \varepsilon \quad \omega \notin \Omega_\varepsilon^s.$$

Then

$$\int_\Omega \left|\frac{f(t+s) - f(t)}{s} - f'(t)\right|d\mu$$

$$< \varepsilon + \int_{\Omega_\varepsilon^s} \left|\frac{f(t+s) - f(t)}{s} - f'(t)\right|d\mu$$

$$< \varepsilon + \mu(\Omega_\varepsilon^s)^{\frac{1}{2}} \left\|\frac{f(t+s) - f(t)}{s} - f'(t)\right\|$$

$$\leq \varepsilon + \sqrt{\varepsilon} \left\|\frac{f(t+s) - f(t)}{s}\right\| + \|f'(t)\|\}. \quad //$$

(B) For $r = 1,2,\ldots$ and M Riemannian we let $BC^r(M;G)$ denote the space of maps $f:M \to G$ for which the first r covariant derivatives exist in the strong Gâteaux sense as continuous maps

$$\nabla^s df: \underbrace{TM \oplus \ldots \oplus TM}_{(2^s \text{ times})} \to G \qquad 0 \leq s \leq r-1$$

(see §5B) and such that the induced s-linear maps of the fibres, $(\nabla^s df)_x \in \mathbb{L}(T_xM,\ldots,T_xM;G)$, for $x \in M$, have norms uniformly bounded over M. If M is finite dimensional this reduces to the usual space of bounded C^r maps. The space $BC^r(M;G)$ will be equipped with the norm

$$|f|_{C^r} = |f|_{C^0} + \sum_{s=0}^{r-1} \sup_{x\in M} |(\nabla^s df)_x|$$

For suitable $h:TM \to G$ measurable we define $\delta P_t(h):TM \to G$ using the S.D.S. $(\delta X,z)$ just as P_t was defined using (X,z). We would like to find general conditions under which P_t maps $BC^r(M;G)$ to itself and under which $dP_t(f) = \delta P_t(df)$. However, we shall have to be content with the following here:

<u>Theorem 8B</u> (i) <u>Let</u> (X,z) <u>be an</u> S.D.S. <u>on the Hilbert space H of class</u> C^2 <u>with</u> $M(X)$ <u>globally Lipschitz</u>. <u>Then</u> P_t <u>restricts to a continuous map of</u> $BC^1(H;G)$ <u>to itself and</u>

$$dP_t(f) = \delta P_t(df)$$

<u>If further</u> X <u>is</u> C^{3+} <u>with</u> X, <u>and its first 3 derivatives bounded on</u> H <u>then</u> $P_t(f)$ <u>is twice Gâteaux differentiable with bounded derivatives whenever</u> $f \in BC^2(M;G)$. <u>In fact</u>

$$d^2P_tf = \delta^2P_t(d^2f).$$

(ii) <u>The corresponding results hold for</u> (X,z) <u>an</u> S.D.S. <u>on a compact</u> C^3 <u>manifold</u> M.

<u>Proof</u> (i) Let $F_t:H \to \mathcal{L}^0(\Omega;H)$, $t \geq a$, be a flow map for (X,z) and $\delta F_t: H \times H \to \mathcal{L}^0(\Omega;H)$ one for $(\delta X,z)$. By Theorem 2D of Chapter VI, F_t is Gâteaux differentiable as a map into $\mathcal{L}^2$ with derivative δF_t whenever X is C^2 and $M(X)$ is globally Lipschitz. Consequently if $f \in BC^1(M;G)$ then P_tf is differentiable with $dP_t(f) = \delta P_t(df)$.

Moreover by Theorem 2B, and Lemma 2C, of Chapter VI the map δF_t is Lipschitz into $\mathcal{L}^2$, and in particular continuous. It follows that P_tf is C^1. Also if $v \in H$

$$|dP_t(f)(v)| = |\delta P_t(df)(v)| = \int_\Omega df(\delta F_t(v))d\mu$$

$$\leq |df| \;\; \|\delta F_t(v)\| \leq \text{const.}\ |df|\ |v|$$

because of the Lipschitz condition. This proves the first part of (i).

If further X is C^{3+} and bounded together with its first three derivatives then δX is C^{2+} and Theorem 8E of Chapter VII implies that δF_t is Gâteaux differentiable in measure with derivative $\delta^2 F_t$. But then we can also use Proposition 5C to see that $\delta^2 F_t$ is bounded into $\mathcal{L}^2$. The previous lemma therefore applies to show that δF_t is Gâteaux differentiable into $\mathcal{L}^1$. Consequently $P_t(f)$ is twice Gâteaux differentiable and

$$d^2 P_t(f) = \delta^2 P_t(d^2 f) \qquad f \in BC^2(M;G).$$

The boundedness of $\delta^2 F_t$ into $\mathcal{L}^2$ and hence into $\mathcal{L}^1$ allows us to conclude that the derivatives of $P_t(f)$ are bounded. This proves (i).

Part (ii) can be deduced from (i) by the standard embedding technique. //

(C) Proposition 8C Let (X,z) be a C^3 S.D.S. on a finite dimensional Riemannian manifold M. Assume

(i) (X,z) is strongly complete and

(ii) the tensors ∇X, $\nabla^2 X$, $R(X,-)(-)$ are uniformly bounded together with $\tilde{\delta}^2 X$ and $\tilde{\delta}^3 X$ as in Proposition 5C.

Then if $f \in BC^2(M;G)$ the map $P_t(f)$ is twice strongly Gâteaux differentiable with bounded derivatives given by

$$dP_t(f) = \delta P_t(df)$$

and

$$\nabla dP_t(f)(u,v) = \int_\Omega \nabla df\Big(\delta F_t(u), \delta F_t(v)\Big) d\mu(\omega) + \int_\Omega df \circ \nabla_1 \delta F_t(u,v) d\mu(\omega)$$

for $u,v \in T_{x_0} M$.

Proof Take a C^2 version F_t as in Theorem 3 with sure derivatives δF_t and $\nabla_1 \delta F_t$. Let σ be a C^1 curve in M with $\sigma(0) = x_0$ and $\dot\sigma(0) = u$. Then

$$f \circ F_t(\sigma(s)) = f \circ F_t(x_0) + \int_0^s df \circ \delta F_t(\dot\sigma(r)) dr.$$

By Proposition 5A we can integrate both sides of this equation over Ω and interchange the order of the repeated integral. This shows that $P_t(f)$ is strongly Gâteaux differentiable with the required first derivative.

For the second derivative let $\sigma_s = \exp_{x_0} su$ and let v_s be the parallel translate along σ_s of $v \in T_{x_0}M$. Then

$$df \circ \delta F_t(v_s) = df \circ \delta F_t(v) + \int_0^s \nabla df\Big(\delta F_t(\dot\sigma_r), \delta F_t(v_r)\Big)dr$$

$$+ \int_0^s df \circ \nabla_1 \delta F_t(\dot\sigma_r, v_r)dr.$$

This time we can use Proposition 5C to integrate both sides and change the order of integration (here we need the fact that the L^1 bound on $\nabla_1 \delta F_t$ in Proposition 5C can be chosen independently of the point x_0). //

(D) Considerably stronger results are available: when z is Brownian motion and X is surjective at each point then P_t will be smoothing operators; see (Azencott 1974).

CHAPTER IX : DIFFUSIONS

Throughout this Chapter we consider an S.D.S. (Y,z) on M which is the direct sum of a (finite dimensional) Brownian motion term and a drift

$$(Y,z) = (X \oplus X_2, B \oplus z_2)$$

where

$$X:M \times \mathbb{R}^n \to TM,$$

B is an $\mathcal{F}_*$-Brownian motion on $\mathbb{R}^n$, $z_2(t,\omega) \equiv t$ and $X_2:M \times \mathbb{R} \to TM$ is given by

$$X_2(m)t = tA(m)$$

for a vector field A on M.

The corresponding stochastic differential equation can be written

$$dx_t = X(x_t)dB_t + A(x_t)dt.$$

We shall prove the semigroup property for the operators P_t defined in Chapter VIII §6, identify the action of their infinitesimal generator on smooth functions, and indicate some applications.

1 THE DIFFERENTIAL GENERATOR

(A) As in the previous chapters we let

$$F(u): [a,\xi^u) \times \Omega \to M$$

denote a maximal l.r. solution of $dx = Ydz$ with $x(a) = u$ a.e. As usual we shall often write $F_t(u)$ for $F(u)(t,-)$.

Let $B\,\mathcal{L}^0(M;G)$ denote the space of all bounded measurable maps of M into a separable Hilbert space G. As in Chapter VIII we define

$$P_t : B\,\mathcal{L}^0(M;G) \to B\,\mathcal{L}^0(M;G) \qquad a < t < \infty$$

by

$$P_t(f)u = \int_{\Omega_t^u} f\Big(F(u)(t,\omega)\Big)d\mu(\omega)$$

where

$$\Omega_t^u = \{\omega : t < \xi^u(\omega)\}.$$

It exists as a map between these spaces by Corollary 7(i) of Chapter VIII.

We shall also define $P_t(f)$ by the same formula whenever the integral exists i.e. even if f is not bounded.

Let $\mathcal{L}_X^2$ be the second order differential operator on M defined by

$$\mathcal{L}_X^2 f = \sum_{i=1}^n X^i(X^i f)$$

for $f:M \to G$, where the vector fields X^i are defined by $X^i = X(-)e_i$ for an orthonormal base $e_1,\ldots,e_n$ of $\mathbb{R}^n$. It follows from the next theorem that the operator is independent of the choice of orthonormal basis. Recall that $A(f):M \to G$ is given by $A(f)(m) = df(A(m))$.

<u>Theorem 1A</u> <u>If</u> $f:M \to G$ <u>is</u> C^2, <u>with both</u>

$$A(f) + \tfrac{1}{2}\mathcal{L}_X^2(f) : M \to G$$

<u>and one of</u> f <u>or</u> $df\circ X : M \to \mathbb{L}(\mathbb{R}^n;G)$ <u>bounded on the image of</u> $F(u)|[0,t] \times \Omega_t^u$ <u>in</u> M <u>for each</u> $u \in M$, <u>then</u> $P_t(f)$ <u>exists and</u>

$$\lim_{t \downarrow a} \frac{P_t f - f}{t-a} = \tfrac{1}{2}\mathcal{L}_X^2(f) + A(f)$$

$$= \tfrac{1}{2}\sum_i \nabla df(X^i,X^i) + \tfrac{1}{2}\sum_i df\circ\nabla X^i(X^i) + A(f)$$

<u>pointwise on</u> M, <u>for any linear connection on</u> M.

Proof By Lemma 9B(ii) of Chapter VII

$$f\Big(F_t(u)\Big) = f(u) + \int_a^{t\wedge\xi} df\circ X(F_s(u))dB_s$$
$$+ \int_a^{t\wedge\xi} df\circ A(F_s(u))ds$$
$$+ \int_a^{t\wedge\xi} \mathcal{L}_X^2(f)\Big(F_s(u)\Big)ds \qquad \text{a.s. on } \Omega_t^u \qquad (1)$$

since, in the notation of that lemma:

$$\int_a^{t\wedge\xi} \frac{d^2}{dt^2} f\circ S(t,x_s)\Big|_{t=0}(dB_s,dB_s)$$
$$= \int_a^{t\wedge\xi} \sum_{i=1}^n \frac{d^2}{dt^2} f\circ S(t,x_s)\Big|_{t=0}(e_i,e_i)ds$$
$$= \int_a^{t\wedge\xi} \sum_{i=1}^n \frac{d}{dt} X^i(f)\Big(S(t,x_s)e_i\Big)\Big|_{t=0} ds$$
$$= \int_a^{t\wedge\xi} \mathcal{L}_X^2(f)(x_s)ds.$$

Since $A(f) + \frac{1}{2}\mathcal{L}_X^2(f)$ is assumed bounded, denoting it by $A(f)$:

$$\lim_{t\downarrow a} \frac{1}{(t-a)} \int_{\Omega_t^u} \int_a^{t\wedge\xi} A(f)(F_s(u))ds\, d\mu$$
$$= \int_\Omega \left(\lim_{t\downarrow a} \frac{1}{(t-a)} \int_a^{t\wedge\xi} A(f)(F_s(u))ds\right)d\mu$$
$$= \int_\Omega A(f)(u)d\mu$$
$$= A(f)(u).$$

For a fixed $u \in M$ take a neighbourhood U of u with $\bar{U}$ contained in the domain of a regular localization for (Y,z) and with $df\circ X$ bounded on $\bar{U}$. Let τ be the first exit time of $F(u)$ from U. Then we have the F_*-process

$$y : [a,\infty) \times \Omega \to M$$

$$y(t,\omega) = F(u)(\tau(\omega) \wedge t,\omega)$$

and by Lemma 5 of Chapter VII

$$\mu\{\omega \in \Omega : \omega \notin \Omega_t^u \text{ or } y(t,\omega) \neq F(u)(t,\omega)\} = o(t-a) \text{ as } t \to a.$$

Then if $A(f)$ and f are bounded, by equation (1)

$$\int_{\Omega_t^u} \int_a^{t\wedge\xi} df\circ X\Big(F_s(u)\Big)dB_s d\mu = \int_\Omega \int_a^{t\wedge\xi} df\circ X(y_s)dB_s d\mu + o(t-a)$$

$$= o(t-a)$$

by Corollary 7C of Chapter III and the martingale property of Brownian motion: Chapter II, Propositiion 3C. The same holds if $df\circ X$ is bounded.

Thus the first equality of the theorem is proved. The second equality follows in exactly the same way using Lemma 9B(iii) of Chapter VII. //

(B) We can strengthen Theorem 1A in certain circumstances:

<u>Theorem 1B</u> <u>Suppose that</u> f, X <u>and</u> A <u>satisfy the conditions of Theorem 1A.</u> <u>If also either</u> $\xi^u \equiv \infty$, <u>or there exists</u> $f(\Delta) \in \mathbb{R}$ <u>with</u> $\lim_{t\to\xi^u} f(F_t(u)) = f(\Delta)$ <u>almost surely on</u> $\{\omega : \xi^u(\omega) < \infty\}$, <u>we then have</u>

$$P_t f(u) = (\mu(\Omega_t^u)-1)f(\Delta) + f(u) + \int_a^t P_s(Af)(u)ds \qquad a < t < \infty$$

where $Af = \tfrac{1}{2}\mathcal{L}_X^2(f) + A(f)$.

<u>Proof</u> Set $g(u) = f(u) - f(\Delta)$ for $u \in M$ and set $g(\Delta) = 0$, Then

$$P_t g(u) = \int_{\Omega_t^u} g(F_t(u))d\mu = \int_\Omega g(F_t(u))d\mu$$

where as usual we have set $F_t(u)(\omega) = \Delta$ if $t > \xi^u(\omega)$.

Take an exhaustion of M by a nested sequence of open domains U_i with $\bar{U}_i$ admitting a uniform cover in M, and set ξ_i equal to the first exit time of $x \equiv F(u)$ from U_i. By Corollary 9B of Chapter VII

$$g(x(t\wedge\xi_i)) = g(u) + \int_a^{t\wedge\xi_i} dg\circ X(x_s)dB_s + \int_a^{t\wedge\xi_i} Ag(x_s)ds.$$

By Corollary 7C of Chapter III this gives

$$\int_\Omega g(x(t\wedge\xi_i))d\mu = g(u) + \int_\Omega \int_a^{t\wedge\xi_i} Ag(x_s)ds\ d\mu.$$

However $t\wedge\xi_i \to t\wedge\xi$ as $i\to\infty$ and so $g(x(t\wedge\xi_i)) \to g(x(t\wedge\xi))$. When g is bounded we can therefore apply the dominated convergence theorem, and the fact that

$$\int_\Omega g(x(t\wedge\xi))d\mu = \int_{\Omega_t} g(x(t\wedge\xi))d\mu$$

to obtain

$$P_t g(u) = g(u) + \int_\Omega \int_a^{t\wedge\xi} Ag(x_s)ds\ d\mu. \tag{2}$$

When $dg\circ X(s)$ is bounded (2) follows from Theorem 7A of Chapter III.

Now define $Ag(\Delta) = 0$. Then $Ag\circ x : [0,\infty) \times \Omega \to G$ remains measurable and so by Fubini's theorem

$$\int_\Omega \int_a^{t\wedge\xi} Ag(x_s)dsd\mu = \int_\Omega \int_a^t Ag(x)dsd\mu = \int_a^t \int_\Omega Ag(x_s)dsd\mu$$

$$= \int_a^t P_s(Ag)(u)ds.$$

Substitution of this in (2) gives the required result after noting that

$$P_t g(u) = P_t f(u) - \mu(\Omega_t^u)f(\Delta). \quad //$$

There is the following intriguing corollary:

<u>Corollary 1B</u> <u>Assuming the conditions of the theorems, setting</u> $x = F(u), \Omega_t = \Omega_t^u$ <u>and</u> $\xi = \xi^u$, <u>if</u> $df\circ X(x_s)$ <u>is bounded on</u> $[0,t\wedge\xi) \times \Omega$ <u>then</u>

$$\int_{\Omega-\Omega_t} \int_a^{t\wedge\xi} df\circ X(x_s)dB_s = (1-\mu(\Omega_t))(f(\Delta) - f(u)).$$

<u>Proof</u> By Lemma 9B of Chapter VII, almost surely on Ω_t

$$f(x_t) = f(u) + \int_a^{t\wedge\xi} df\circ X(x_s)dB_s + \int_a^{t\wedge\xi} A(f)(x_s)ds.$$

Therefore

$$\int_{\Omega_t}\int_a^{t\wedge\xi} df\circ X(x_s)dB_s = P_t f(u) - \mu(\Omega_t^u)f(u)$$

$$- \int_{\Omega_t}\int_a^{t\wedge\xi} A(f)(x_s)ds$$

$$= P_t f(u) - \mu(\Omega_t^u)f(u) - \int_a^t P_s f(u)ds.$$

The result follows from the theorem using Proposition 7C of Chapter III to replace the integral over Ω_t by one over $\Omega-\Omega_t$. //

(C) The operator A will be called the *differential generator* of the system (Y,z). We have

$$A(f) = \tfrac{1}{2}\,\mathcal{L}_X^2(f) + A(f)$$

$$= \tfrac{1}{2}\sum_i \nabla df(X^i,X^i) + \tfrac{1}{2}\sum_i df\circ\nabla X^i(X^i) + A(f).$$

In local coordinates, if dim M = m, using the summation convention

$$A(f)(x) = \tfrac{1}{2}\, a^{\alpha\beta}(x)\,\frac{\partial^2 f}{\partial x^\alpha \partial x^\beta} + \left(X^{i\beta}(x)\,\frac{\partial X^{i\alpha}}{\partial x^\beta} + A^\alpha(x)\right)\frac{\partial f}{\partial x^\alpha}$$

where the indices α,β are summed from 1 to m, and i is summed from 1 to n; $\{X^{i\beta}\}_{\beta=1}^m$ are the components of X^i and $\{A^\alpha\}_{\alpha=1}^m$ those of A and

$$a^{\alpha\beta}(x) = \sum_i X^{i\alpha}(x)X^{i\beta}(x).$$

Note that since the matrix $[a^{\alpha\beta}(x)]$ is the product of the adjoint of X(x) with X(x) itself, it is positive semi-definite. That is to say that A is *semi-elliptic*. It is *elliptic* i.e. $[a^{\alpha\beta}(x)]$ is positive definite if and only if $X(x):\mathbb{R}^n \to T_xM$ is surjective for each $x \in M$. If so we say that the system (Y,z), or the diffusion F(x), is *non-degenerate*. The extreme case of a degenerate diffusion is given by an ordinary dynamical system i.e. $X \equiv 0$. We shall not prove it here, but in the non-degenerate case any solution instantaneously spreads over all of M for M connected: if $f \geq 0$, continuous, and positive at some point then $P_t f(x) > 0$ for all $x \in M$ and $t > 0$ see §8 Chapter VI of (Ikeda & Watanabe 1981) for a discussion of Stroock and Varadhan's results on the "support" of degenerate diffusions (piecewise linear approximation can be useful here).

§2 UNIQUENESS OF SOLUTIONS OF THE DIFFUSION EQUATION

(A) As before define $A:C^2(M;G) \to C^0(M;G)$ by

$$A(f) = \tfrac{1}{2}\mathcal{L}_X^2(f) + A(f).$$

The *diffusion equation* $\frac{\partial f_t}{\partial t} = Af_t$ for $t \geq a$ and f_a given is often called the *Kolmogorov equation* or *backward parabolic equation* when written as $\frac{\partial g_t}{\partial t} + Ag_t = 0$ for $t \leq T$ with g_T given. There are some easy results, using essentially only the Itô formula, as in (Friedman 1975) Chapter 6. They will be extended in §7 below to allow A to have a zero order (i.e. potential) term. The Markov process theory of §§3-5 is not needed for these extensions either. We take $0 \leq a < b < \infty$ throughout.

Theorem 2A Suppose $f:[a,b] \times M \to G$ is C^2 and satisfies $\frac{\partial f}{\partial t} = Af$ with initial condition f_a. Let $u \in M$. Assume that (Y,z) is complete and that f_a is bounded on the image of $[a,b] \times \Omega$ under $F(u)$. Then

$$f(t,u) = P_t(f_a)(u) \qquad a \leq t \leq b.$$

Proof We can assume $a = 0$. For $\tau \in [a,b]$ define $g:[0,\tau] \times M \to G$ by $g(t,u) = f(\tau-t,u)$. Then $\frac{\partial g}{\partial t} = -Ag$. Set $x_t = F_t(u)$ for F as in §1. By Itô's formula, (applied to the process (t,x_t) on $[0,\tau] \times M$, or using Remark 3 Chapter V extended to the manifold case).

$$\begin{aligned} g(t,x_t) &= g(0,u) + \int_0^t D_2g(s,x_s)(X(x_s)dB_s) + \int_0^t \frac{\partial g}{\partial s}(s,x_s)ds \\ &\quad + \int_0^t A(g)(s,x_s)ds \qquad 0 \leq t \leq \tau \\ &= g(0,u) + \int_0^t D_2g(s,x_s)(X(x_s)dB_s), \end{aligned}$$

where D_2 denotes the partial derivative in the second variable. Setting $t = \tau$ this gives

$$f(a,x_t) = f(\tau,u) + \int_0^\tau D_2f(\tau-s,x_s)(X(x_s)dB_s) \qquad a \leq \tau \leq b .$$

Integrating over Ω, using Corollary 7C of Chapter III we get

$P_\tau f_a(u) = f(\tau,u)$ as required. //

(B) Next we have a minor modification of Theorem 2A. It can be used when M is the interior of a manifold with boundary and we are imposing zero boundary values on our problem; when M is the interior of a domain with compact closúre the explosion time ξ would be the exit time from M. To obtain maximum generality we topologize $M^+ = M \cup \{\Delta\}$ as in Remark 6 of Chapter VII.

Theorem 2B Suppose that $f:[a,b] \times M \to G$ is C^2 and satisfies $\frac{\partial f}{\partial t} = Af$ with initial condition f_a. Assume that

(i) f_a is bounded on $F(u)[[a,b\wedge\xi) \times \Omega]$

(ii) $f(t,m) \to 0$ as $m \to \Delta$, uniformly in $t \in [a,b]$

Then $f(t,u) = P_t f_a(u), \qquad a \leq t \leq b.$

Proof Assume a = 0, and take t in [a,b]. Choose an increasing sequence $\{U_i\}$ of open domains of M whose union is M, with each $\bar{U}_i$, admitting a uniform cover in M, and such that $|f(s,m)| < 2^{-i}$ if $m \notin U_i$ for all $0 \leq s \leq b$ and $i = 1,2,\ldots$. Let ξ_i be the first exit time of $x \equiv F(u)$ from U_i and set $\tau_i = t\wedge\xi_i$. Proceed as in the proof of the previous theorem, using Remark 7B of Chapter III to obtain

$$f(t-\tau_i,x(\tau_i)) = f(t,u) + \int_0^{\tau_i} D_2f(t-s,x_s)(X(x_s)dB_s).$$

Now on $\{\omega:t < \xi_i\}$ we have $f(t-\tau_i,x(\tau_i)) = f(0,x_t)$, while on its complement $|f(t-\tau_i,x(\tau_i))| < \frac{1}{2^i}$. Therefore we can apply Corollary 7C of Chapter III to get

$$\int_\Omega f(t-\tau_i,x(\tau_i))d\mu = f(t,u).$$

From the proof of Lemma 9F of Chapter VII we know that $\xi_i \uparrow \xi$ so that

$$\int_\Omega f(t-\tau_i,x(\tau_i))d\mu = \int_{\{\omega:t<\xi_i\}} f(0,x_t)d\mu$$

$$+ \int_{\{\omega:t\geq\xi_i\}} f(t-\xi_i,x(\xi_i))d\mu \to \int_{\Omega_t} f(0,x_t)d\mu$$

as $i \to \infty$. Thus $f(t,u) = P_t f(0,u)$ as required. //

(C) Now suppose $M = O(N)$ where N is finite dimensional and Riemannian and go back to the case of Brownian motion on N using the notation of §12, Chapter VII. The heat flow on 1-forms will be considered again later.

Theorem 2C Suppose $\phi:[0,b] \times TN \to \mathbb{R}$ is a C^2 time dependent 1-form on N. Assume

(i) ϕ satisfies $\frac{\partial\phi}{\partial t} = \frac{1}{2}\Delta\phi$ with initial condition ϕ_0

(ii) ϕ is closed

(iii) Both ϕ and $\nabla_2\phi$ are bounded on $[a,b] \times N$, where ∇_2 is covariant differentiation with respect to the second variable.

(iv) Both the curvature R and the covariant derivative ∇K of the Ricci curvature of N are uniformly bounded on N (or more generally the process ξ, A of §12D of Chapter VII lie in L^2 on $[a,b]$).

(v) N is stochastically complete.

Then $\phi(t,v_0) = \delta P_t\phi_0(v_0) \qquad (t,v_0) \in [0,b] \times TN$

Proof Given $t \in [0,b]$ define $\psi:[0,t] \times TN \to E$ by

$$\psi_s \equiv \psi(s,-) = \phi(\tau-s,-).$$

Then $\psi(s,-)$ is a closed 1-form on N satisfying $\frac{\partial\psi}{\partial s} = -\frac{1}{2}\Delta\psi$. Consequently by Theorem 12D of Chapter VII for $v_t = \delta F(v_0)t$ and A as in that theorem,

$$\psi(t,v_t) = \psi(0,v_0) + \int_0^t \nabla\psi_s(u_s dz_s)(v_s) + \int_0^t \psi_s(u_s A_s dz_s) \qquad (1)$$

Now by assumption (iv) using Corollary 12F of Chapter VII both v_s and A_s lie in L^2 for $0 \leq s \leq t$, and so if $|\nabla\phi_s|$ and $|\phi_s|$ are the supremum norms of $\nabla\phi_s$ and ϕ_s where $\phi_s = \phi(s,-)$ then

$$\| \nabla\phi_s(u_s-)(v_s) \| \leqslant |\nabla\phi| \quad \|v_s\|$$

while

$$\| \phi_s(u_s A_s-) \| \leqslant |\phi_s| \quad \|A_s\| .$$

Consequently integrating (1) over Ω we get

$$\int_\Omega \phi(0,v_t)d\mu = \int_\Omega \phi(t,v_0)d\mu$$

i.e. $\delta P_t\phi_0(v_0) = \phi(t,v_0)$ as required. //

(D) Of course other results than these are available using partial differential equation theory, especially when A is elliptic. For uniqueness theorems for the heat flow on 1-forms see (Vauthier 1979; Dodziuk 1981) and Theorem 7C below.

§3 THE SEMIGROUP PROPERTY OF FLOWS

(A) For $u \in \mathcal{L}^0(\Omega,F_r;M)$ and $r \geqslant 0$ let

$$F^r(u):[r,\xi_r^u) \times \Omega \to M$$

be the maximal solution of $dx = Ydz$ with $F^r(u)(r,\omega) = u(\omega)$ for $\omega \in \Omega$.

Without extra conditions it is *not necessarily true* that

$$F^s\Big(F^r(m)(s,\omega)\Big)(t,\omega) = F^r(m)(t,\omega)$$

a.s., for $m \in M$ and $0 \leqslant r \leqslant s \leqslant t$, and $t < \xi_r^m(\omega)$. In fact the equation does not really make sense since $F^s(x)(t)$ is only defined up to a set of measure zero depending on $x \in M$:

<u>Example 3A</u> Consider the case $M = \mathbb{R}^n$, $n \geqslant 2$ with $X(m):\mathbb{R}^n \to \mathbb{R}^n$ the identity for each m. For the S.D.S. (X,B) on $\mathbb{R}^n$ we can choose

$$F^0(m)(s,\omega) = B_t(\omega) + m \qquad m \in M$$

and for fixed $m_0 \in M$, $s > 0$ we can choose, for $x \in \mathbb{R}^n$,

$$F^s(x)(t,\omega) = \begin{cases} x \text{ if } x = F^0(m_0)(s,\omega) \\ x + B_t(\omega) - B_s(\omega) \text{ otherwise.} \end{cases}$$

This is possible since $\mu\{\omega:F^0(m_0)(s,\omega) = x\} = 0$, each $x \in \mathbb{R}^n$. But then $F^s(F^0(m_0)(s,\omega))(t,\omega) = F^0(m_0)(s,\omega)$ all $t \geq s$. //

(B) To get over this difficulty we will choose special versions of our flows. Since we wish to include the infinite dimensional case we cannot apply Chapter VIII to get continuous versions.

Let $M^+ = M \cup \{\Delta\}$ for a coffin state Δ, topologized as a disjoint union for simplicity. For a flow F^r we shall adopt the convention that $F^r(\Delta) = \Delta$, and so can consider F^r as a map

$$F^r: M^+ \times \Omega \to C^\Delta([r,\infty);M^+)$$

where the target space is the space of maps $\sigma:[r,\infty) \to M^+$ such that (i) if $\sigma(s) = \Delta$ then $\sigma(t) = \Delta$ for all $t \geq s \geq r$, and (ii) σ is continuous until it first leaves M. This path space will be given the σ-algebra $\bar{C}^r$ generated by the evaluation maps ρ_s as in Chapter II §1A. Note that we are taking a different viewpoint now from that of Chapter VIII: see Warning 2C of Chapter VIII.

Let G^r be the σ-algebra generated by $\{B_t - B_r: t \geq r\}$ on Ω, and let G^r_t, $t \geq r$, be the σ-algebra generated by $\{B_s - B_r: r \leq s \leq t\}$ together with all sets of measure zero in G^r, as in §2F of Chapter VII.

First we have a key lemma:

<u>Lemma 3B</u> <u>Let</u> G <u>be a</u> σ-<u>subalgebra of</u> F. <u>Give</u> $M \times \Omega$ <u>the</u> σ-<u>algebra</u> Borel $(M) * G$ <u>and suppose</u> $V \subset M \times \Omega$ <u>is measurable.</u> <u>For</u> $i = 1,2$ <u>let</u> $\theta_i: V \to Z$ be measurable maps into a measurable space $\{Z,C\}$ such that for <u>each</u> $m \in M$, <u>for almost all</u> $\omega \in \Omega$ <u>with</u> $(m,\omega) \in V$ <u>we have</u> $\theta_1(m,\omega) = \theta_2(m,\omega)$. <u>Suppose</u> $u: \Omega \to M$ <u>is measurable and independent of</u> G. <u>Then</u>

$$\theta_1(u(\omega),\omega) = \theta_2(u(\omega),\omega)$$

<u>for almost all</u> ω <u>with</u> $(u(\omega),\omega) \in V$.

<u>Proof</u>. Let ν be the image measure of μ on $M \times \Omega$ determined by the map

$$\Omega \to M \times \tilde{\Omega}$$
$$\omega \longmapsto (u(\omega),\omega)$$

where we have let $\tilde{\Omega}$ denote Ω when considered with the σ-algebra G. By

Exercise 3(ii) of Chapter I it is the product measure:

$$\nu = u(\mu) \otimes (\mu|G).$$

Now for each m there is a subset A(m) in G with $\mu(A(m)) = 0$ such that $\theta_1(m,\omega) = \theta_2(m,\omega)$ when $(m,\omega) \in V$ and $\omega \notin A(m)$. It follows by Fubini's theorem that $\theta_1(m,\omega) = \theta_2(m,\omega)$ for ν-almost all point (m,ω) of V, and this is equivalent to what we had to prove. //

Theorem 3B For each $r > 0$ there is a version of F^r which is Borel $M^+ * G^r$ measurable and adapted to $\{G_t^r : t > r\}$. For such versions

(i) for each $s > r > 0$ and each $m \in M$

$$F^s(F^r(m,\omega)(s),\omega)(t) = F^r(m,\omega)(t) \text{ all } t > s, \qquad \text{a.s.}$$

and (ii) if $u \in \mathcal{L}^0(\Omega,F,M)$ is independent of G^r then

$F^r(u(\omega),\omega)$ is a version of $F^r(u)(-,\omega)$.

Proof As in §2F of Chapter VII there is no difficulty in obtaining versions adapted to $\{G_t^r : t > r\}$. Some additional general theory is needed to get the Borel $M^+ * G^r$ measurable version, and that is left to Appendix C.

Assuming we have chosen such versions we shall first prove (ii). Suppose therefore that $u \in \mathcal{L}^0(\Omega,F;M)$ is independent of G^r. Take $t > r$ and consider a partition Π of $[r,t]$, letting z_π be the corresponding piecewise linear approximation to z. Assume first that (Y,z_π) and (Y,z) are complete for all such Π. Let F_π be the flow of (Y,z_π) from time r.

Write $M \times \Omega^r$ for $M \times \Omega$ furnished with the σ-algebra Borel $M * G^r$ and as in the lemma equip it with the image measure ν of μ under $\omega \mapsto (u(\omega),\omega)$.

By Theorem 10 of Chapter VII

$$F_\pi(u)(t) \to F^r(u)(t) \tag{1}$$

in measure as mesh $\Pi \to 0$. Since, for all $\omega \in \Omega$

$$F_\pi(u)(t,\omega) = F_\pi(u(\omega))(t,\omega) \tag{2}$$

this implies that

$$F_\pi(-)(t,-):M \times \Omega^r \to M$$

is Cauchy in measure. There is therefore a sequence $\{\Pi_j\}_{j=1}^\infty$ with mesh $\Pi_j \to 0$ such that writing

$$^jF = F_{\pi_j}(-)(t,-):M \times \Omega^r \to M$$

$\{^jF\}_j$ converges on some measurable V in $M \times \Omega^r$ with $\nu(V) = 1$, to some $G:V \to M$. Moreover $G(u(\omega),\omega) = F^r(u)(t,\omega)$, a.s., by (1) and (2).

On the other hand, for each $m \in M$, Theorem 10 of Chapter VII assures us that there is a subsequence $\{j_k\}_{k=1}^\infty$ such that $^{j_k}F(m)(\omega)$ converges to $F^r(m,\omega)(t)$ almost surely. Therefore we can apply the lemma with $\theta_1 = F^r(-,-)(t)|V$ and $\theta_2 = G|V$ to conclude that $F^r(u(\omega),\omega)(t) = G(u(\omega),\omega) = F^r(u)(t,\omega)$ almost surely. This proves (ii) under our completeness assumptions.

For incomplete systems let ξ^u be the explosion time of $F^r(u)$ and ξ^m that of $F^r(m,-)$. Set $\Omega_t^u = \{\omega:t < \xi^u(\omega)\}$ as usual. Given $\varepsilon > 0$ and $t \geqslant r$ we can apply the technique of §7B Chapter VII to obtain an open set U_ε of M and a system (Y_ε,z) such that

(a) $Y_\varepsilon|U_\varepsilon = Y|U_\varepsilon$

(b) (Y_ε,z) and (Y_ε,z_π) are complete for all partitions Π of $[r,t]$.

(c) if $F_\varepsilon^r(u)$ denotes the flow of (Y_ε,z) with initial distribution u and τ_ε its first exit time from U_ε then $\tau_\varepsilon \leqslant \xi^u$ almost surely (by (i) and (ii))and

$$\mu\{\omega \in \Omega_t^u:\tau_\varepsilon(\omega) \leqslant t\} < \varepsilon.$$

We now know, because of (b), that almost surely

$$F_\varepsilon^r(u(\omega),\omega)(s) = F_\varepsilon^r\ (u)(s,\omega) \qquad 0 \leqslant s \leqslant t. \tag{3}$$

For $m \in M$ let τ_ε^m be the first exit time of $F_\varepsilon^r(m,-)$ from U_i. From (a)

$$F_\varepsilon^r(m,\omega)(s) = F^r(m,\omega)(s) \qquad r \leqslant s < \tau_\varepsilon^m \tag{4}$$

a.s. for each $m \in M$ and

$$F^r_\varepsilon(u)(s,\omega) = F^r(u)(s,\omega) \qquad r \leqslant s < \tau_\varepsilon, \text{ a.s.} \tag{5}$$

Applying the lemma with $V = \{(m,\omega):t < \tau^m_\varepsilon(\omega)\}$, for fixed t, we have

$$F^r_\varepsilon(u(\omega),\omega)(t) = F^r(u(\omega),\omega)(t)$$

almost surely on $\{\omega:t < \tau^{u(\omega)}_\varepsilon(\omega)\}$.

Now $\tau^{u(\omega)}_\varepsilon(\omega) = \tau_\varepsilon(\omega)$ a.s. by (3), and so by (c), (3), and sample continuity we see that for $(s,\omega) \in [r,\xi^u) \times \Omega$

$$F^r(u)(s,\omega) = F^r(u(\omega),\omega)(s) \qquad \text{a.s.}$$

In particular $\xi^{u(\omega)}(\omega) \geqslant \xi^u(\omega)$ a.s.. However for any t with $r \leqslant t < \xi^{u(\omega)}(\omega)$ we have $F^r(u(\omega),\omega)$ continuous on [r,t],whereas if $\xi^u(\omega) \leqslant t$ we have $F^r(u)(s,\omega) \to \Delta$ as $s \uparrow \xi^u(\omega)$ as in §6 of Chapter VII. Therefore $\xi^u(\omega) = \xi^{u(\omega)}(\omega)$ with probability one and $F^r(u)(s,\omega) = F^r(u(\omega),\omega)(s)$, for all $s \geqslant r$. Thus (ii) is proved.

To prove (i) first observe that (ii) remains true even when u has values in M^+. This can be seen simply by replacing (Ω,F,μ) by (Ω',F',μ') where $\Omega' = u^{-1}(M)$, F' is the induced σ-algebra on Ω', and $\mu = \mu(\Omega')^{-1}\mu|F'$. By independence $B_t - B_r$ is still a Brownian motion on (Ω',F',μ') and we can apply (ii) as it stands. Having observed this (i) follows immediately by taking $u = F^r(m,-)(s)$, since G^o_r is independent of G^r by Proposition 3B of Chapter II and by uniqueness of solutions

$$F^s(F^r(m,-)(s))(t,\omega) = F^r(m,\omega)(t) \qquad \text{a.s.} \quad //$$

When $u \in \mathcal{L}^0(\Omega,F;M)$ is independent of G^r we can, and will always, take a version of $F^r(u)$ adapted to $\{\sigma(u) \wedge G^r_t:t \geqslant r\}$. See §2F of Chapter VII. From now on we will also always choose versions of our flows as in Theorem 3B. Therefore we have immediately:

Corollary 3B If $u \in \mathcal{L}^0(\Omega,F;M)$ is independent of G^r then for each $s \geqslant r$

$$F^s(F^r(u)(s,\omega),\omega)(t) = F^r(u)(t,\omega) \qquad s \leqslant t < \infty \quad \text{a.s.} \ //$$

We shall also not be particular in what way we write the arguments of our flow maps: so $F^r_t(m)(\omega) = F^r(m,\omega)(t) = F(m)(t,\omega)$ etc.

§4 TIME HOMOGENEITY

It is important to remember here that Y is assumed time independent.

(A) Proposition 4A Let $F = F^0$ then for each $r \geq 0$ and $u \in \mathcal{L}^0(\Omega,\mathcal{F};M)$ independent of $\mathcal{G}^0$

$$F^r(u)(t,\omega) \overset{\sim}{\cdot} F(u)(t-r,\omega)$$

as processes in $t \geq r$.

Proof Set $z^r(t) = z(t+r) - z(r)$ $\qquad t \geq r.$

Then if $\hat{F}$ is the flow from time 0 of (X,z^r) we have

$$F^r(u)(t,\omega) = \hat{F}(u)(t-r,\omega) \qquad t \geq r \qquad \text{a.s.}$$

However $z^r \overset{\sim}{\cdot} z$ by Corollary 3A of Chapter II. Consequently $\hat{F}(u) \overset{\sim}{\cdot} F(u)$ by Theorem 6B of Chapter VI extended to the manifold case by the usual localization procedure (the joint isonomy required in Theorem 6B comes from the independence of u and z). //

(B) Let $u \in \mathcal{L}^0(\Omega,\mathcal{F};M)$ be independent of $\mathcal{G}^s$ where $s \geq 0$. For $t \geq s$ let $p(s,u;t,-)$ be the measure on M determined by the distribution of $F_t^s(u)$: for $B \in$ Borel M

$$p(s,u;t,B) = \mu\{\omega:F_t^s(u)(\omega) \in B\}.$$

Theorem 4B For $t \geq s$ and $r \geq 0$

$$p(s+r,u;t+r,-) = p(s,u;t,-)$$

Proof By Proposition 4A, as processes in $t \geq s$

$$F^{s+r}(u)(t+r) \overset{\sim}{\cdot} F^s(u)(t). \quad //$$

Now let ξ_r^u be the explosion time of $F^r(u)$ and set

$$\Omega_t^{r,u} = \{\omega:t < \xi_r^u(\omega)\} \qquad t \geq r \geq 0.$$

For suitable $f:M \to G$ and $0 \leq s \leq t < \infty$ and u independent of $\mathcal{G}^s$ define

$$P_t^s(f)(u) = \int_{\Omega_t^{s,u}} f(F_t^s(u)(\omega))d\mu(\omega)$$

$$= \int_M f(x)p(s,u;t,dx).$$

As before we write $P_t(f)(u)$ for $P_t^0(t)(u)$. We then have immediately:

Corollary 4B For $r > 0$ and $t > s$

$$P_{t+r}^{s+r}(f)(u) = P_t^s(f)(u).$$

In particular

$$P_t^s(f)(u) = P_{t-s}(f)(u). \quad //$$

§5 CHAPMAN-KOLMOGOROV IDENTITY; MARKOV PROCESSES

We continue to use the notation of the previous sections.

(A) Theorem 5A Suppose $0 < r < s$. Let $A \in \text{Borel}(M)$ and let E be a measurable subset of $C^\Delta([s,\infty);M^+)$. Suppose that u is independent of G^r. Then

$$\mu\{\omega:F^r(u)(s,\omega) \in A \ \& \ F^r(u)(-,\omega)|[s,\infty) \in E\}$$

$$= \int_A \mu\{\omega:F^s(x,\omega) \in E\}\ p(r,u;s,dx).$$

Proof Let N denote the space of measurable maps $f:M \to C^\Delta([s,\infty);M^+)$ furnished with the usual σ-algebra generated by evaluations, and define ν to be the measure on $M \times N$ which is the image of μ restricted to $\Omega_s^{r,u}$ under the map

$$\omega \mapsto (F^r(u)(s,\omega),F^s(-,\omega)).$$

By independence $\nu = p(r,u;s,-) \otimes F^s(-,\mu)$.

By Theorem 3B

$$\mu\{\omega:F^r(u)(s,\omega) \in A \ \& \ F^r(u)(-,\omega)|[s,\infty) \in E\}$$

$$= \nu\{(m,f) \in A \times N:f(m) \in E\}$$

and the result follows by Fubini's theorem. //

Iteration of the conclusion of the theorem yields the first corollary:

Corollary 5A1 *Suppose* $0 \leqslant r \leqslant t_1 \leqslant \ldots \leqslant t_k < \infty$ *and* $A_1, \ldots, A_k$ *are Borel subsets of* M. *Then if* u *is independent of* G^r

(i) $\mu\{\omega : F^r(u)(t_j,\omega) \in A_j \text{ for } j = 1 \text{ to } k\}$

$$= \int_{A_1} \ldots \int_{A_k} p(t_{k-1},x_{k-1};t_k,dx_k)p(t_{k-2},x_{k-2};t_{k-1},dx_{k-1}) \ldots$$

$$\ldots\ p(r,u;t_1,dx_1)$$

and, in particular, taking $A_1 = M$ *and* $k = 2$.

(ii) (*Chapman-Kolmogorov identity*). *For* $m \in M$

$$\int_M p(t_1,x_1;t_2,A_2)p(r,m;t_1,dx_1) = p(r,m;t_2,A_2). \quad //$$

Corollary 5A2 *Suppose* $f: M \to G$ *is measurable and* $P_t f$ *exists for all* $t \geqslant 0$. *Then* $P_s(P_t f)$ *exists and*

$$P_s P_t f = P_{s+t} f.$$

Proof Note first that $P_t f$ is measurable by Corollary 7 of Chapter VIII. Now suppose $G = \mathbb{R}$. When f is the characteristic function of some Borel set A of M the result is just the Chapman-Kolmogorov identity above. It is therefore true for any non-negative f, since then we can approximate f by a sequence $\{f_i\}$ of non-negative simple functions with $f_i \uparrow f$. Then $P_t f_i \uparrow P_t f$ and so by the monotone convergence theorem $P_{s+t} f_i = P_s P_t f_i \uparrow P_s P_t f$, whence $P_{s+t} f = P_s P_t f$.

For general G let $\mathcal{L}^1_{t,m}$ denote $\mathcal{L}^1(M, \text{Borel } M, p(0,m;t,-);G)$. Then if $f \in \mathcal{L}^1_{t,m}$ for all $t \geqslant 0$ and $m \in M$ we have, for all $m \in M$

$$P_s(|P_t f(\cdot)|)(m) \leqslant P_s(P_t|f(\cdot)|)(m) = P_{s+t}(|f(\cdot)|)(m)$$

$$= \|f\|_{\mathcal{L}^1_{s+t,m}} .$$

Therefore $P_s P_t f$ exists, and moreover $f \mapsto P_s(P_t f)(m)$ is continuous on

$\mathcal{L}^1_{s+t,m}$. Since our assertion is clear on the dense subset of simple functions in $\mathcal{L}^1_{s+t,m}$ by the Chapman-Kolmogorov identity, it follows for general f. //

We shall not need to use the next corollary.

Corollary 5A3 Let z' be independent of z but with $z' \overset{\bullet}{\underset{\bullet}{\sim}} z$. Let F and F' be flows of (Y,z) and (Y,z') from time zero. Then if u is independent of z and z' and $0 \leq r \leq s$

$$F^r_t(u) \overset{\bullet}{\underset{\bullet}{\sim}} F'_{t-s}(F_{s-r}(u))$$

as processes in $t \in [s,\infty)$.

Proof Set $\tilde{F}^r_t(u) = F'_{t-s}(F_{s-r}(u))$ for $t \geq s$. As in Corollary 5A1(i) we can write down an expression for the finite dimensional distributions of $\tilde{F}^r(u)(t)$, $t \geq s$. Comparison of that formula with the given one for $F^r(u)$ using Theorem 4B and isonomy invariance shows that they are the same. //

(B) If $D \in F$ and $\mathcal{B}$ is a sub-σ-algebra of F the *conditional probability* $\mathbb{P}(D|\mathcal{B})$ is by definition the conditional expectation $\mathbb{E}(\chi_D|\mathcal{B}):\Omega \to \mathbb{R}$ of the characteristic function of D.

Proposition 5B With the notation of Theorem 5A:

(i) (*Markov Property of* $F^r(u)$). For each $s \geq r$

$$\mathbb{P}(\{\omega:F^r(u)(-,\omega)|[s,\infty) \in E\}\big|\sigma\{F^r(u)(q):r \leq q \leq s\})$$

$$= \mathbb{P}(\{\omega:F^r(u)(-,\omega)|[s,\infty) \in E\}\big|\sigma\{F^r(u)(s)\}) \qquad \text{a.s..}$$

In particular: (ii) If $0 \leq r \leq s \leq t$ and $A \in$ Borel (M)

$$\mathbb{P}(\{\omega:F^r(u)(t,\omega) \in A\}\big|\sigma\{F^r(u)(q):r \leq q \leq s\})(\omega)$$

$$= p(s,F^r(u)(s,\omega);t,A) \qquad \text{a.s.}$$

Moreover: (iii) For $0 \leq r \leq s \leq t < \infty$ "the past" $\sigma\{F^r(u)(q):r \leq q \leq s\}$ is conditionally independent of "the future" $\sigma\{F^r(u)(t):s \leq t < \infty\}$ given "the present" $\sigma\{F^r(u)(s)\}$.

Proof To verify (i) it is enough to check equality of both sides after they have been integrated over an event $P = \{\omega : F^r(u)(q_j,\omega) \in B_j \text{ for } j = 1 \text{ to } k\}$ where $B_j \in \text{Borel}(M)$ and $r < q_j < s$ for $j = 1$ to k. From Theorem 5A we see that the right hand side of (i) is just $G(F^r(u)(s,\omega))$ where $G:M \to \mathbb{R}$ is given by

$$G(x) = \mu\{\omega : F^s(x)(-,\omega) \in E\}.$$

By Corollary 5A1, assuming that $q_k = s$ without loss of generality,

$$\int_P G(F^r(u)(s,\omega))d\mu(\omega) = \int_{B_1}\cdots\int_{B_k} G(x_k)p(q_{k-1},x_{k-1};q_k,dx_k) \cdots \cdots p(r,u;q_1,dx_1)$$
$$= \mu\{\omega : \omega \in P \;\&\; F^r(u)(-,\omega)|[s,\infty) \in E\}$$

by Theorem 5A iterated. But this is just the integral of the left hand side of (i) over P. Thus (i) and (ii) are proved.

Part (iii) is automatic from (i). Indeed let A, N, B represent the 'past', 'present', and 'future' σ-algebras and suppose $P \in A$ and $Q \in B$. We must show that

$$\mathbb{E}(\chi_P \cdot \chi_Q | N) = \mathbb{E}(\chi_P | N)\mathbb{E}(\chi_Q | N).$$

It suffices to take Q to be of the form $Q = \{\omega : F^r(u)(-,\omega)|[s,\omega) \in E\})$. Then

$$\mathbb{E}(\chi_P \cdot \chi_Q | N) = \mathbb{E}(\mathbb{E}(\chi_P\chi_Q | A)|N)$$
$$= \mathbb{E}(\chi_P\, \mathbb{E}(\chi_Q | A)|N)$$
$$= \mathbb{E}(\chi_P\, \mathbb{E}(\chi_Q | N)|N)$$

by (i), and the result follows. //

The Chapman-Kolmogorov identity of Corollary 5A1(ii) ensures that the specification of part (i) of Corollary 5A1 satisfies the Kolmogorov consistency conditions of Chapter II §1 for processes with values in M^+. In particular that specification determines $F^r(u)$ up to isonomy, given the family $\{p(s,x,t,-) : r < s < t,\ x \in M\}$. The processes $\{F^r(m) : m \in M, r > 0\}$ form the *Markov process* with transition (sub)-probabilities $p(s,x,t,-)$. Assertions (i) and (ii) are equivalent formulations of the *Markov property* of the particular process $F^r(u)$.

(C) Example: The Ornstein-Uhlenbeck position process x described in the Introduction does *not* have the Markov property. Indeed its behaviour is specified after time s by knowledge of its position x_s and velocity v_s at time s. The velocity v_s can be determined by knowledge of x throughout any interval $(s-\varepsilon,s]$. On the other hand the Markov property would mean that no better estimate of x in $[s,\infty)$ could be obtained from knowledge previous to s than is obtainable from knowledge of x at the instant s. However the pair (x,v) considered as a process on $\mathbb{R}^n \times \mathbb{R}^n$ is Markov being a solution of the equation

$$dv_t = -\beta v_t\, dt + \gamma dB_t$$

$$dx_t = v_t\, dt.$$

The velocity process v by itself is also Markov.

There are corresponding remarks about solutions y to stochastic dynamical systems of the type we are considering but with the *white* noise dB_t replace by *coloured* or *real noise* dx_t, for x the O-U position process. Then

$$dy_t = X(y_t)(v_t)dt + V(y_t)dt$$

and to obtain a Markov process the state space has to be increased: for example by considering $(y,\frac{dy}{dt})$ or (y,v). The point of taking such a "Markovianization" is that the resulting process is completely determined by its transition probabilities. For a Markovianization of delay equations see (Mohammed 1981).

The derivative process v of a Brownian motion on N, as described in §12D of Chapter VII will not be Markov in general, but it is a projection of the Markov process δF_t on TO(N). However Brownian motion on N is Markov. In fact using the notation of §12 of Chapter VII we can define transition probabilities p(s,m;t,-) on N by letting

$$p(s,m;t,B) = \mu\{\omega : x^s(t,\omega) \in B\} \qquad B \in \text{Borel}\ (N)$$

for x^s a Brownian motion starting from $m \in N$ at time s. The argument at the beginning of §11 Chapter VII together with the fact that Brownian motion on $\mathbb{R}^n$ is invariant, up to isonomy, by the action of O(n) shows that this definition is independent of the choice of $u_s \in \pi^{-1}(m) \subset O(N)$.

Theorem 5C *Let* $x:[0,\xi) \times \Omega \to M$ *be a Brownian motion on* N *with* $x(0,\omega) = m$. *Then if* $0 < t_1 < \ldots < t_k < \infty$ *and* $A_1,\ldots,A_k \in$ Borel M

$$\mu\{\omega : x(t_j,\omega) \in A_j \text{ for } j = 1 \text{ to } k\}$$

$$= \int_{A_1} \ldots \int_{A_k} p(t_{k-1},x_{k-1}:t_k,dx_k) \ldots p(0,m;t_1,dx_1).$$

In particular x *has the Markov property.*

Proof Let $\tilde{p}$ denote the transition probabilities of the system (X,B) on O(N). Set $\tilde{A}_j = \pi^{-1}(A_j)$. We have $x(t,\omega) = \pi(u(t,\omega))$ for some solution u of (X,B) with $u(0) = u_0 \in \pi^{-1}(m)$ say. Using Corollary 5A1

$$\mu\{\omega : x(t_j,\omega) \in A_j \text{ for } j - \text{ to } k\} = \mu\{\omega : u(t_j,\omega) \in \tilde{A}_j \text{ for } j = 1 \text{ to } k\}$$

$$= \int_{\tilde{A}_1} \ldots \int_{\tilde{A}_k} \tilde{p}(t_{k-1},u_{k-1};t_k,du_k) \ldots \tilde{p}(0,u_0;t_1,du_1).$$

However π maps $\tilde{p}(t_{j-1},u_{j-1};t_j,-)$ to $p(t_{j-1},\pi(u_{j-1});t_j,-)$ and so the result follows. //

It makes sense to call what we have defined to be Brownian motions on N *standard models* of Brownian motion on N and to allow any process with continuous sample paths which is isonomous to one of these standard models to be called a *Brownian motion*.

§6 CRITERIA FOR NON-EXPLOSION AND STOCHASTIC COMPLETENESS

(A) The Khasminskii tests for non-explosion are described in detail in (Azencott 1974), (as also are criteria for recurrence and transience). We give a simple, but useful result from (Elworthy 1981). The notation is as before with A the differential generator for the system (Y,z).

Theorem 6A *Let* M *be finite dimensional and suppose there exists a* C^2 *map* $\alpha: M \to [0,\infty)$ *such that*

(i) $\alpha(y) \to \infty$ *as* $y \to \infty$ *in* M

(ii) *there exists a sequence* $\{r_n\}_{n=1}^{\infty}$ *in* $[0,\infty)$, *increasing to* ∞ *so that for*

$$A_n = \{y \in M : \alpha(y) < r_n\}$$

we have $\lim\limits_{n\to\infty} \frac{1}{r_n} \sup \{A(\alpha)(y) : y \in A_n\} \leq 0.$

Then (Y,z) is complete.

Proof Set $k_n = \sup \{A(\alpha)(y) : y \in A_n\}$. Let $\{x_t : t \geq 0\}$ be a solution starting from a point x_0 of M, and let τ_n denote its first exit time from A_n. Set $\Omega_t^n = \{\omega \in \Omega : t < \tau_n(\omega)\}$. By Itô's formula, for $t \geq 0$,

$$\alpha(x_{t\wedge\tau_n}) = \alpha(x_0) + \int_0^{t\wedge\tau_n} d\alpha\Big(X(x_s)dB_s\Big) + \frac{1}{2}\int_0^{t\wedge\tau_n} A(\alpha)(x_s)ds$$

where $x_{t\wedge\tau_n}$ is the map $\omega \mapsto x(t\wedge\tau_n(\omega),\omega)$. Therefore by Corollary 7C of Chapter III

$$\int_\Omega \alpha(x_{t\wedge\tau_n})d\mu \leq \alpha(x_0) + \frac{1}{2}tk_n.$$

However

$$\int_\Omega \alpha(x_{t\wedge\tau_n})d\mu \geq r_n(1 - \mu(\Omega_t^n))$$

since if $\omega \notin \Omega_t^n$ we have $\alpha(x_{t\wedge\tau_n}(\omega)) = r_n$. Thus

$$1 - \mu(\Omega_t^n) \leq \frac{\alpha(x_0)}{r_n} + \frac{1}{2}t\,\frac{k_n}{r_n}$$

whence by (ii)

$$1 - \mu(\bigcup_n \Omega_t^n) \leq 0.$$

The result follows since a path must leave each A_n before it explodes by Corollary 6.2 of Chapter VII. //

Remarks 6A (i) We have been a bit imprecise in the definition of completeness. At least we mean that solutions starting from all points of M go on for all time. In the past we have taken it to include solutions with initial functions $u:\Omega \to M$. Luckily Theorem 3B, and its extensions to more general systems, shows that completeness for initial functions is implied by that for initial points, at least for the situations we are mainly interested in. This is because the measurable version of the flow in Theorem 3B can be chosen so that $F^r(M \times \Omega)$ lies in $C([0,\infty);M)$ given completeness from each point of M.

(ii) Theorem 6A remains true when M is infinite dimensional, with the same proof, if condition (i) is replaced by the assumption that each A_n admits a uniform cover and that $M = \bigcup_n A_n$.

Corollary 6A A finite dimensional Riemannian manifold M is stochastically complete if there exists $\alpha:M \to [0,\infty)$ satisfying conditions (i) and (ii) of the theorem with A replaced by $\frac{1}{2}\Delta$.

Proof It is enough to show that the canonical S.D.S. (X,B) on O(M) is complete. This follows from the theorem by considering the map $\alpha\circ\pi:O(M) \to [0,\infty)$.

(B) Let M be a complete connected Riemannian manifold with corresponding distance function d. An obvious choice of α in Theorem 6A or its corollary is

$$\alpha(x) = d(x,p)$$

for some fixed p in M. Unfortunately this is not differentiable in general, and not just at the point p where it can easily be modified. For example on the circle S^1 it is not differentiable at the antipodal point to p. The bad points are points of the *cut locus* Cut (p) of p, see (Cheeger & Ebin 1975; Kobayashi 1967). The point p is a *pole* of M if the exponential map $\exp_p:T_pM \to M$ is a diffeomorphism, or equivalently if Cut (p) is empty. This is so for all points p of any simply connected complete manifold M with non-positive sectional curvatures by the Cartan-Hadamard Theorem e.g. (Milnor 1963) Theorem 19.2. See Appendix B, §4C.

In general, for $\alpha(x) = d(x,p)$, there is an upper estimate in (Yau 1975) for $\Delta\alpha(x)$ when $x \notin$ Cut (p) in terms of the Ricci tensor of M. In particular if the Ricci is bounded below on M then for each $\delta > 0$ $\Delta\alpha(x)$ is bounded above on $\{x \in M:\alpha(x) > \delta \;\&\; x \notin \text{Cut}(p)\}$. See also (Greene & Wu 1979) where also, by the Hessian comparison theorem, the second covariant derivative $\nabla^2\alpha$ is shown to be bounded above on the same sets provided all the sectional curvatures have a common lower bound.

Another special class of points of M are the *conjugate points* to p. These are the critical values of $\exp_p:T_pM \to M$, i.e. the images under

$\exp_p$ of points v in T_pM at which $T_v \exp_p: T_vT_pM \to T_{\exp_p(v)}M$ fails to be invertible. If p has no conjugate points then $\exp_p$ is a covering map (see Appendix A) and determines a complete Riemannian metric on T_pM for which each $T_v \exp_p$ is an isometry, i.e. such that $\exp_p$ is a *Riemannian cover*. This holds for all points p when M has non-positive sectional curvatures.

Proposition 6B(i) Let $f:\tilde{M} \to M$ be a C^2 covering map. Given any S.D.S. (X,z) on M let $(\tilde{X},z)$ be the lift of (X,z) to $\tilde{M}$. For an F_a-measurable map $\tilde{x}_a:\Omega \to \tilde{M}$ the solution to $d\tilde{x} = \tilde{X}(\tilde{x})dz$ starting from $\tilde{x}_a$ is mapped by f to the solution from $f(\tilde{x}_a)$ to $dx = X(x)dz$, and both solutions have the same explosion time.

(ii) Let $f:\tilde{N} \to N$ be a Riemannian cover of complete Riemannian manifolds. Then f maps Brownian motion on $\tilde{N}$ to Brownian motion on N, and they both have the same explosion time.

Proof Part (i) is a simple special case of Theorem 13C of Chapter VII, and (ii) follows from it since $f:\tilde{N} \to N$ determines a covering

$$O(f):O(\tilde{N}) \to O(N)$$

such that the canonical S.D.S. of $\tilde{N}$ is the lift of that of N, and such that $O(f)$ lies over f. //

(C) When p has no conjugate points the above considerations can easily be used to find criteria for non-explosion by lifting via the exponential map to T_pM where the distance function from the origin is just the Euclidean distance in the inner product of T_pM. However note that Theorem 6A is not strong enough to give the linear growth criterion of Example (ii) §6 Chapter VII when M is $\mathbb{R}^n$.

In any case there is the result of Greene and Wu, see (Wu 1979) Theorem 4, that if the Ricci curvature is bounded below there exists a C^∞ map $\alpha:M \to \mathbb{R}$ with $\alpha(x) \to \infty$ as $x \to \infty$ and such that $\Delta\alpha$ is bounded on M. Their map is even Lipschitz. The construction seems technically very complicated but assuming its existence Proposition 6A immediately yields the following result proved analytically by S.-T. Yau:

Theorem 6C (Yau 1978) Every complete Riemannian manifold

with Ricci curvature bounded below is stochastically complete. //

For an alternative analytic proof see (Dodziuk 1980). For a strengthened version proved probabilistically and for a criterion for stochastic incompleteness see (Ichihara 1980). A non-explosion criterion for submanifolds of $\mathbb{R}^n$ is given in (Baxendale 1980).

§7 VERTICAL DRIFT: THE FEYNMAN - KAC FORMULA, HEAT FLOW ON 1-FORMS

(A) The differential generators A of the semi-groups we have looked at so far have no zero order term. The Feynman-Kac formula gives a method of obtaining more general semi-groups.

Suppose that $V:M \to \mathbb{R}$ is continuous and $v_0 \in \mathbb{R}$. Define

$$\bar{v}:[a,\xi) \times \Omega \to M \times \mathbb{R}$$

by

$$\bar{v}_t(\omega) = (x_t(\omega), v_t(\omega))$$

where

$$\frac{dv}{dt} = V(x_t)v_t$$

and

$$x_t = F_t(u),\ \xi = \xi^u,\ \text{for } u \in M.$$

This is a special case of §13A Chapter VII; in fact now we have an explicit solution on Ω_t^u:

$$v_t = (\exp \int_0^t V(x_s)ds)v_0$$

For $f \in B\mathcal{L}^0(M;\mathbb{R})$ define $\tilde{f}:M \times \mathbb{R} \to \mathbb{R}$ by $\tilde{f}(m,r) = f(m)r$. Then, if $v_0 = 1$, almost surely on Ω_t^u, $0 \leqslant t < \infty$

$$\begin{aligned}\tilde{f}(\bar{v}_t) &= f(x_t)v_t \\ &= f(u) + \int_0^{t\wedge\xi} df\circ X(x_s)(dB_s)v_s \\ &\quad + \int_0^{t\wedge\xi} A(f)(x_s)v_s ds + \int_0^{t\wedge\xi} f(x_s)V(x_s)ds.\end{aligned}$$

If V is bounded above clearly v_t is in $\mathcal{L}^1$ for each t and we can define

$$Q_t:B\mathcal{L}^0(M;\mathbb{R}) \to B\mathcal{L}^0(M;\mathbb{R})$$

by

$$Q_t(f) = \int_{\Omega_t^u} f(x_s)(\exp \int_0^t V(x_s)ds)d\mu$$

to obtain a semi-group (by Corollary 5A2) satisfying (by Theorem 1A)

$$\lim_{t \downarrow 0} \frac{Q_t(f)(u)-f(u)}{t} = Af(u) + V(u)f(u) \qquad u \in M$$

whenever $f:M \to \mathbb{R}$ is C^2 and f and $Af + Vf$ are bounded.

The same formula (the 'Feynman-Kac' formula) clearly works when x is Brownian motion on a Riemannian manifold N, so that $A = \frac{1}{2}\Delta$. In particular, (saying that a function f of two variables is $C^{1,2}$ if the partial derivatives $D_1 f$ and $D_2 D_2 f$ exist and are continuous):

Theorem 7A Let N be stochastically complete and $V:N \to \mathbb{R}$ continuous and bounded above. If $f_0:N \to \mathbb{R}$ is bounded and C^2 the equation for $f:[0,\infty) \times N \to \mathbb{R}$

$$\frac{\partial f}{\partial t} = \tfrac{1}{2}\Delta f + Vf$$

$$f(0,x) = f_0(x) \qquad\qquad x \in N$$

has at most one solution which is $C^{1,2}$. It is given by

$$f(t,x_0) = \int_\Omega f_0(x_t) \exp [\int_0^t V(x_s)ds]d\mu \qquad x_0 \in N$$

where x_t is Brownian motion on N starting from x_0.

Proof Apply Theorem 2A to the relevant S.D.S. on $O(M) \times \mathbb{R}$ constructed as above. The Itô formula used in the proof of Theorem 2A only requires the function g to be $C^{1,2}$. //

Remarks 7A(i) When V is locally Hölder continuous a $C^{1,2}$ solution does exist: see (Azencott 1974).

(ii) In partial differential equation theory the zero order term V is often assumed non-positive. Note that if λ is an upper bound for V and f a solution as above then $e^{-\lambda t}f(t,x)$ satisfies the same equation with V replaced by $\bar{V}$, for $\bar{V}(x) = V(x) - \lambda \leqslant 0$, all $x \in N$.

(iii) For a detailed discussion of the Feynman-Kac formula in $\mathbb{R}^n$, and applications to quantum physics, see (Simon 1979).

(B) A more sophisticated Feynman-Kac formula can be used to obtain a semi-group for the heat flow of 1-forms on a Riemannian manifold N. The approach described here was given in (Airault 1976). We shall use the results of §13, Chapter VII with B = TN, together with the usual notation we have assembled for Brownian motion on N.

Consider the covariant equation on B

$$(\text{It\hat{o}})\quad Dv(t) = -\tfrac{1}{2}K(v(t),-)^{\#}dt$$

where K is the Ricci tensor, and $K(v(t),-)^{\#} \in T_{x(t)}N$ corresponds to $K(v(t),-) \in T^*_{x(t)}N$. Let

$$v:[0,\xi) \times \Omega \to TN, \quad v(o) = v_0 \in T_{x_0}N$$

be a maximal solution. Then by Proposition 13B of Chapter VII if ϕ is a C^2 1-form on N, almost surely on Ω_t,

$$\phi(v_t) = \phi(v_0) + \int_0^{t\wedge\xi} \nabla\phi(u_s dB_s)v_s - \tfrac{1}{2}\int_0^{t\wedge\xi} K(v_s,\phi^{\#}(x_s))ds$$

$$+ \tfrac{1}{2}\int_0^{t\wedge\xi} \text{trace}\,(\nabla^2\phi)(v_s)ds$$

$$= \phi(v_0) + \int_0^{t\wedge\xi} \nabla\phi(u_s dB_s)v_s + \tfrac{1}{2}\int_0^{t\wedge\xi} \Delta\phi(v_s)ds \qquad (1)$$

by the Weitzenböck formula §12D, Chapter VII). Here we have used the fact that, in the notation of Proposition 13B, for $e \in \mathbb{R}^n$

$$\frac{D}{\partial t}\, T\pi \circ \tilde{X}(\tilde{S}(t,u_s)e)e = \frac{D}{\partial t}\,(\tilde{S}(t,u_s)e)e$$

$$= 0.$$

Now writing $\tilde{v}_t = u_t^{-1}v_t$, the defining equation for v gives

$$\frac{d}{dt}\tilde{v}_t = -\tfrac{1}{2}\,u_t^{-1}\,K(v_t,-)^{\#} \qquad 0 < t < \xi(\omega).$$

Consequently (by elementary calculus)

$$|\tilde{v}_t|^2 = |\tilde{v}_0|^2 + \langle 2\tilde{v}_t, -\tfrac{1}{2}\,u_t^{-1}K(v_t,-)^{\#}\rangle$$

giving

$$|v_t|^2 = |v_0|^2 - \int_0^t K(v_s,v_s)ds\ . \qquad (2)$$

It follows that if K is bounded below then $|v(t)|$ is bounded on $[0,\tau\wedge\xi)\times\Omega$ for each $\tau \in [0,\infty)$. This, together with Corollary 5A2 and the proof of Theorem 1A applied to equation (1) gives:

Proposition 7B Suppose the Ricci curvature K of N is bounded below, then defining

$$Q_t(\phi)(v_0) = \int_{\Omega_t} \phi(v_t)d\mu$$

we obtain a semi-group on the space of bounded measurable 1-forms on N with the property that whenever ϕ is C^2 and ϕ and $\Delta\phi$ are bounded then

$$\lim_{t\downarrow 0} \frac{Q_t(\phi)(v_0) - \phi(v_0)}{t} = \tfrac{1}{2}\Delta\phi(v_0) \quad \text{all} \quad v_0 \in TN. \quad // \qquad (3)$$

When N is complete as well as having K bounded below we know that N is also stochastically complete by Yau's theorem: Theorem 6C. We can then therefore strengthen (3) to the equality

$$Q_t(\phi)(v_0) = \phi(v_0) + \tfrac{1}{2}\int_0^t Q_s\Delta(\phi)(v_0)ds.$$

The above proposition together with equation (2) give a good idea of the effect of positive Ricci curvature on the heat flow of 1-forms. For example Bochner's theorem in the form: there exists no non-zero harmonic 1-form on a compact manifold of strictly positive Ricci curvature, is an immediate consequence. Using the fact that Brownian motion on a compact manifold is recurrent one can obtain the corresponding result when the Ricci curvature is non-negative everywhere and strictly positive at at least one point of N. See also (Malliavin 1974; Debiard et al. 1976). For an analytical discussion of a generalization of Bochner's theorem to non-compact N see (Yau 1976). For a general discussion of heat flows of 1-forms including the *geodesic deviation* of Dohrn & Guerra see (Meyer 1981b).

Note that by Theorem 12D of Chapter VII, the other 'heat flow' $\mathcal{D}P_t(\phi)$ obtained from δF_t seems to have a more restricted domain of definition.

(C) We have another uniqueness theorem on the lines of Theorem 2C; it is taken from (Elworthy 1981):

Theorem 7C Suppose that ϕ_t is a C^2 1-form on N for $0\le t\le T<\infty$, some $T > 0$, which satisfies the heat equation $\frac{\partial\phi_t}{\partial t} = \frac{1}{2}\Delta\phi_t$. Assume that N is complete and has Ricci curvature bounded below. Then if ϕ_t is uniformly bounded on M for $0 \leq t \leq T$ we have

$$\phi_t = Q_t(\phi_0) \qquad\qquad 0 \leq t \leq T$$

for Q_t as above. In particular such a solution is uniquely determined by its initial value ϕ_0.

Proof As for Theorem 2C, using the stochastic completeness of N from Yau's theorem: Theorem 6C. //

The uniqueness part of Theorem 7C was proved analytically in (Dodziuk 1980).

§8 THE DIRICHLET PROBLEM: GREENS MEASURE

(A) Consider an open subset U of M with boundary ∂U. Suppose $V:\bar{U} \to \mathbb{R}$ is continuous and bounded above. Let $\tau \equiv \tau^{x_0}(U)$ be the first exit time of $x \equiv F(x_0)$ from U, where $x_0 \in U$. For simplicity we shall assume that $\tau^{x_0} < \xi^{x_0}$ almost surely.

Proposition 8A Suppose $h:\bar{U} \to \mathbb{R}$ is C^2 on U and continuous on $\bar{U}$, and satisfies

$$A(h) + Vh = 0 \qquad \text{on } U.$$

Assume that h is bounded and that $\exp\left(\int_0^\tau V^+(x_s)ds\right)$ is in $\mathcal{L}^1(\Omega,F,\mu;\mathbb{R})$ where $V^+ = \frac{1}{2}(V + |V|)$. Then

$$h(x_0) = \int_\Omega h\Big(x\,(\tau(\omega),\omega)\Big)\Big(\exp\int_0^{\tau(\omega)} V(x_s(\omega))ds\Big)d\mu(\omega). \qquad (1)$$

Proof As in §7A the Itô formula yields

$$h(x_\tau)\exp\int_0^\tau V(x_s)ds = h(x_0) + \int_0^\tau dh\circ X(x_s)(dB_s)(\exp\int_0^s V(x_r)dr). \qquad (2)$$

Now if τ_R is any stopping time less than or equal to τ then, a.s.,

$$h(x_{\tau_R})\exp\int_0^{\tau_R} V(x_s)ds \leq (\sup_{u\in\bar{U}} |h(u)|)\exp\int_0^\tau V^+(x_s)ds.$$

It follows that the use of the dominated convergence theorem in the proof of Corollary 7C of Chapter III is valid in our situation and integration of (2) yields (1) as required. //

Note that the right hand side of (1) depends only on the restriction of h to ∂U since $x(\tau(\omega),\omega) \in \partial U$. Thus, under good conditions (1) gives the solution to the Dirichlet problem: $A(h) + Vh = 0$ with $h|\partial U$ given. For a more careful discussion see (Friedman 1975).

The map $\Omega \to \partial U$ defined by $\omega \mapsto x(\tau(\omega),\omega)$ is measurable and so determines a measure $g(x_0,-)$ on ∂U. This is the *Greens measure*. When $V \equiv 0$ equation (1) becomes

$$h(x_0) = \int_{\partial U} h(u)g(x_0,du) \qquad\qquad x_0 \in U.$$

(B) The sort of argument used above can be used to estimate exit times. The following was used to compare exit times from balls in spaces of different constant curvature by Debiard et al. (1976). See also (Pinsky 1978). A similar method is used in (Ichihara 1980). For an application to the estimation of eigenfunctions see (Carmona & Simon 1981).

Proposition 8B Assume that $\tau^{x_0} < \xi^{x_0}$ almost surely: for example that $\bar{U}$ is compact or admits a uniform cover. Let $h:\bar{U} \to \mathbb{R}$ be a C^2 and bounded solution to the Dirichlet problem

$$\begin{aligned} A(h)(u) &= -1 \qquad\qquad u \in U. \\ h|\partial U &= 0 \end{aligned}$$

Assume also that

$$\lim_{t\to\infty} h(x(t\wedge\tau^{x_0}(\omega),\omega)) = 0 \quad \text{a.s. for all } x_0 \text{ in } U.$$

Then

$$\int_\Omega \tau^{x_0}(\omega)d\mu(\omega) = h(x_0).$$

Proof By the Itô formula, for $\tau = \tau^{x_0}$,

$$h(x_{t\wedge\tau}) = h(x_0) + \int_0^{t\wedge\tau} dh\circ X(x_s)dB_s - \int_0^{t\wedge\tau} 1\, ds.$$

Therefore

$$h(x_0) = \int_\Omega [h(x_{t\wedge\tau}) + t\wedge\tau]d\mu.$$

The result follows by letting $t \to \infty$, using the dominated and monotone convergence theorems. //

§9 $P_t f$ AS A SOLUTION

(A) So far we have given various results expressing solutions of the diffusion equation in the form $P_t f$. It is time to examine when $P_t f$ *is* a solution. We have shown that $\{P_t\}_{t\geq 0}$ is a semigroup on $B\mathcal{L}^0(M;\mathbb{R})$ with $||P_r|| \leq 1$ for each t. Moreover

(i) (positivity) $P_t f \geq 0$ when $f \geq 0$

(ii) $P_t f(x)$ is measurable in t for all x in M, and continuous in t if f is continuous

(iii) if f is C^2, bounded, and with A(f) bounded then

$$\lim_{t\downarrow 0} t^{-1}(P_t f(x) - f(x)) = Af(x)$$

all $x \in M$.

Of these (i) is clear, (ii) comes immediately from the sample continuity of our solution process, and (ii) is Theorem 1A. We gave some continuity results in Theorem 6C of Chapter VIII.

If we knew that $P_t f$ was C^2 with $A(P_t f)$ bounded, for all $t \geq 0$, we would have, for $x \in M$,

$$\lim_{h\downarrow 0} h^{-1}[P_{t+h} f(x) - P_t f(x)] = \lim_{h\downarrow 0} h^{-1}[P_h - 1]P_t f(x) = A(P_t f)(x)$$

by (iii), and so $P_t f$ is a (classical) solution of the diffusion equation

$$\frac{\partial}{\partial t} P_t f = A(P_t f).$$

Strictly speaking we have only shown that the right hand time derivative of $P_t f$ behaves as required. However if also (Y,z) is complete, or if f converges to 0 at ∞, then $P_t f$ is differentiable in t by Theorem 1B. Alternatively use Theorem 2, §3 Chapter IX of (Yosida 1968) where there is also a simple argument to show that under general conditions

$$\frac{\partial}{\partial t} P_t f = \bar{A}(P_t f)$$

where $\bar{A}$ is the infinitesimal generator of $\{P_t\}_t$.

Differentiability results were given in Theorems 8B and 8C of Chapter VIII. Unfortunately there remains a gap between being twice strongly Gâteaux differentiable and being C^2. To get over this some more work would be needed, especially in the infinite dimensional case. However here we shall content ourselves with a result which at least is reasonably definitive for compact manifolds:

Theorem 9A Suppose that Y is C^3 and that M is finite dimensional with (Y,z) strongly complete. Assume also:

(i) the tensors ∇Y, $\nabla^2 Y$, $R(Y,-)(-)$ are uniformly bounded together with $\tilde{\delta}^2 Y$ and $\tilde{\delta}^3 Y$ as in Proposition 5C of Chapter VIII;

(ii) $f: M \to \mathbb{R}$ is C^2 and bounded; and

(iii) $A(f \circ F_t): M \to \mathbb{R}$ is almost surely bounded for each $t > 0$, where F_t is a C^2 version of the flow of (X,z).

Then $P_t f$ is twice strongly Gâteaux differentiable and satisfies

$$\frac{\partial}{\partial t} P_t f = A(P_t f) \qquad t > 0.$$

Proof Using Theorem 3 of Chapter VIII we can take our flow maps to be C^2. By Proposition 8C of Chapter VIII, $P_t f$ is twice strongly Gâteaux differentiable and, as a map of M,

$$AP_t f = \int_\Omega A(f \circ F_t) d\mu \tag{1}$$

By Theorem 3B, and then using independence and time homogeneity (Proposition 4A):

$$\begin{aligned} P_{t+s} f - P_t f &= \int_\Omega \{f(F^s_{t+s} \cdot F_s) - f(F_t)\} d\mu \\ &= \int_\Omega \{P_s(f \circ F_t) - f \circ F_t\} d\mu \\ &= \int_\Omega \int_0^s P_r\, A(f \circ F_t) dr\, d\mu \end{aligned}$$

by Theorem 1B and (ii). Therefore Fubini's theorem and (1) give:

$$P_{t+s} f - P_t f = \int_0^s P_r\, AP_t f\, dr$$

and the result follows. //

Remarks (a) All the hypotheses are satisfied by a C^3 system (Y,z) on a compact manifold M with f a C^2 map.

(b) Condition (iii) is satisfied if f has compact support and F_t is a diffeomorphism of M onto M

(c) Strong completeness should not be necessary: it was used so as to be able to apply the Itô formula to $f\circ F_t(-,\omega):M \to \mathbb{R}$. However the Itô formula can be applied to $f\circ F_t:M \to L^2(\Omega,F,\mu;\mathbb{R})$ if that map is C^2.

(B) When A is elliptic i.e. when $X(u):\mathbb{R}^n \to T_uM$ is surjective for each u in M, the situation is particularly nice and there is a very clear discussion in (Azencott 1974). We shall not go into details but the key point is that if $\{U_n\}_{n=1}^\infty$ is a family of open sets covering M with $\bar{U}_n$ compact and in U_{n+1} for each n, then

$$P_t f(x) = \sup_n P_{n,t} f(x) \qquad (1)$$

for $f \in B\mathcal{L}^0(M;\mathbb{R})$ where $\{P_{n,t}\}_{t>0}$ is the semigroup associated to (Y,z) restricted to U_n i.e.

$$P_{n,t} f(x) = \int_{\tau(U_n)(\omega)<t} f(F_t(x,\omega))d\mu(\omega).$$

This is quite obvious. On the other hand if each U_n has smooth boundary ∂U_n the problem for $g_t:\bar{U}_n \to \mathbb{R}$, $t \geq 0$,

$$\frac{\partial g_t}{\partial t} = Ag_t \quad \text{on } U_n$$

$$g_t|\partial U_n \equiv 0,$$

with g_t continuous on $\bar{U}_n$ and g_0 a given C^2 function with compact support in U_n is well known (by partial differential equation theory) to have a C^2 solution, continuous on $[0,\infty) \times \bar{U}_n$. By Theorem 2B this must be $g_t = P_{n,t}\, g_0$.

Assuming that f is C^2 with compact support this shows that each point of M has a neighbourhood where, for sufficiently large n, $P_{n,t}f$ is $C^{1,2}$, and satisfies $\partial g_t/\partial t = Ag_t$. Azencott then applies Schauder's interior estimates for parabolic equations to (1) to see that the same holds for P_tf. Thus P_tf is a solution. On the other hand using (1) and the maximal principle, if f is also non-negative, then $P_tf(x)$ is less than or equal to any other such solution. In fact Azencott shows that in the elliptic case then $\{P_t\}_t$ is the unique *minimal* semigroup associated to A, in the sense that if $f \in B\mathcal{L}^0(M,[0,\infty))$ then

$$P_t f \leq Q_t f$$

for any other semigroup $\{Q_t\}$ associated to A.

There is also a *density, heat kernel, fundamental solution,* or *propagator*,

$$p_t : M \times M \to \mathbb{R}(\geq 0),\ t > 0$$

such that

$$P_t f(x) = \int_M f(y) p_t(x,y) dy$$

where the integration is with respect to the volume element of a Riemannian metric on M. Azencott also gives the proof that $P_t f$ is C^2 for all f in $B\mathcal{L}^0(M;\mathbb{R})$.

(C) For a general discussion of the semigroups associated to Markov processes see (Dynkin 1965). The powerful *martingale method* gives a very neat way of expressing the relationship between our processes and their differential generator A, as well as a general method of constructing the processes with very weak conditions on the coefficients: see (Williams 1980a) for a very helpful review of (Stroock & Varadhan 1979).

§10. OTHER GEOMETRICAL CONSTRUCTIONS OF BROWNIAN MOTION AND RELATED PROCESSES

(A) The construction of Brownian motion via the development map gives a canonical construction for any C^3 Riemannian manifold N. To construct a Markov process on such N with differential generator $\mathcal{A}f = \frac{1}{2}\Delta f + A(f)$ for A a C^1 vector field on N we can simply take the horizontal lift $\tilde{A}$ of A to O(N) to give a horizontal vector field $\tilde{A}$ on O(N). For $x_0 \in N$ choose $u_0 \in \pi^{-1}(x_0)$ and let x_t be the projection of the maximal solution u_t to $du_t = X(u_t)dB_t + \tilde{A}(u_t)dt$ from u_0, where (X,B) is the canonical S.D.S. If $f: N \to \mathbb{R}$ is C^2 we have, a.s. on Ω_t

$$f(x_t) = f(x_0) + \int_0^{t\wedge\xi} df(u_s dB_s) + \int_0^{t\wedge\xi} \mathcal{A}f(x_s) ds.$$

Also x_t has the Markov property, as in Theorem 5C, and up to isonomy is independent of the choice of u_0 in $\pi^{-1}(x_0)$. It will be called *standard Brownian motion on* N *with drift* A. If we set

$$P_t f(x_0) = \int_{\Omega_t} f(x_t)d\mu$$

the considerations of §9B show that $P_t f$ is the minimal solution of

$$\frac{\partial g_t}{\partial t} = \tfrac{1}{2}\Delta g_t + A(g_t)$$

with $g_0 = f$. This can be boosted up by the Feynman-Kac formula, as in §7, to include a potential V.

(B) As discussed in §4F of Appendix B, if M is a C^2 finite dimensional Riemannian manifold there exists a C^2 vector bundle map $X:\underline{\mathbb{R}}^m \to TM$, for some m, which restricts to an orthogonal projection of $\mathbb{R}^m$ onto T_xM over each x of M. The solutions of the Itô equation (see §14 Chapter VII)

$$(\text{Itô}) \quad dx_t = X(x_t)dB_t$$

for B an m-dimensional Brownian motion, will then be Brownian motions on M, since this equation is equivalent to

$$dx_t = X(x_t)dB_t - \tfrac{1}{2}\text{ trace } \nabla X(X(x_t)-)(-).$$

In fact we only need the Riemannian structure to be C^1 for this to work.

In this construction the X is not canonical, and not closely related to the Riemannian geometry of M. It is therefore less likely to be useful in tying in the geometry of M with its stochastic analysis.

(C) Now suppose that M is a C^2 submanifold of $\mathbb{R}^m$, some m, with induced Riemannian metric i.e. as a Riemannian manifold it is *isometrically embedded*, e.g. S^n in $\mathbb{R}^{n+1}$. As in Example 1A(iv) we can take $X:\underline{\mathbb{R}}^m \to TM$ to be the orthogonal projection of $\mathbb{R}^m$ onto T_xM for each point x of M. Fix x in M and take an orthonormal base $e_1,\ldots,e_m$ of $\mathbb{R}^m$ with $e_1,\ldots,e_n$ in T_xM and $e_{n+1},\ldots,e_m$ in $T_xM^\perp$. Let $E_1,\ldots,E_m$ be the corresponding constant vector fields on $\mathbb{R}^m$, and set $Z(x) \equiv I-X(x):\mathbb{R}^m \to \mathbb{R}^m$. Then E_i has derivative zero and

$$E_i(x) = X(x)E_i(x) + Z(x)E_i(x).$$

By the formulae of Gauss and Weingarten, §5 Appendix B, if we differentiate and take the component in T_xM,

$$0 = \nabla X_i(v) - A_x(v,Z_i(x))$$

for some $A_x: T_xM \times T_xM^{\perp} \to T_xM$, where $v \in T_xM$ and $X_i \equiv X(\cdot)e_i$, $Z_i(\cdot) = Z(\cdot)e_i$. From this we see

$$\nabla X_i(v) = 0 \qquad i = 1,\ldots,n \quad \text{all } v \in T_xM$$

while

$$\nabla X_i(X_i(x)) = 0 \; i = n+1,\ldots,m$$

because $X_i(x) = 0$ for these i. Thus trace $\nabla X(X(x)(-))(-) \equiv 0$ for all x in M, and the equation

$$(\text{Itô})\; dx_t = X(x_t)dB_t$$

is equivalent to the equation $dx_t = X(x_t)dB_t$, so that the solutions of either are Brownian motions on M. See Examples 3(i) of Chapter VII for a special case.

(D) If M is a finite dimensional Lie group with a left invariant Riemannian metric we can take a left invariant trivialization $X:\mathbb{R}^n \to TM$ as in §5B of Appendix A, with X(1) an isometry. Again this gives a left invariant stochastic differential equation

$$dx_t = X(x_t)dB_t - \tfrac{1}{2}\,\text{trace}\,\nabla X(X(x_t)(-))(-)$$

whose solutions are Brownian motions on M. This is complete, even strongly complete by §2F of Chapter VIII. Thus M is stochastically complete.

For each e in $\mathbb{R}^n$, $\nabla X(X(\cdot)e)(e)$ is identically zero if and only if all the integral curves of $X(\cdot)e$ are geodesics: just differentiate an integral curve twice. This will be so for all e in $\mathbb{R}^n$ if and only if the Lie group exponential map (§5C Appendix A) coincides with the exponential map of the Levi-Civita connection (§4 Appendix B). This is true when the metric is *bi-invariant* (i.e. both right and left invariant): see (Milnor 1963) §21, or (Arnold 1978) Appendix 2, for a more general discusssion. A detailed discussion of the curvature properties of Lie groups with left invariant metrics is in (Milnor 1976). Brownian motion on Lie groups is discussed in (McKean 1969).

(E) Let $\pi: B \to M$ be a C^r *Riemannian submersion*, $r \geqslant 3$. This means by definition, that π is surjective and $T_b\pi: T_bB \to T_{\pi(b)}M$ is an orthogonal projection for each b in B, so we can write

$$T_bB = VT_bB \oplus HT_bB \qquad\qquad b \in B$$

where VT_bB is the null space of $T_b\pi$ and has orthogonal complement the 'horizontal' subspace HT_bB which is mapped isometrically by $T_b\pi$ onto $T_{\pi(b)}M$. For a vector field V on M let $\tilde{V}$ denote its horizontal lift to B, i.e. the vector field on B such that $\tilde{V}(b) \in HT_bB$ for each b in B and $\tilde{V}$ and V are π-related: $T_b\pi(\tilde{V}(b)) = V(\pi(b))$.

Let ∇ and $\tilde{\nabla}$ denote covariant differentiation with respect to the Levi-Civita connections of B and M respectively. The following comes from (O'Neill 1966):

<u>Lemma 10E</u> <u>If</u> V <u>and</u> W <u>are vector fields on</u> M <u>then the vector field</u> $\widetilde{\nabla V}(\tilde{W}(\cdot))$ <u>is</u> π-<u>related to</u> $\nabla V(W(\cdot))$.

<u>Proof</u> For any x in M, $b \in \pi^{-1}(x)$, and C^1 vector field U on M:

$$\tilde{W}(\langle\tilde{V},\tilde{U}\rangle)(b) = d(\langle\tilde{V},\tilde{U}\rangle)(\tilde{W}(b))$$

$$= \langle\widetilde{\nabla V}(\tilde{W}(b)),\tilde{U}(b)\rangle_b + \langle\tilde{V}(b),\widetilde{\nabla U}(\tilde{W}(b))\rangle_b$$

while $[\tilde{V},\tilde{W}](b) = \widetilde{\nabla W}(\tilde{V}(b)) - \widetilde{\nabla V}(\tilde{W}(b))$

(see Appendix B, equation (5)). Therefore

$$2\langle\widetilde{\nabla V}(\tilde{W}),\tilde{U}\rangle = \tilde{W}(\langle\tilde{V},\tilde{U}\rangle) + \tilde{V}(\langle\tilde{U},\tilde{W}\rangle) - \tilde{U}(\langle\tilde{W},\tilde{V}\rangle)$$

$$- \langle\tilde{W},[\tilde{V},\tilde{U}]\rangle + \langle\tilde{V},[\tilde{U},\tilde{W}]\rangle + \langle\tilde{U},[\tilde{W},\tilde{V}]\rangle \qquad (*)$$

However, for example $\tilde{W}(\langle\tilde{V},\tilde{U}\rangle)(b) = W(\langle V,U\rangle)(x)$ and

$$\langle\tilde{W}(b),[\tilde{V},\tilde{U}](b)\rangle_b = \langle W(x),[V,U](x)\rangle_x$$

by differential calculus and the isometric property of $T\pi$. Thus

$$2\langle\widetilde{\nabla V}(\tilde{W}(b)),\tilde{U}(b)\rangle_b = 2\ \langle\nabla V(W(x)),U(x)\rangle_x$$

and the result follows. //

The following is an extension of Proposition 6B(ii) on Riemannian covers. However we can no longer assert that the explosion times of the Brownian motions are the same (since B could equally well be replaced by any open subset B_0 of B with $\pi(B_0) = M$). It also fits into a series of results relating various classes of maps e.g. harmonic maps, harmonic morphisms, to their effect on Brownian motions: see (Meyer 1981; Darling 1982), and also (Blanchard & Sirugue 1981) for application to a problem from quantum mechanics. It could also be proved using the martingale problem characterization of Brownian motion and the fact that under our assumptions if $f:M \to \mathbb{R}$ is C^2 then $\Delta(f\circ\pi) = (\Delta f)\circ\pi$, see (Wallach 1972). We say π has *minimal fibres* if $\pi^{-1}(x)$ is a minimal submanifold (§5 Appendix B) of B for all x in M.

Theorem 10E Let $\pi:B \to M$ be a Riemannian submersion with minimal fibres. If $\{u_t:0 \leq t < \xi^{u_0}\}$ is a Brownian motion on B from $u_0 \in B$ then $\{\pi(u_t):0 \leq t < \xi^{u_0}\}$ is isonomous to Brownian motion on M from $\pi(u_0)$ restricted to $[0,\xi^{u_0}) \times \Omega$.

Proof Take any C^2 section X of $\mathbb{L}(\mathbb{R}^m;TM)$, some m, which is an orthogonal projection on each fibre, as in §B above. Take its horizontal lift $\tilde{X}:\mathbb{R}^m \to TB$ i.e. for each e in $\mathbb{R}^m$, $\tilde{X}(\cdot)e$ is the horizontal lift of $X(\cdot)e$. Now take any C^2 section Y of $\mathbb{L}(\mathbb{R}^p;VTB)$ which is an orthogonal projection on each fibre, some p. Let $\{B_t:t \geq 0\}$ and $\{B_t^v:t \geq 0\}$ be Brownian motions on $\mathbb{R}^m$ and $\mathbb{R}^p$ respectively. Then Brownian motions on M and B can be represented as solutions of

$$dx_t = X(x_t)dB_t - \tfrac{1}{2}\,\mathrm{trace}\,\nabla X(X(x_t)(\cdot))(\cdot)dt$$

and

$$du_t = \tilde{X}(u_t)dB_t + Y(u_t)dB_t^v - \tfrac{1}{2}\,\mathrm{trace}\,\tilde{\nabla}\tilde{X}(\tilde{X}(u_t)(\cdot))(\cdot)dt$$

$$- \tfrac{1}{2}\,\mathrm{trace}\,\nabla Y(Y(u_t)(\cdot))(\cdot)dt$$

respectively. It therefore suffices by Proposition 9E of Chapter VII to show that the two systems are π-related.

Now the field Y is tangent to the fibres $\pi^{-1}(x)$, all x in M. By Gauss's formula (§5 Appendix B) the horizontal component $\nabla Y(Y(b)e)(e)$ is therefore given by the second fundamental form of the fibre through b. The minimality of that submanifold therefore ensures that $\mathrm{trace}\,\nabla Y(Y(b)(\cdot))(\cdot)$ is in the null space of $T_b\pi$ for all $b \in B$ and $e \in \mathbb{R}^p$. The previous lemma takes care of the covariant derivatives of X and $\tilde{X}$, and the results follows. //

Example 10E (Homogeneous spaces) Let H be a closed subgroup of a finite dimensional Lie group G and set M = G/H. Let G have a left invariant

Riemannian metric and M a Riemannian metric invariant under the left action $(g,g_0H) \mapsto (gg_0)H$ of G on M, such that the projection $p:G \to G/H$ is a Riemannian submersion. Let $\underline{h}$ and $\underline{g}$ denote the Lie algebras of H and G, see (Appendix A, §5B). Let $\underline{h}^{\perp}$ be the orthogonal complement of $\underline{h}$ in $\underline{g} \approx T_eG$ using the metric. Assume also that $[h,h'] \in \underline{h}^{\perp}$ whenever $h \in \underline{h}$ and $h' \in \underline{h}^{\perp}$. We will show that p has minimal fibres (having been guided by J. Rawnsley).

In fact if U,V,W are left invariant vector fields on G it follows directly from the identity (*) in the proof of Lemma 10E that, for the Levi-Civita connection on G:

$$2\langle\nabla V(W),U\rangle = -\langle W,[V,U]\rangle + \langle V,[U,W]\rangle + \langle U,[W,V]\rangle \qquad (**)$$

Therefore if U and V are left invariant with $V(e) \in \underline{h}$ and $U(e) \in \underline{h}^{\perp}$ we have

$$\langle\nabla V(V),U\rangle = 2\langle V,[U,V]\rangle = 0$$

since $[\underline{h},\underline{h}^{\perp}] \subset h^{\perp}$. Since U(e) and V(e) were arbitrary in T_eH and $(T_eH)^{\perp}$ this shows that the second fundamental form α of H in G has $\alpha_e(v,v) = 0$ all $v \in T_eH$. Its trace is therefore zero at e and therefore at all points of H, and similarly for all left cosets of H, by left invariance, as required. (In fact $\alpha \equiv 0$ since it is symmetric so the cosets are 'totally geodesic').

If M = G/H is a reductive homogeneous space with G-invariant Riemannian metric, see (Kobayashi & Nomizu 1963) Volume II, in particular a Riemannian symmetric space, then a Riemannian metric exists on G which satisfies our conditions. The standard example is $M = S^{n-1} = SO(n)/SO(n-1)$.

It follows from Theorem 10E that p maps Brownian motion on G to Brownian motion on M. In particular if $\{g_t : t \geq 0\}$ is Brownian motion on G with $g_0 = e$ then for each x_0 in M

$$x_t = g_t \cdot x_0$$

is Brownian motion from x_0 in X. In particular we have a nice 'Brownian flow' on M by isometries. For an analytical discussion see (Yosida 1968).

Another consequence is that if $p_t : M \times M \to \mathbb{R}$ is the heat kernel of M and $\tilde{p}_t : G \times G \to \mathbb{R}$ that of G then $p_t(x,y)$ can be expressed as an integral of $\tilde{p}_t(\tilde{x},-)$ over the coset $\pi^{-1}(y)$, for $\tilde{x} \in \pi^{-1}(x)$.

§11. VERTICAL NOISE: GIRSANOV-CAMERON-MARTIN FORMULA

(A) Let W be a vector field on M of the form

$$W(u) = X(u)(W_0(u)) \qquad u \in M$$

where $W_0: M \to \mathbb{R}^n$ is continuous. For u in M let

$$M_t: [0,\xi) \times \Omega \to \mathbb{R}$$

be the solution, almost surely on Ω_t, of

$$M_t = 1 + \int_0^t \langle W_o(x_s), dB_s \rangle_{\mathbb{R}^n} M_s \tag{1}$$

i.e.

$$M_t = \exp\{\int_0^{t\wedge\xi} \langle W_o(x_s), dB_s \rangle_{\mathbb{R}^n} - \tfrac{1}{2}\int_0^{t\wedge\xi} |W_0(x_s)|^2 ds\} \tag{2}$$

a.s. on Ω_t, where $x_t = F_t(x_0)$ and $\xi = \xi^{x_0}$.

Alternatively this is a special case of §13 Chapter VII with B the trivial bundle $\underline{\mathbb{R}}$ over M with trivial connection: if

$$\bar{v}: [0,\xi) \times \Omega \to M \times \mathbb{R}$$

is

$$\bar{v}_t = (x_t, v_t)$$

with $v_t = M_t v_0$ then $\bar{v}$ is a solution to

$$(\text{Itô})\ D\bar{v}_t = J(x_t)(dB_t)\bar{v}_t$$

where

$$J: M \times \mathbb{R}^n \to \mathbb{L}(\mathbb{R};\mathbb{R}) \simeq \mathbb{R}$$

is

$$J(u)(e) = \langle W_o(u), e \rangle_{\mathbb{R}^n}.$$

For $f \in C^2(M;\mathbb{R})$, as in §7 consider $\tilde{f}: M \times \mathbb{R} \to \mathbb{R}$ given by $\tilde{f}(m,r) = f(m)r$. Then, by Proposition 13B of Chapter VII or, directly, from Lemma 9B of Chapter VII, a.s. on Ω_t:

$$f(x_t)v_t = \tilde{f}(x_t, v_t) = f(x_0) + \int_0^{t\wedge\xi} (df\circ X(x_s)dB_s)v_s$$
$$+ \int_0^{t\wedge\xi} f(x_s)\langle W_o(x_s), dB_s \rangle v_s + \int_0^{t\wedge\xi} A(f)(x_s)v_s ds + \int_0^{t\wedge\xi} df(W(x_s))ds.$$

For the last term we have used

$$\sum_i (df \circ X(x_s)e_i) \langle W_0(x_s), e_i \rangle = df(W(x_s)).$$

Let $\{\tau_n : n = 1,2,\ldots\}$ be a family of stopping times with $\tau_n \uparrow \xi$ but $\tau_n < \xi$ for each n, such that $|W_0(x_\cdot)|$ and $M_\cdot$ are bounded on each $[0,\tau_n] \times \Omega$. Then $M_t = \lim_{n\to\infty} M_{t\wedge\tau_n}$ almost surely on Ω_t, and each $\{M_{t\wedge\tau_n} : t \geq 0\}$ is a martingale by (1) and Proposition 7C of Chapter III. Therefore by Fatou's lemma

$$\int_{\Omega_t} M_t \, d\mu \leq \underline{\lim}_{n\to\infty} \int_{\Omega} M_{t\wedge\tau_n} \, d\mu = 1.$$

In particular $M_t \in \mathcal{L}^1(\Omega_t, F, \mu; \mathbb{R})$ for all t, and we can define

$$Q_t : B\mathcal{L}^0(M;\mathbb{R}) \to B\mathcal{L}^0(M;\mathbb{R})$$

by

$$Q_t(f)(x_0) = \int_{\Omega_t} f(x_t) M_t \, d\mu$$

to obtain a semi-group (by Corollary 5A2) which satisfies (by the proof of Theorem 1A):

$$\lim_{t\downarrow 0} \frac{Q_t f(u) - f(u)}{t} = A(f)(u) + W(f)(u)$$

whenever $f : M \to \mathbb{R}$ is C^2 with f and $A(f) + W(f)$ both bounded on M.

If W_0 is bounded, Theorem 13A of Chapter VII assures us that M_t lies in $\mathcal{L}^2$ for each t. If also (Y,z) is complete then $\{M_t : t \geq 0\}$ is a martingale because of (1) and so for $T > 0$ we can write, for $0 \leq t \leq T$:

$$Q_t f(u) = \int_{\Omega} \exp\{\int_0^T \langle W_0(x_s), dB_s \rangle_{\mathbb{R}^n} - \tfrac{1}{2} \int_0^T |W_0(x_s)|^2 ds\} f(x_t) d\mu. \tag{3}$$

In general in the elliptic case the same argument as in §9B shows that $\{Q_t : t \geq 0\}$ is the minimal semigroup with differential generator $f \mapsto A(f) + W(f)$. We can also combine this 'vertical noise' with a 'vertical drift' as in §7, and work on the frame bundle of our Riemannian manifold N as in §10A. This gives the 'Girsanov-Feynman-Kac' formula:

<u>Theorem 11A</u> <u>Let</u> A <u>be a locally Lipschitz vector field on</u> N

and let $\{x_t : t \geq 0\}$ be the standard Brownian motion from x_0 with drift A, assumed non-explosive. Suppose $V : N \to \mathbb{R}$ is bounded above and W is a vector field on N with both V and W locally Hölder continuous, then the unique minimal solution to

$$\frac{\partial g_t}{\partial t} = \tfrac{1}{2}\Delta g_t + A(g_t) + W(g_t) + Vg_t \qquad t \geq 0$$

$$g_0 = f$$

for given $f \in BC^0(N;\mathbb{R})$ can be written as

$$g_t(x_0) = \int_\Omega M_t \exp\left(\int_0^t V(x_s)ds\right) f(x_t) d\mu$$

where

$$M_t = \exp\left\{ \int_0^t \langle W(x_s), u_s dB_s \rangle_{x_s} - \tfrac{1}{2} \int_0^t |W(x_s)|^2_{x_s} ds \right\}$$

for $\{u_t : t \geq 0\}$ the horizontal lift of $\{x_t : t \geq 0\}$ in O(N). //

Remarks 11A(i) The Hölder continuity of V and W is so that Azencott's arguments in §9B can be applied.

(ii) If W were locally Lipschitz we could of course have combined W with A: however §12 below shows that even when W = -A it can be useful to write the solution in a form like the one given.

(iii) From Proposition 13D of Chapter VII we obtain generalizations of the Girsanov-Feynman-Kac formula giving semigroups on spaces of sections of vector bundles. In particular §13F gives an expression for the diffusion semigroup associated to a quantum particle in an electromagnetic field c.f. (Simon 1979; Streater 1981). Non-abelian gauge fields can be treated the same way provided the Schrödinger operator has the gauge invariant Laplacian as its highest order term. See also (Gaveau & Vauthier 1981).

If we take $V \equiv 0$ and $f \equiv 1$ in the theorem, and use the uniqueness of minimal semigroups we have the following corollary. See also (McKean 1969).

Corollary 11A With the hypotheses of the theorem suppose also that W is locally Lipschitz. Let ξ^{x_0} be the explosion time of the Brownian motion on N with drift A + W. Then

$$\mu\{\omega : t < \xi^{x_0}(\omega)\} = \int_\Omega M_t \, d\mu. \qquad //$$

(B) We now go rather more deeply. As with Theorem 11A the results which follow are valid in greater generality. In particular the hypothesis that W_0 is bounded in the next theorem is too restrictive: for example the conclusions also hold if it is replaced by

$$\exp(\tfrac{1}{2}\int_0^t |W_0(x_s)|^2 ds)d\mu < \infty. \qquad t > 0,$$

see (Ikeda & Watanabe 1981) or (Friedman 1975; Stroock & Varadhan 1979). If the conclusions do hold we shall say that W is a *Girsanov drift* for x_s. The notation is as in §A. The following is a version of the *Girsanov theorem;* it is essentially a probabilistic version of Theorem 11A but applies to degenerate systems.

Theorem 11B Assume the solution x to (Y,z) from x_0 is non-explosive and that W_0 is locally Lipschitz and bounded above. Then the solution y with $y_0 = x_0$ to

$$dy_t = Y(y_t)dz_t + W(y_t)dt \tag{4}$$

is non-explosive. For fixed T > 0 let μ_T be the measure on Ω with $\mu_T \sim \mu$ and

$$\frac{d\mu_T}{d\mu} = \exp\{\int_0^T \langle W_0(x_s), dB_s\rangle_{\mathbb{R}^n} - \tfrac{1}{2}\int_0^T |W_0(x_s)|^2 ds\}$$

Then μ_T is a probability measure and $\{y_t : 0 \le t \le T\}$ is isonomous to $\{x_t : 0 \le t \le T\}$ when the latter is considered on the probability space (Ω, F, μ_T).

Proof Since "$d\mu_T = M_T d\mu$" we see $\mu_T(\Omega) = 1$ by the martingale property of $\{M_t : t \ge 0\}$. Now x is a solution of

$$dx_t = X(x_t)dB_t + A(x_t)dt$$

which by substitution (Proposition 4B of Chapter III) is equivalent to

$$dx_t = X(x_t)d\tilde{B}_t + A(x_t)dt + W(x_t)dt$$

where

$$\tilde{B}_t = B_t - \int_0^t W_0(x_s)ds.$$

We will show that $\{\tilde{B}_t : 0 \le t \le T\}$ is a Brownian motion when considered on

$\{\Omega,F,\mu_T\}$. Both the completeness of (4) and the stated isonomy will then follow from the isonomy uniqueness of solutions (we are leaving the extension from the globally Lipschitz case of §6 Chapter VI as an exercise).

In fact $\{\tilde{B}_t : 0 \le t \le T\}$ is a G_*-Brownian motion on (Ω,F,μ_T) (see §5B of Chapter V) by Theorem 5C of Chapter V. To see this first observe that $\{M_t\tilde{B}_t : t \ge 0\}$ is a G_*-martingale on (Ω,F,μ) since, using *Itô* differentials (to save writing the integral signs): by the Itô formula and equations (1) for M_t

$$(\text{Itô})\ d(M_t\tilde{B}_t) = M_t\, d\tilde{B}_t + (dM_t)\tilde{B}_t + dM_t\, dB_t$$

$$= M_t\, dB_t - M_tW_0(x_t)dt + (dM_t)\tilde{B}_t + M_t\langle W_0(x_t), dB_t\rangle_{\mathbf{R}^n}\, dB_t$$

$$= M_t\, dB_t + (dM_t)\tilde{B}_t$$

and both B_t and M_t are G_*-martingales and so therefore are the stochastic integrals with respect to them. Now suppose $G \in G_s$ then for $0 \le s \le t \le T$

$$\int_G \tilde{B}_t M_T\, d\mu = \int_G \tilde{B}_t M_t\, d\mu = \int_G \tilde{B}_s M_s\, d\mu = \int_G \tilde{B}_s M_T\, d\mu$$

using, in turn, the martingale properties of M_t, $\tilde{B}_tM_t$, and M_t again. Thus $\{\tilde{B}_t : 0 \le t \le T\}$ is a G_*-martingale (Ω,F,μ_T).

If $n = 1$ the result follows by Theorem 5C of Chapter V since then

$$(\tilde{B}_t)^2 = 2\int_0^t B_s d\tilde{B}_s + \int_0^t dB_s\, dB_s = 2\int_0^t B_s d\tilde{B}_s + t.$$

For general n we can simply extend the proof of that theorem by applying the Itô formula to $\theta(\tilde{B}_t, t)$ where

$$\theta(v,t) = \exp\left(i\langle\xi, v\rangle + \tfrac{1}{2}|\xi|^2 t\right)$$

for arbitrary ξ in $\mathbf{R}^n$. The essential point is that if $\tilde{B}_t = (\tilde{B}_t^1, \ldots, \tilde{B}_t^2) \in \mathbf{R}^n$ then

$$d\tilde{B}_t^j\, d\tilde{B}_t^k = \delta^{jk} t. \quad //$$

<u>Corollary 11B1</u> <u>The semigroup associated to equation</u> (4) <u>is the same as the semigroup</u> $\{Q_t\}_t$ <u>of</u> §A. //

<u>Corollary 11B2</u> <u>The corresponding results to Theorem 11B hold</u>

for locally Lipschitz vector fields A and W on a Riemannian manifold N, with W bounded, for x and y Brownian motions on N with drifts A and A + W respectively, and with

$$\frac{d\mu_T}{d\mu} = \exp\left\{\int_0^T \langle W(x_s), u_s dB_s\rangle_{x_s} - \tfrac{1}{2}\int_0^T |W(x_s)|^2 ds\right\}$$

where $\{u_t : t \geq 0\}$ is the horizontal lift of x.

Proof Apply the theorem to the relevant processes on O(N). //

(C) Under the conditions of Theorem 11B we have measurable maps $x_\cdot : \Omega \to C_{x_0}([0,T],M)$ and $y_\cdot : \Omega \to C_{x_0}([0,T];M)$. These induce measures $P \equiv x_\cdot(\mu)$ and $P_W \equiv y_\cdot(\mu)$ on $C_{x_0}([0,T];M)$. The theorem implies that $x_\cdot(\mu_T) = P_W$. Using the terminology of the end of §3A Chapter I:

Theorem 11C (Girsanov-Cameron-Martin formula) Under the conditions of Theorem 11B we have $P \approx P_W$ with

$$\frac{dP_W}{dP}(\rho) = \mathbb{E}\{M_T | x_\cdot = \rho\} \qquad \rho \in C_{x_0}([0,T];M)$$

where

$$M_T = \exp\left\{\int_0^T \langle W_0(x_s), dB_s\rangle_{\mathbb{R}^n} - \tfrac{1}{2}\int_0^T |W_0(x_s)|^2 ds\right\}.$$

Proof Let B be a Borel subset of $C_{x_0}([0,T],M)$. Then

$$\begin{aligned}\int_B E\{M_T | x_\cdot = \rho\}\, dP(\rho) &= \int_{x_\cdot^{-1}(B)} \mathbb{E}\{M_T | \sigma\{x_\cdot\}\} d\mu \\ &= \int_{x_\cdot^{-1}(B)} M_T\, d\mu = \mu_T(x_\cdot^{-1}(B)) \\ &= P_W(B)\end{aligned}$$

as required. //

Remarks 11C(i) The corresponding formula holds for Brownian motions with drifts on N as in Corollary 11B2.

(ii) The same formulae, and Theorem 11B and its corollaries, remain true when A and W are allowed to be time dependent: All we have to do is to replace M by M × [0,T] and consider the equation for

$(x_t, r_t) \in M \times [0,T]$:

$$\begin{cases} dx_t = X(x_t)dB_t + A(x_t, r_t)dt \\ dr_t = dt \end{cases}$$

with corresponding equation for (y_t, r_t). (However this method does require unnecessary regularity in the time variable.)

§12 HEAT KERNEL FOR A MANIFOLD WITH A POLE: THE B.R. BRIDGE

(A) The following is given as an application, particularly of the Girsanov-Feynman-Kac formula. It is taken from (Elworthy & Truman 1982). We will find an explicit path integral representation for the fundamental solution, or 'propagator' $p_t(x,y)$ of the equation

$$\frac{\partial g_t}{\partial t} = \tfrac{1}{2} \Delta g_t + V g_t \tag{1}$$

on a complete C^∞ Riemannian manifold N under certain conditions on N and V. The conditions will not be examined too carefully here.

(B) Assume that $V: N \to \mathbb{R}$ is continuous and bounded above, and locally Hölder continuous. Let $\{P_t : t \geq 0\}$ be the minimal semigroup on $B\mathcal{L}^0(N;\mathbb{R})$ associated to (1) and let

$$p_t : N \times N \to \mathbb{R}$$

be its density:

$$P_t f(x_0) = \int_N p_t(x_0, y) f(y) d(\mathrm{vol})(y) \tag{2}$$

where d(vol) refers to the measure from the volume element of the Riemannian metric of N. Then, for x_0 and y_0 in N

$$p_t(x_0, y_0) = \lim_{\lambda \downarrow 0} \; (2\pi\lambda)^{-n/2} \, P_t f^\lambda(x_0) \tag{3}$$

where

$$f^\lambda(y) = T_0(y) \exp\{-\frac{d(y, y_0)^2}{2\lambda}\} \tag{4}$$

for d the Riemannian distance on N and T_0 any bounded continuous function,

$T_0:N \to \mathbb{R}$, with $T_0(y_0) = 1$ and support in the domain of normal coordinates about y_0 (to make it more obvious). This follows from (2) and the fact that in normal coordinates about y_0 the distance $d(x,y_0)$ is just the Euclidean distance, using the inner product of $T_{y_0}N$, and the volume element has value 1 at o (see below).

(C) In fact assume to start with that y_0 is a pole for N so that $\exp_{y_0}:T_{y_0}N \to N$ is a C^∞ diffeomorphism. We therefore have a global system of normal coordinates from y_0 (using the chart $\exp_{y_0}^{-1}:N \to T_{y_0}N$) in which the geodesics from y_0 are represented as straight lines from the origin. We shall work in these coordinates (but ∇,Δ, div will refer to the Riemannian ∇,Δ, and div). Let G be the local representative of the metric. The volume element d(vol) is then given by $\sqrt{g(v)}dv$ where $g(v) = \det G(v)$ and v is Lebesgue measure (on $T_{y_0}N$ using $<-,>_{y_0}$). In fact $g(v) = \theta_{y_0}(v)$ where $\theta_{y_0}(v)$ is the Jacobian determinant of $\exp_{y_0}$ at v,('Ruses invariant'):

$$\theta_{y_0}(v) = |\det{}_N T_v \exp_{y_0}|$$

where $\det_N$ indicates that the inner products of $T_{y_0}N$ and $T\exp_{y_0}(v)N$ are used to define the determinant. Thus $\theta_{y_0}:T_{y_o}N \approx N \to [0,\infty)$. We shall often write θ for θ_{y_0}.

Fix $T > 0$ and $\lambda > 0$ and let $\{A_t^\lambda:0 \le t \le T\}$ be the time dependent vector field $A_t^\lambda = \nabla Y_t^\lambda$ where $Y_t^\lambda:N \to \mathbb{R}$ is given in our coordinates by

$$Y_t^\lambda(v) = -\tfrac{1}{2}\frac{|v|^2}{\lambda+T-t} - \tfrac{1}{2}\log\theta(v) + \tfrac{1}{2}\log\theta\left(\frac{\lambda v}{\lambda+T-t}\right). \qquad (5)$$

The general philosophy leading to this choice of A_t^λ is given in (Elworthy & Truman 1982). It may be apparent from the proof of part (ii) of the following lemma. The first part depends only on the fact that A_t^λ is the gradient of some function Y_t^λ:

Lemma 12C(i) Let $\{x_t^\lambda:0 \le t \le T)$ be the standard Brownian motion on N with drift $\{A_t^\lambda:0 \le t \le T\}$, from x_0 (as in §10A). Assume that it is non-explosive and for $0 \le t \le T$ define $M_t^\lambda:\Omega \to \mathbb{R}$ by

$$M_t^\lambda = \exp\{-\int_0^t <A_s^\lambda(x_s^\lambda),u_s^\lambda\, dB_s> - \tfrac{1}{2}\int_0^t |A_s^\lambda|^2 ds\} \qquad (6)$$

<u>where</u> u^λ <u>is the horizontal lift of</u> x^λ. <u>Then</u>

$$M_t^\lambda = \exp\{Y_0^\lambda(x_0) - Y_t^\lambda(x_t^\lambda) + \int_0^t \left(\frac{\partial Y_s}{\partial s}(x_s^\lambda) + \tfrac{1}{2}|A_s^\lambda(x_s^\lambda)|^2\right)ds + \tfrac{1}{2}\int_0^t \Delta Y_s^\lambda(x_s^\lambda)ds\}. \quad (7)$$

(ii) $$\frac{\partial Y_s}{\partial s} + \tfrac{1}{2}|A_s^\lambda|^2 + \tfrac{1}{2}\Delta Y_s^\lambda = -\frac{n}{2}\frac{\partial}{\partial s}\log\left(\frac{\lambda}{\lambda+T-s}\right) + \tfrac{1}{4}\Delta\log\phi_{T-s}^\lambda + \frac{1}{8}|\nabla\log\phi_{T-s}^\lambda|^2 \quad (8)$$

<u>where</u>

$$\phi_t^\lambda(v) = \left(\frac{\lambda}{\lambda+t}\right)^n \theta(v)^{-1}\,\theta\left(\frac{\lambda}{\lambda+t}v\right) \quad (9)$$

<u>Proof</u> We shall often neglect to write the superscript λ in what follows. By Itô's formula

$$Y_t(x_t) = Y_0(x_0) + \int_0^t \frac{\partial Y_s}{\partial s}(x_s)ds + \int_0^t \langle \nabla Y_s(x_s), u_s\, dB_s\rangle + \int_0^t \left(\tfrac{1}{2}\Delta Y_s(x_s) + \langle A_s(x_s), \nabla Y_s(x_s)\rangle\right)ds$$

and (7) follows by using this to eliminate

$$\int_0^t \langle A_s(x_s),\, u_s\, dB_s\rangle$$

from (6), since $A_s = \nabla Y_s$.

(ii) The point here is that if $S_t: N \to \mathbb{R}$ and $\Phi_t: N \to N$ are defined (in normal coordinates) by

$$S_t(v) = \tfrac{1}{2}\frac{|v|^2}{\lambda+t} \quad (10)$$

and

$$\Phi_t(v) = \frac{\lambda+t}{\lambda}v \quad (11)$$

then

$$\frac{d}{dt}\Phi_t(v) = \nabla S_t(\Phi_t(v)) \qquad t \geq 0. \quad (12)$$

Also

$$\phi_t(v) = |\det_N T_v(\Phi_t^{-1})|$$

and so by general principles, e.g. see (Elworthy & Truman 1981) §6, ϕ_t satisfies the continuity equation

$$\frac{\partial \phi_t}{\partial t} + \operatorname{div}(\phi_t \nabla S_t) = 0. \tag{13}$$

From this, since $\Delta = \operatorname{div} \nabla$, we have

$$\Delta S_t = -\frac{\partial}{\partial t} \log \phi_t - \langle \nabla \log \phi_t, \nabla S_t \rangle . \tag{14}$$

On the other hand by definition (9) of ϕ_t we see that

$$Y_t(v) = -S_{T-t}(v) + \tfrac{1}{2} \log \phi_{T-t}(v) - \frac{n}{2} \log \left(\frac{\lambda}{\lambda+T-t}\right). \tag{15}$$

Therefore

$$\Delta Y_t = -\frac{\partial}{\partial t} \log \phi_{T-t} + \langle \nabla \log \phi_{T-t}, \nabla S_{T-t} \rangle + \frac{1}{2} \Delta \log \phi_{T-t}$$

$$= -\frac{\partial}{\partial t} \log \phi_{T-t} - |A_t|^2 + |\nabla S_{T-t}|^2 + \tfrac{1}{4} |\nabla \log \phi^\lambda_{T-t}|^2 + \tfrac{1}{2} \Delta \log \phi^\lambda_{T-t}. \tag{16}$$

The result follows on observing that S_t satisfies the Hamilton-Jacobi equation

$$\frac{\partial S_t}{\partial t} + \tfrac{1}{2} |\nabla S_t|^2 = 0. \quad //$$

(D) From Lemma 12C we see that

$$M^\lambda_t = \exp \{Y^\lambda_0(x_0) - Y^\lambda_t(x^\lambda_t) - \frac{n}{2} \log \left(\frac{\lambda+T}{\lambda+T-t}\right) + I^\lambda_t\} \tag{17}$$

where

$$I^\lambda_t = \int_0^t \{\tfrac{1}{4} \Delta \log \phi^\lambda_{T-s}(x^\lambda_s) + \frac{1}{8} |\nabla \log \phi^\lambda_{T-s}(x^\lambda_s)|^2\} ds. \tag{18}$$

Therefore

$$M^\lambda_T = \left(\frac{\lambda}{\lambda+T}\right)^{n/2} \theta\left(\frac{\lambda x_0}{\lambda+T}\right)^{\frac{1}{2}} \theta(x_0)^{-\frac{1}{2}} \exp\{-\tfrac{1}{2} \frac{|x_0|^2}{\lambda+T} + \frac{|x_0|^2}{2\lambda} + I^\lambda_T\}. \tag{19}$$

Applying the Girsanov-Feynman-Kac formula, Theorem 11A, under the assumption that $\{x_t^\lambda : 0 \le t \le T\}$ is non-explosive, we have, by (4) and (19)

$$P_T f^\lambda(x_0) = \left(\frac{\lambda}{\lambda+T}\right)^{n/2} \theta_{y_0}\left(\frac{\lambda x_0}{\lambda+T}\right)^{-\frac{1}{2}} \theta_{y_0}(x_0)^{\frac{1}{2}} \exp\left\{-\tfrac{1}{2}\frac{|x_0|^2}{\lambda+T}\right\}$$

$$\times \int_\Omega \exp\left\{I_T^\lambda + \int_0^T V(x_s^\lambda)ds\right\}d\mu. \qquad (20)$$

Therefore, by (3) and the fact that $\theta_{y_0}(0) = 1$:

$$p_T(x_0, y_0) = (2\pi T)^{-n/2}\, \theta_{y_0}(x_0)^{-\frac{1}{2}} \exp\left\{-\frac{|x_0|^2}{2T}\right\} \lim_{\lambda \downarrow 0} \int_\Omega \exp\left\{I_T^\lambda\right.$$

$$\left. + \int_0^T V(x_s^\lambda)ds\right\}d\mu \qquad (21)$$

Now from the definition (9) of ϕ_t^λ

$$\frac{1}{4}\Delta(\log \phi_{T-s}^\lambda)(v) + \frac{1}{8}|\nabla(\log \phi_{T-s}^\lambda)(v)|^2$$

$$= -\frac{1}{4}\Delta(\log\theta)(v) + \frac{1}{4}\left(\frac{\lambda}{\lambda+T-s}\right)^2 \Delta(\log\theta)\left(\frac{\lambda}{\lambda+T-s}v\right)$$

$$+ \frac{1}{8}\left|\nabla(\log\theta)(v) - \frac{\lambda}{\lambda+T-s}\nabla(\log\theta)\left(\frac{\lambda}{\lambda+T-s}v\right)\right|^2$$

$$\to -\frac{1}{4}\Delta\log\theta(v) + \frac{1}{8}|\nabla\log\theta(v)|^2$$

$$= \frac{1}{2}\sqrt{\theta}(v)\Delta\sqrt{\theta}^{-1}(v) \qquad (22)$$

as $\lambda \downarrow 0$, (where θ^{-1} is the *reciprocal* of θ), provided $0 \le s < T$.

Also for $0 \le t < T$ as $\lambda \downarrow 0$ so $A_t^\lambda \to Z_t$ where

$$Z_t = -\frac{v}{T-t} - \frac{1}{2}\nabla\log\theta_{y_0}(v) \qquad 0 \le t < T \qquad (23)$$

It follows from §4 Chapter VIII and Theorem 8C of Chapter VII that as $\lambda \downarrow 0$ so x_t^λ converges in probability, for $0 \le t < T$, to a standard Brownian motion y on N with drift Z_t. To see what happens at time T define $r: N \to \mathbb{R}$ by $r(v) = d(v, y_0)$, so $r(v) = |v|$ in our coordinates. Then r is C^2 on $N - \{y_0\}$ while, if $n \ge 2$, and $x_0 \ne y_0$, with probability one x_t^λ avoids y_0 for $0 \le t < T$ as does y_t for $0 \le t < T$: see (Friedman 1975) Vol. II. Therefore by the Itô formula, for $0 \le t < T$:

$$r(y_t) = r(x_0) + \int_0^t dr(u_s dB_s) + \int_0^t dr\,(Z_s(y_s))ds + \frac{1}{2}\int_0^t \Delta r(y_s)ds \tag{24}$$

where now $\{u_s : 0 \le s < T\}$ denotes the horizontal lift process of y. In our coordinates

$$\nabla r(v) = \frac{v}{|v|} \tag{25}$$

and it is easily seen from equation (12)' of Appendix B that

$$\Delta r = \frac{n-1}{r} + \frac{\partial}{\partial r} \log \theta \tag{26}$$

where $\frac{\partial}{\partial r}$ denotes differentiation in the radial direction. Also by (25)

$$dr(Z_t(v)) = -\frac{r(v)}{T-t} - \frac{1}{2}\;\frac{\partial}{\partial r} \log \theta(v).$$

Therefore by (24) setting $r_t = r(y_t)$ and $\tilde{B}_t = \int_0^t dr(u_s dB_s)$:

$$r_t = r_0 + \tilde{B}_t + \frac{1}{2}(n-1)\int_0^t \frac{ds}{r_s} - \int_0^t \frac{r_s}{T-s}\,ds \qquad 0 \le t < T. \tag{27}$$

Now $\{\tilde{B}_t ; 0 \le t < T\}$ is a one-dimensional Brownian motion by Corollary 5C of Chapter V. It follows that this equation for r_t is essentially independent of N! In fact, for example by comparison with the formulae in (Simon 1979) and (Ikeda & Watanabe 1981) we see that it is just the radial component of the *Brownian bridge* or *tied down Brownian motion* from x_0 to the origin, in $\mathbb{R}^n$. In particular with probability one r_t converges to y_0 as $t \uparrow T$, so we may define $Z_T = 0 \equiv y_0$ to obtain a sample continuous process $\{Z_t : 0 \le t \le T\}$. This process will be called the *Brownian-Riemannian bridge* (B.R. bridge) *from* x_0 *to* y_0.

Theorem 12D (Elworthy & Truman 1982). Suppose that N is complete and $V : N \to \mathbb{R}$ is locally Hölder continuous and bounded above. Let y_0 be a pole for N. Assume that Brownian motion x^λ on N with drift $\{A_t^\lambda : 0 \le t \le T\}$, determined by (5), is non-explosive for sufficiently small λ, and that $\Delta \log \theta_{y_0}$ is bounded above and $|\nabla \log \theta_{y_0}|$ is bounded. Then if the minimal semigroup associated to

$$\frac{\partial g_t}{\partial t} = \frac{1}{2}\Delta g_t + V g_t$$

has density $\{p_t : 0 \le t < \infty\}$, for all x_0 in N we have

$$p_t(x_0, y_0) = (2\pi t)^{-n/2}\, \theta_{y_0}(x_0)^{-\frac{1}{2}} \exp\{-\frac{d(x_0,y_0)^2}{2t}\}$$

$$\int_\Omega \exp\{\int_0^t (\tfrac{1}{2}\sqrt{\theta_{y_0}}(y_s)\Delta\sqrt{\theta_{y_0}^{-1}}(y_s) + V(y_s))ds\}d\mu$$

where $\{y_s : 0 \le s \le t\}$ is the B.R. bridge from x_0 to y_0 in time t.

The radial component of this B.R. bridge is isonomous to that of the corresponding Brownian bridge in $\mathbb{R}^n$.

Proof This follows from the previous discussion and the dominated convergence theorem. //

Remarks 12D (i) When y_0 is not necessarily a pole but nevertheless has no conjugate points, so the exponential map is a covering map but not a diffeomorphism, we can replace N by $\tilde{N} = T_{y_0}N$ with its induced metric and use Proposition 6B(ii) to see that $p_t(x_0,y_0) = \sum_i \tilde{p}_t(x_0^i, y_0)$ where $\{x_0^i\}_i$ are the points of $\exp_{y_0}^{-1}(x_0)$, $\tilde{p}_t$ is the density for $\tilde{N}$ and $\tilde{y}_0 = 0$ in $T_{y_0}N = \tilde{N}$. Since $\tilde{y}_0$ will now be a pole for $\tilde{N}$ we can apply the theorem to $\tilde{N}$. Then $p_t(x_0,y_0)$ is expressed as a sum of integrals over different B.R. bridges, one for each geodesic from y_0 to x_0, c.f. (Colin de Verdière 1973; Sunada 1982).

(ii) The formula makes it very easy to see the behaviour of p_t as $t \downarrow 0$, see for example (Pinsky 1978). This is made even clearer by introducing a parameter μ as in (Elworthy & Truman 1982), c.f. (Debiard et al. 1976) where a 'tied down Brownian motion on N' is also used.

Example 12D For hyperbolic n-space N with constant sectional curvatures $-\frac{1}{R^2}$ we have

$$\theta_{y_0}(x) = (\frac{R}{r}\sinh\frac{r}{R})^{n-1}$$

for $r = d(x,y_0)$. From this, using the extension of (26) that Δf, when f is a function of r, is given by

$$\Delta f = \frac{d^2 f}{dr^2} + \left(\frac{n-1}{r} + \frac{\partial}{\partial r} \log \theta_{y_0}\right)\frac{df}{dr}$$

we see

$$\frac{1}{2} \sqrt{\theta_{y_0}}\ \Delta\sqrt{\theta_{y_0}}^{-1} = \frac{-(n-1)^2}{8R^2} + \frac{(n-1)(n-3)}{8}\left(r^{-2} - \left(R^2 \sinh^2\left(\frac{r}{R}\right)\right)^{-1}\right).$$

For n = 3 our formula then gives the known result for the heat kernel of N (i.e. with $V \equiv 0$):

$$p_t(x_0,y_0) = (2\pi t)^{-3/2}\, e^{\frac{-t}{2R^2}}\, \theta_{y_0}(x_0)^{-\frac{1}{2}}\, e^{\frac{-d(x_0,y_0)^2}{2t}}$$

(If ρ is the sectional curvature then $\rho = -\frac{n(n-1)}{R^2}$.)

(E) For a non-probabilistic treatment of related questions, and more differential geometric background see (Berger et al. 1971). For a probabilistic treatment see (Azencott et al. 1981), and for a discussion (of a much more difficult problem) from the point of view of theoretical physics see (DeWitt 1965) equations (17.53) to (17.59).

APPENDIX A: MANIFOLDS AND FIBRE BUNDLES

§1 NOTATION FOR DIFFERENTIALS ETC

(A) Let E and F be Banach spaces and U an open subset of E. If $f:U \to F$ is differentiable let $Df(x)$ denote the (Fréchet) derivative of f at the point x of U. Then $Df(x) \in \mathbb{L}(E;F)$. As x varies in U we have $Df:U \to \mathbb{L}(E;F)$. If Df is differentiable, i.e. if f is twice differentiable we write D^2f for

$$D(Df):U \to \mathbb{L}(E;\mathbb{L}(E;F)) \cong \mathbb{L}(E,E;F).$$

Thus $D^2f(x):E \times E \to F$ is a continuous bilinear map. Inductively $D^rf \equiv D(D^{r-1})f$, so that $D^rf(x)$ is an r-linear map of $E \times \ldots \times E$ to F for $r = 1,2,\ldots$. The map is C^r if D^rf exists and is continuous, and is C^∞ if it is C^r for all r.

When $E = \mathbb{R}^p$ and $F = \mathbb{R}^q$ writing

$$x = (x^1,\ldots,x^p) \text{ and } f(x) = (f^1(x),\ldots,f^q(x))$$

we have the matrix representation for $Df(x)$ as $[\frac{\partial f^i}{\partial x^j}(x)]_{i,j}$ so that for $u = (u^1,\ldots,u^p)$

$$(Df(x)u)^i = \sum_{j=1}^{p} \frac{\partial f^i}{\partial x^j}(x)u^j, \qquad i = 1 \text{ to } q.$$

The second derivative $D^2f(x)$ is represented by $[\frac{\partial f^i}{\partial x^j \partial x^k}]_{i,j,k}$ with

$$(D^2f(x)(u,v))^i = \sum_{j,k=1}^{p} \frac{\partial^2 f^i}{\partial x^j \partial x^k}(x)u^jv^k.$$

The basic properties of these derivatives are concisely summarized in (Lang 1962). However we do need to indicate the relationship with the Gâteaux derivative GDf. By definition

$$GDf(x)u = \lim_{t \downarrow 0} \frac{f(x + tu) - f(x)}{t} \qquad x \in U, \quad u \in E$$

whenever the limit exists. When f is Fréchet differentiable it is Gâteaux differentiable and

$$GDf(x)u = Df(x)u \qquad x \in U, \quad u \in E.$$

The converse is not true. However there is the following well known exercise on the mean value theorem, see e.g. (Vainberg 1964; Schwartz 1969): Suppose $f: U \to F$ has a Gâteaux derivative $GDf(x)u$ for all $x \in U$ and $u \in E$, and that $u \mapsto GDf(x)u$ is continuous and linear for each x in U. If the resulting map $GDf: U \to \mathbb{L}(E;F)$ is continuous on U then f is C^1 on U.

§2 DIFFERENTIABLE MANIFOLDS, MAPS, SUBMANIFOLDS, AND LIE GROUPS

(A) Let N be a metrizable space and H a Banach space, not necessarily Hilbert here. A C^r *atlas* $\{(U_\alpha, \phi_\alpha): \alpha \in A\}$ for N *modelled on* H consists of an open cover $\{U_\alpha: \alpha \in A\}$ of N with homeomorphisms ϕ_α of U_α onto some open subset of H

$$\phi_\alpha : U_\alpha \to H$$

such that each

$$\phi_\alpha \circ \phi_\beta^{-1} : \phi_\beta(U_\alpha \cap U_\beta) \to \phi_\alpha(U_\alpha \cap U_\beta)$$

is C^r. A C^r *structure* on N, modelled on H, is a maximal such atlas. The space N together with a given C^r structure is a C^r *manifold*. It is usually simply denoted by N. Each element (U,ϕ) of a C^r structure is a C^r *chart*, or *local coordinate system* (especially for $H = \mathbb{R}^n$ when $\mathbb{R}^n$ is equipped with its usual basis).

Each C^r atlas determines a C^r structure. Open subsets U of H, in particular U = H, are considered with the C^∞ structure determined by the trivial atlas $\{(U,i)\}$ where i is the inclusion. If M and N are C^r manifolds modelled on Banach spaces G and H then $M \times N$ is given the C^r structure determined by the atlas with charts $(U \times V, \phi \times \psi)$ for each C^r chart (U,ϕ) of M and (V,ψ) of N, where $\phi \times \psi : U \times V \to G \times H$ is $(x,y) \mapsto (\phi(x), \psi(y))$.

Note that a C^r structure determines a C^p structure for each $0 \leq p \leq r$.

(B) For M and N as above, a continuous map $f:M \to N$ is a C^r *map* if for each C^r chart (U,ϕ) of M and (V,ψ) of N the *local representative*

$$\psi o f o \phi^{-1}:\phi(f^{-1}V) \to \psi(V)$$

is a C^r map (of open sets of Banach spaces). It is a C^r *diffeomorphism* if it is C^r and bijective with a C^r inverse.

When M is separable and H is infinite dimensional and Hilbert there exists a C^r diffeomorphism f of M onto an open subset of H, see (Eells & Elworthy 1970). Equivalently, M has an atlas consisting of one chart (M,f). Also when M is finite dimensional its C^r structure can be refined to a C^∞ structure, if $r \geq 1$: see (Hirsch 1976).

(C) A subset Y of a C^r manifold N is a C^r *submanifold* if for each point y of Y there is a C^r chart $\phi:U \to H$ *about* y (i.e. $y \in U$) and a linear isomorphism (splitting) $j:H \to H_1 \times H_2$ onto a product of Banach spaces such that

$$Y \cap U = (j o \phi)^{-1}(H_1 \times \{0\}).$$

Then $(U, j o \phi)$ is a *submanifold chart* for Y at y. If Y is connected it is easy to see that the isomorphism class of H_1 does not depend on the point y or the submanifold chart, and a C^r atlas for Y can be obtained from $\{(U \cap Y, j o \phi|Y):(U,\phi)$ is a C^r submanifold chart$\}$. Thus Y has a natural C^r structure. Any open subset of M is a C^r submanifold, and can be given its induced C^r structure: it is called an *open submanifold*.

The standard example of a C^∞ submanifold is the sphere S^{n-1} in $\mathbb{R}^n$ defined by $S^{n-1} = \{x \in \mathbb{R}^n : |x| = 1\}$ where we use the Euclidean norm. Another C^∞ example is the group $O(n)$ of orthogonal matrices in the open subset $GL(n)$ of $\mathbb{L}(\mathbb{R}^n;\mathbb{R}^n)$ consisting of invertible elements. The same holds when $\mathbb{R}^n$ is replaced by an arbitrary Hilbert space H, see (Lang 1962), in which case we use the notation $O(H)$ and $GL(H)$.

A map $f:M \to N$ is a C^r *embedding* if $f(M)$ is a C^r submanifold of N and f gives a C^r diffeomorphism onto $f(M)$ with its natural C^r structure. It is a *closed embedding* if $f(M)$ is a closed subset of N, i.e. if $f(M)$ is a *closed submanifold* of N. Whitney's embedding theorem states that any C^r manifold, $r \geq 1$, *of dimension* n, i.e. modelled on $\mathbb{R}^n$, where $n < \infty$, has a closed C^r embedding into $\mathbb{R}^{2n+1}$, e.g. see (Hirsch 1976). The corresponding result holds for separable, infinite dimensional Hilbert manifolds, with similar proof, $\mathbb{R}^{2n+1}$ being replaced by a separable Hilbert space, see (Kuiper 1971).

(D) A C^r *Lie group* G is a topological group with a C^r structure such that all the group operations are C^r maps. In particular left and right multiplication by a fixed element g of G determine C^r diffeomorphisms L_g and R_g of G with itself, viz. $L_g(x) = g \cdot x$ and $R_g(x) = x \cdot g$. Standard examples are O(n), GL(n), and O(H) as above; and GL(H) for H an arbitrary Banach space.

§3 TANGENT VECTORS AND THE TANGENT BUNDLE

(A) For a point x of our C^r manifold N, $r \geq 1$, a *tangent vector* at x can be considered as an equivalence class of C^r curves

$$\sigma:(-\varepsilon_\sigma,\varepsilon_\sigma) \to N$$

some $\varepsilon_\sigma > 0$, such that $\sigma(0) = x$, under the equivalence relation $\sigma \sim \sigma'$ if for some (and hence any) C^r chart (U,ϕ) about x we have

$$D(\phi o \sigma)(0) = D(\phi o \sigma')(0):\mathbb{R} \to H.$$

Let T_xN denote the set of such tangent vectors: it is the *tangent space* to N at x. For each chart (U,ϕ) about x there is then a natural bijection

$$d\phi_x:T_xN \to H$$

given by

$$d\phi_x(v) = D(\phi o \sigma)(0)(1)$$

where σ is a representative of v. If (U',ϕ') is another chart about x we see that

$$d\phi'_x \, o(d\phi_x)^{-1} = D(\phi' o \phi^{-1})(\phi(x)):H \to H \tag{1}$$

which is a linear isomorphism, inverse $D(\phi o \phi'^{-1})(\phi'(x))$. Thus T_xN has a unique topological vector space structure isomorphic to H by $d\phi_x$ for any chart (U,ϕ) about x. The *local reprsentative of* $v \in T_xN$ *in the chart* (U,ϕ) is $d\phi_x(v)$. Equation (1) shows how these change under change of chart.

(B) For a C^r map $f:M \to N$, $r \geq 1$, there is a unique continuous linear map $T_x f:M \to N$ for each $x \in M$ such that we have the commutative diagram

$$\begin{array}{ccc} T_x M & \xrightarrow{T_x f} & T_x N \\ {\scriptstyle d\psi_x}\downarrow & & \downarrow{\scriptstyle d\phi_x} \\ G & \xrightarrow[D(\phi f \psi^{-1})(\psi(x))]{} & H \end{array}$$

for all charts (V,ψ) and (U,ϕ) of M and N about x and f(x). Thus the *local representatives* of $T_x f$ are the derivatives of the local representatives of f. Alternatively $T_x f$ could be defined using the map $\sigma \mapsto f \circ \sigma$ for σ a curve in M.

For any open subset U of H there is a natural isomorphism of $T_x U$ with H, and this is considered as an identification. Correspondingly it is normal to abuse notation and write $T_x \psi$ for $d\psi_x$ when (V,ψ) is a chart at x. For a C^1 curve $\sigma:(a,b) \to N$ we write $\frac{d\sigma}{dt}$ or $\dot{\sigma}(t)$ for $T_t \sigma(1) \in T_{\sigma(t)} N$ having identified $T_t(a,b)$ with $\mathbb{R}$.

The tangent space $T_y Y$ at a point y of a submanifold Y of N has a natural inclusion in $T_y N$ as a closed linear subspace, as can be seen using submanifold charts. This is also often used as an identification, especially when N is a Banach space.

(C) Let TN be the disjoint union $\cup\{T_x N : x \in N\}$ and define $p:TN \to N$ by $p^{-1}(x) = T_x N$. Then TN has a unique topology, and C^{r-1} structure, with charts $(p^{-1}(U), T\phi)$ for each C^r chart (U,ϕ) of M where

$$T\phi : p^{-1}(U) \to H \times H$$

is given by

$$T\phi(v) = (\phi(x), T_x\phi(v)) \text{ for } x = p(v).$$

This manifold TN (with its additional structure abstracted in §3 below) is the *tangent bundle* of N and p is the *projection*. For a C^r map $f:M \to N$, $r \geq 1$ we obtain a C^{r-1} map $Tf:TM \to TN$ defined by $Tf(v) = T_x f(v)$ for $v \in T_x M$. When $f:M \to F$ is a map into a Banach space we write

A

$df : TM \to F$

for the map given by $df(v) = T_x f(v)$, $v \in T_x M$.

The tangent bundle to an open set U of H is just $U \times H$ with projection $p : U \times H \to U$ the projection on the first factor.

(C) A C^{r-1} *vector field* on M is a C^{r-1} map $A : M \to TM$ with $A(x) \in T_x M$ for each x in M. A C^r curve $\sigma : (a,b) \to M$ determines $\dot{\sigma} : (a,b) \to TM$ as described above, and σ is an *integral curve* of A if $\dot{\sigma}(t) = A(\sigma(t))$ for $a < t < b$. In the notation of the text it is a solution of $d\sigma = A(\sigma)dt$.

§4 VECTOR BUNDLES, PRINCIPAL BUNDLES, AND FIBRE BUNDLES

(A) Let F be a Banach space and G a subgroup of GL(F) with a differentiable structure which makes it a C^r Lie group such that the inclusion $i : G \to GL(F)$ is C^r. A C^r *vector bundle* over M, *fibre* (or *model*) F and *structure group* G consists of a C^r manifold B, the *total space*, with a C^r map $p : B \to M$, the *projection* together with a maximal family, $\{(U_\alpha, \tau_\alpha) : \alpha \in A\}$ of *local trivializations*

$$\tau_\alpha : p^{-1}(U_\alpha) \to U_\alpha \times F$$

such that

(i) $\{U_\alpha : \alpha \in A\}$ is an open cover of M (not necessarily related to an atlas)

(ii) each τ_α is a C^r diffeomorphism and has the form

$$\tau_\alpha(v) = (x, \tau_{\alpha,x}(v))$$

for $x = p(v)$, where $\tau_{\alpha,x} : p^{-1}(x) \to F$.

(iii) there is a C^r map $g_{\alpha\beta} : U_\alpha \cap U_\beta \to G$, each α, β, such that

$$\tau_{\alpha,x} \circ \tau_{\beta,x}^{-1}(w) = g_{\alpha\beta}(x)w$$

for all $x \in U_\alpha \cap U_\beta$ and $w \in F$.

It follows that each $p^{-1}(x)$ has a unique topological vector space structure such that $\tau_{\alpha,x} : p^{-1}(x) \to F$ is a linear isomorphism, for each local trivialization. In fact any structure on F preserved by the group G can be transferred to give a corresponding structure on $p^{-1}(x)$ independent of the trivialization e.g. if F is Hilbert and each element of G is orthogonal then $p^{-1}(x)$ has a natural inner product such that each $\tau_{\alpha,x}$ is orthogonal.

(B) If G is contained in another suitable subgroup G' of GL(F) and the inclusion is C^r we can relax the structure of our vector bundle to that of a vector bundle group G'. This relaxes condition (iii) above and has the effect of allowing more local trivializations.

Conversely we may be able to find a *trivializing cover* (i.e. a family satisfying (i), (ii), (iii) above) such that each $g_{\alpha\beta}$ is C^r into some smaller group G" than G. This then determines a vector bundle with group G": a *reduction* of the structure group to G".

When the structure group is not specified it is taken to be GL(F). For a Hilbert space F, a reduction to O(F) corresponds to a Riemannian metric, described in Appendix B, see (Lang 1962).

(C) The bundle is *trivializable* if it has a local trivialization (U,τ) with $U = M$. Then $\tau:B \to M \times F$ is a C^r diffeomorphism. This gives, in particular, a reduction of the structure group to the identity subgroup $\{I\}$ of GL(F). The *trivial* or *product* bundle $\underline{F}$ is $M \times F \to M$ with group $\{I\}$ and trivialization (M,τ) for τ the identity map.

(D) The *tangent bundle* $p:TN \to N$ is as described in §2 with trivializing cover $\{(U_\alpha,\tau_\alpha):\alpha \in A\}$ for any atlas $\{(U_\alpha,\phi_\alpha):\alpha \in A\}$ where

$$\tau_\alpha:p^{-1}(U_\alpha) \to U_\alpha \times H$$

is given by

$$\tau_\alpha(v) = (x,T_x\phi_\alpha(v)) \qquad v \in p^{-1}(x),\ x \in U_\alpha.$$

It is a C^{r-1} vector bundle, group GL(H), for N a C^r manifold modelled on H.

(E) As with Banach spaces new vector bundles may be constructed from old. From $p:B \to M$ we obtain $p^*:B^* \to M$ with $p^{*-1}(x) = p^{-1}(x)^* = \mathbb{L}(p^{-1}(x);\mathbb{R})$, the *dual bundle*, called the cotangent bundle and written T*M when B is the tangent bundle to M. The model is F*. More generally if $p':B' \to M$ is another vector bundle, model F' say, we can form $\mathbb{L}(p;p') : \mathbb{L}(B;B') \to M$ defined so that $\mathbb{L}(p;p')^{-1}(x) = \mathbb{L}(p^{-1}(x);p'^{-1}(x))$. Thus B* =

$\mathbb{L}(B;\mathbb{R})$.

A C^r *section* of $p:B \to M$ is a C^r map $s:M \to B$ such that $s(x) \in p^{-1}(x)$ for all x in M. A *vector field* on M is a section of the tangent bundle to M; a 1-*form* is a section of the cotangent bundle, and a C^r map $f:B \to B'$ is a C^r *vector bundle map* if it arises as a C^r section s of $\mathbb{L}(p;p')$, i.e. $f(v) = s(p(x))v$. In particular f restricts to a continuous linear map $f_x:p^{-1}(x) \to p'^{-1}(x)$ which is just s(x).

A 1-form is often considered as a map $\theta:TM \to \mathbb{R}$ linear on each fibre T_xM. Thus a C^r map $f:M \to \mathbb{R}$ determines a C^{r-1} 1-form df, by this abuse of notation. Here $\mathbb{R}$ may be replaced by any Banach space E to obtain a 1-*form with values in* E.

Since each fibre $p^{-1}(x)$ of a vector bundle is a vector space the space of C^r sections of it also forms a real vector space under pointwise operations. The zero in this space is called the *zero-section* of the bundle.

In a local trivialization $\tau_\alpha:p^{-1}(U_\alpha) \to U_\alpha \times F$ our section s has the local representation $x \mapsto (x,s_\alpha(x))$ where $s_\alpha:U_\alpha \to F$ is $s_\alpha(x) = \tau_{\alpha,x}(s(x))$. In the case of the tangent bundle, and bundles derived from it, it is convenient to work over a chart $(U,\phi),\phi:U \to H$. Then a vector field A, for example, is represented by

$$\begin{aligned}\phi(U) &\to \phi(U) \times H\\ x &\mapsto (x,\phi_*(A)(x))\end{aligned}$$

where $\phi_*(A)(x) = T_{\phi^{-1}(x)}A(\phi^{-1}(x))$, and is called the *principal part* of the local representative of A.

(F) When M is separable and modelled on a Hilbert space, for any C^r vector bundle $p:B \to M$ there exists a Banach space $\tilde{F}$ and a C^r vector bundle map $f:\underline{\tilde{F}} \to B$ such that for all x in M, $f_*:\tilde{F} \to p^{-1}(x)$ is surjective (with kernel a direct summand of $\tilde{F}$, even if F is infinite dimensional). When M and the model $\tilde{F}$ of $p:B \to M$ are finite dimensional $\tilde{F}$ can also be chosen finite dimensional. These facts can be seen from the proof in (Lang 1962) Proposition 9 Chapter III, modifying it when M is infinite dimensional to allow $\tilde{F}$ to be an infinite sum of copies of F, e.g. $\ell_2(F)$, and when M is finite dimensional using the fact that M has finite covering dimension.

If the model F is Hilbert and p has a fixed reduction to O(F), or equivalently a Riemannian metric (i.e. an inner product $\langle\ ,\ \rangle_x$ on each fibre $p^{-1}(x)$ which is C^r in x: see Appendix B) then $\tilde{F}$ can be chosen Hilbert, with dim $\tilde{F} < \infty$ if dim $F < \infty$ and dim $M < \infty$ and $f:\tilde{F} \to B$ can be chosen so that each $f_x:\tilde{F} \to p^{-1}(x)$, $< >_x$ is an orthogonal projection.

To prove this, first take a Hilbert space $\tilde{F}$, finite dimensional if both M and F are, with a C^r vector bundle map $\tilde{f}:\underline{\tilde{F}} \to B$ with each $\tilde{f}_x$ surjective. Now, because of the surjectivity, if $\ker \tilde{f} = \bigcup_{x \in M} \ker \tilde{f}_x$ then ker f has a natural vector bundle structure and we can decompose $\underline{\tilde{F}}$ as $\ker \tilde{f} \oplus (\ker \tilde{f})^{\perp}$ where $(\ker \tilde{f})^{\perp}$ is the union of the orthogonal complements of $\{\ker f_x : x \in M\}$: see (Lang 1962). Each $(\ker \tilde{f}_x)^{\perp}$ can be given an inner product such that the restriction to it of $\tilde{f}_x$ is an isometry, and each $\ker \tilde{f}_x$ has an inner product induced on it from $\tilde{F}$. These give Riemannian metrics to $\ker \tilde{f}$ and $(\ker \tilde{f})^{\perp}$ and so to $\underline{\tilde{F}}$. Thus we have a C^r map $G:M \to Pos(\tilde{F})$, into the positive definite symmetric elements of $GL(\tilde{F})$, with $\langle u,v\rangle_x = \langle G(x)u,v\rangle$ for $u,v \in \tilde{F}$ and $x \in M$. As in (Lang 1962) Chapter VII Theorem 1 we can let B(x) be the square root of G(x) and define $\tau:\underline{\tilde{F}} \to \underline{\tilde{F}}$ by $\tau_x(v) = B(x)v$. Then $f = \tilde{f}\circ\tau^{-1}:\underline{\tilde{F}} \to B$ is as required.

When F is infinite dimensional we could have applied Kuiper's theorem to show that p itself was trivializable, even with group O(F). Then f could simply be such a trivialization: see (Kuiper 1971).

An example is $f:\underline{\mathbb{R}^n} \to TS^{n-1}$ where f_x is the projection of $\mathbb{R}^n$ onto the tangent space to the sphere in $\mathbb{R}^n$ at the point x, taken with its usual induced Riemannian metric. Any other submanifold of $\mathbb{R}^n$ would do equally well.

(G) If p has group G, for each x in M let G_x denote the set of linear isomorphisms $u:F \to p^{-1}(x)$ such that for each local trivialization (U_α,τ_α) the composition $\tau_{\alpha,x}\circ u:F \to F$ is in G. The disjoint union $G(B) = \bigcup_{x \in M} G_x$ can be given a topology and C^r structure, locally diffeomorphic to $U_\alpha \times G$ over each U_α, and with these local trivializations and the obvious projection $G(B) \to M$ forms the *principal bundle* associated to p. When $G = GL(F)$ the space G(B) is just that subset of $\mathbb{L}(\underline{F};B)$ whose elements are linear isomorphisms $F \to p^{-1}(x)$, some x in M. The case $G = O(H)$, which is the Riemannian metric case is considered in more detail in Appendix B, at least for the tangent bundle. For the general case see (Steenrod 1951) Part §8.

(H) More generally a C^r *left action* of a Lie group G on a manifold M is a C^r map $\alpha:G \times M \to M$ such that the map $g \mapsto \alpha(g,-)$ is a group homomorphism of G into the group of C^r diffeomorphisms of M. A *right action* is defined similarly, but the corresponding map $g \mapsto \alpha(-,g)$ is an anti-homomorphism. Usually $\alpha(g,x)$ is written as gx or $g\cdot x$. Typical examples are the C^∞ action of GL(F) on F:

$$GL(F) \times F \to F$$

$$(T,v) \mapsto Tv,$$

and the natural left action of G on itself: $\alpha(g,x) = L_g(x)$.

For any C^r manifold F and C^r left action of a Lie group G on F the definition of a *fibre bundle* group G, fibre F, action as given, with base space M, is the same as the definition of a vector bundle given in §4A but taking into account the fact that we have a more general action. It is convenient here to assume that the map $g \mapsto \alpha(-,g)$ is injective, i.e. the action is *effective*.

A *principal bundle* is just a fibre bundle whose fibre and group are the same, with the left action of the group on itself. We have seen examples of these in the previous section. When the group G is discrete, for example G = Z, we arrive at the notion of a covering space. A C^r map $p:B \to M$ is a *covering space* if there exists an open cover $\{U_\alpha : \alpha \in A\}$ of M such that on each component of $p^{-1}(U_\alpha)$ the restriction of p is a C^r diffeomorphism onto U_α, for each α in A. The standard example is the map

$$p:\mathbb{R} \to S^1$$

$$t \mapsto e^{it}.$$

Covering spaces have the path lifting property: if $\sigma:[a,b) \to M$ is a C^p path $0 \leqslant p \leqslant r$ and $v_0 \in p^{-1}(\sigma(a))$ then there is a unique C^p map $\tilde{\sigma}:[a,b) \to B$ with $\tilde{\sigma}(a) = v_0$ and $p\circ\tilde{\sigma} = \sigma$. Moreover $\tilde{\sigma}$ depends continuously in σ: the details are given in greater generality in (Dugundji 1966) Chapter XX. In particular this means that a sample continuous process with values in M has a natural lift to a B-valued process. Covering spaces are discussed in (Steenrod 1951; Kobayashi & Nomizu 1963).

(I) Given a C^r vector bundle $p:B \to M$ and a C^r map $f:M \to N$ we can form the *pull-back bundle* $f^*(p):f^*(B) \to N$. This is a vector bundle with the same group as p and with a map $p^*(f):f^*(B) \to B$ with commutative diagram

$$\begin{array}{ccc} f^*(B) & \xrightarrow{p^*(f)} & B \\ {\scriptstyle f^*(p)}\downarrow & & \downarrow{\scriptstyle p} \\ N & \xrightarrow{f} & M \end{array}$$

such that $p^*(f)$ is a linear isomorphism on the fibres. In fact we can consider $f^*(B)$, as a topological space, as

$$f^*(B) = \{(x,b) \in N \times B : f(x) = p(b)\}$$

and then $f^*(p)$ and $p^*(t)$ are just the restrictions of the projections onto N and B. Pull backs of fibre bundles can be constructed in the same way.

§5 LIE BRACKETS, LIE GROUPS, AND LIE ALGEBRAS

(A) A C^p vector field A on a C^r manifold N, $0 < p < r-1$, acts as a first order differential operator on functions $f:N \to F$ into a Banach space F, by letting $A(f):N \to F$ be given by

$$A(f)(x) = T_x f(A(x)) \qquad x \in N.$$

Locally

$$A(f)(\phi^{-1}(x)) = D(f \circ \phi^{-1})(x)(\phi_*(A)(x)) \qquad x \in \phi(U)$$

for a chart (U,ϕ). If f is C^r then $A(f)$ is C^p.

Given C^p vector fields A_1 and A_2, with $1 < p < r-1$, there is a unique C^{p-1} vector field, denoted by $[A_1,A_2]$ with

$$[A_1,A_2](f) = A_1(A_2(f)) - A_2(A_1(f))$$

for all C^p maps $f:N \to F$. Locally, in (U,ϕ), if we write A_1^ϕ for $\phi_*(A_1)$ etc, $[A_1,A_2]$ is represented by $[A_1,A_2]^\phi$ where

$$[A_1,A_2]^\phi(x) = DA_2^\phi(x)(A_1(x)) - DA_1^\phi(x)(A_2^\phi(x)) \qquad x \in \phi(U).$$

This is the *Lie bracket* of A_1 and A_2.

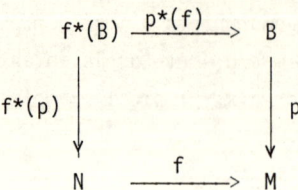

(B) For a Lie group G with identity element denoted by e, each v in T_eG determines a *left invariant vector field* A_v, i.e. a vector field such that for all g in G

$$A_v(gx) = TL_g(A_v(x)) \qquad x \in G$$

where L_g is left translation by g, see §2D. In fact we can simply set

$$A_v(x) = TL_x(v) \qquad x \in G.$$

This determines a bijection between T_eG and the set of all left invariant vector fields. Moreover the Lie bracket of left invariant vector fields is left invariant, and so we can define a Lie bracket on T_eG by

$$[v,w] = A(e) \text{ where } A = [A_v,A_w] \qquad v,w \in T_eG.$$

This way T_eG becomes a Lie algebra: the *Lie algebra* of G. The standard examples are GL(F) when the Lie algebra can be identified with $\mathbb{L}(F;F)$, and O(H) for H a Hilbert space with Lie algebra the space of skew-symmetric maps in $\mathbb{L}(H;H)$. In both cases the Lie bracket is given by $[v,w] = vw-wv$.

The assignment $v \mapsto A_v$ determines a trivialization

$$X:\underline{E} \to TG$$
$$X(x,v) = A_v(x)$$

for $E = T_eG$. The corresponding statements hold for right invariant vector fields.

(C) For each $v \in T_eM$ let $\sigma^v:[0,\infty) \to G$ be the integral curve of the vector field A_v with $\sigma^v(0) = e$: see §3C. It is defined for all time by Examples (i) §6 of Chapter VII, or by a more elementary proof. The *exponential map* of G is the map

$$\exp:T_eG \to G$$
$$v \mapsto \sigma^v(1).$$

It should not be confused with the exponential map of a connection, although it does agree with it in certain circumstances: see Appendix B.

§6 MANIFOLDS OF MAPS

(A) Let M and N be manifolds, for simplicity assume that both are C^∞ and that M is compact, although this is not essential. Let N be modelled on the Banach space H and suppose that to each C^∞ vector bundle $p:B \to M$ with fibre H we have assigned a Banach space S(p) of sections. For example S(p) could be the space $C^r(p)$ of C^r sections for some fixed $0 < r < \infty$, or the space of H^s sections as in Chapter VIII. However it is important to assume that S(p) consists of continuous sections and that the inclusion $S(p) \to C^0(p)$ is continuous into the usual (uniform) topology.

Under suitable conditions on $\{S(p)\}_p$ we will construct a C^∞ structure on a corresponding space of maps $f:M \to N$, to be denoted by S(M;N). To do this we take any C^∞ connection on N, with exponential map $\exp_x : T_xN \to N$, maybe only partially defined; see Appendix B. For each C^∞ map $f:M \to N$ we can find an open neighbourhood U_f of f in the C^0 topology of the space $C^0(M;N)$ of continuous maps such that

$$\phi_f(g)(x) \equiv \exp^{-1}_{f(x)} g(x) \in T_{f(x)}N$$

is defined for all $g \in U_f$ and $x \in M$ and gives a homeomorphism ϕ_f of U_f onto a neighbourhood of the zero-section in the pull-back $f^*(\pi)$ of the tangent bundle $\pi:TN \to N$ of N. Here we have identified a map $\tilde{g}:M \to TN$ such that $\tilde{g}(x) \in T_{f(x)}N$ for all x with a section of $f^*(\pi)$.

A map $g:M \to N$ is defined to be in S(M;N) if when it lies in U_f for some C^∞ function f then $\phi_f(g)$ lies in $S(f^*(\pi))$. For a wide class of assignments $\{S(p)\}_p$ this last condition is independent of the function f for which $g \in U_f$ and moreover each map

$$\phi_f \circ \phi_h^{-1} : \phi_h(U_h \cap U_f) \cap S(h^*(\pi)) \to \phi_f(U_f \cap U_h) \cap S(f^*(\pi))$$

will be C^∞, for f and h both C^∞. This defines a topology on S(M;N) and on each component the spaces $S(f^*(\pi))$ will be linearly isomorphic, so that $\{(U_f, \phi_f)\}_f$ determines a C^∞ atlas (strictly speaking on each component, since in our definition the whole manifold had to have the same model space, but this is not really important). For the details see (Eliasson 1967; Palais 1968). However we should remark that the tangent space $T_gS(M;N)$ at g can be identified, naturally, with the space of maps v in S(M;TN) such that $v(x) \in T_{g(x)}N$ for all $x \in M$, or equivalently, with $S(g^*(\pi))$.

APPENDIX B: SOME DIFFERENTIAL GEOMETRY USING THE FRAME BUNDLE

§1 CONNECTIONS, CURVATURE FORMS AND TORSION

(A) Let N be a C^{p+1} manifold modelled on a Hilbert space H (e.g. $H = \mathbb{R}^n$), $p \geq 2$. A C^p *Riemannian metric* on N is an assignment of an inner product $\langle\ ,\ \rangle_x$ to the tangent space T_xN for each x in N, such that if (U,ϕ) is a chart of N, with $\phi:U \to H$ then there is a C^p map

$$G:\phi(U) \to \mathrm{Pos}(H)$$

into the positive definite symmetric elements of GL(H) with

$$\langle v_1,v_2\rangle_x = \langle G(\phi(x))T_x\phi(v_1),\ T_x\phi(v_2)\rangle \qquad v_1,v_2 \in T_xN,\ x \in U$$

where $\langle\ ,\ \rangle$ is a fixed inner product for H. In finite dimensions G(x) can be represented by a matrix, components $g_{ij}(x)$.

Given such a metric, for each $x \in N$ let $O(N)_x$ be the space of all isometries

$$u:H,\ \langle\ ,\ \rangle \longrightarrow T_xN,\ \ \langle\ \ \rangle_x.$$

Set $$O(N) = \bigcup_{x\in N} O(N)_x$$

and define $\pi:O(N) \to N$ by $\pi^{-1}(x) = O(N)_x$.

For a chart (U,ϕ) as above define

$$\tilde{\phi}:\pi^{-1}(U) \to U \times O(H)$$

where O(H) is the orthogonal group of H, by

$$\tilde{\phi}(U) = (x,G(\phi(x))^{\frac{1}{2}}\circ T_x\phi\circ u) \qquad u \in O(N)_x,\ \ x \in U.$$

Then O(N) has a C^p manifold structure such that each $\tilde{\phi}$ is a C^p diffeomorphism and π is C^p. There is also a natural right action of O(H) on O(N):

$$O(N) \times O(H) \to O(N)$$

given by $(u,a) \longmapsto u \cdot a$.

We will denote this by ua or u.a.
(Note: "O(H)" refers to the group, and is not the same as O(N) even when N = H!).

This action commutes with each $\tilde{\phi}$, using the natural right action of O(H) on U × O(H). The map $\pi : O(N) \to N$ is called the *orthonormal frame bundle* of N. (We have now shown that it is a principal bundle with group O(H).)

(B) The map π can be differentiated to give $T\pi : TO(N) \to N$. For the chart (U,ϕ) let $O(N)_U = \pi^{-1}(U)$. We have commutative diagrams

$$\begin{array}{ccc} TO(N)_U & \xrightarrow{T\tilde{\phi}} & TU \times TO(H) \\ \downarrow & & \downarrow \\ O(N)_U & \xrightarrow{\tilde{\phi}} & U \times O(H) \\ & \pi \searrow \quad \swarrow \text{proj.} & \\ & U & \end{array}$$

and

$$\begin{array}{ccc} TO(N)_U & \xrightarrow{T\tilde{\phi}} & TU \times TO(H) \\ & T\pi \searrow \quad \swarrow \text{proj} & \\ & TU & \end{array}$$

For each $u \in O(N)$ we have the subspace $\ker T_u\pi$ in $T_uO(N)$. The *vertical tangent bundle* $VTO(N) = \cup \ker \{T_u\pi : u \in O(N)\}$ is a subvector bundle of TO(N). It follows that there will be a complementary "horizontal" tangent bundle HTO(N) i.e. a sub-bundle of TO(N) with

$$TO(N) = VTO(N) \oplus HTO(N)$$

equivalently

$$T_uO(N) = VT_uO(N) \oplus HT_uO(N) \quad \text{each } u \in O(N).$$

In particular $T_u\pi|HT_uO(N)$ will be an isomorphism onto T_xN for each $u \in \pi^{-1}(x)$ with $x \in N$. However there is no unique choice of such a horizontal bundle.

A C^{p-1} *Riemannian connection* on the Riemannian manifold N is a choice of such a C^{p-1} horizontal bundle HTO(N) in TO(N) with the additional property that the decomposition of TO(N) is invariant under the action of O(H) i.e. if $a \in O(H)$ and $R_a:O(N) \to O(N)$ denotes right multiplication by a then we require

$$HT_{ua}O(N) = T_uR_a[HT_u(O(N))] \qquad u \in O(N).$$

This still does not give a unique choice. The basic theorem of Riemannian geometry says only that there is a unique Riemannian connection with vanishing "torsion" (we discuss torsion below). This is the *Levi-Civita connection.*

(C) A connection can equally well be specified by a *connection form* ϖ. This is an o(H)-valued 1-form on O(N), where o(H) is the Lie algebra of O(H), viz. the skew adjoint operators in $\mathbb{L}(H;H)$. Thus

$$\varpi:TO(N) \to o(H)$$

and ϖ is linear on $T_uO(N)$ for each $u \in O(N)$. To be a connection form it must satisfy

$$\text{(i)} \quad \varpi \circ TR_a = ad(a^{-1})\circ\varpi \qquad a \in o(H)$$

and $$\text{(ii)} \quad \varpi(A^*(u)) = A \qquad u \in O(N), \quad A \in O(H)$$

where $ad(a^{-1});o(H) \to o(H)$ is the adjoint action:

$$ad(a^{-1})(A) = a^{-1}Aa$$

and A^* is the vertical vector field on O(N) given by

$$A^*(u) = \frac{d}{dt}(u.\text{exp}tA)\Big|_{t=0}.$$

From a connection form ϖ we can obtain a horizontal bundle by setting

$HT_uO(N) = \ker \varpi|T_uO(N)$. Conversely given $HTO(N)$ we can obtain a unique connection form ϖ using condition (ii). From now on we suppose we have chosen a connection.

Let $h:TO(N) \to TO(N)$ be the projection along $VTO(N)$ onto $HTO(N)$. Define the *curvature form* $\underset{\sim}{\Omega}$ as the $o(H)$ valued 2-form $\underset{\sim}{\Omega}$ on $O(N)$ given by

$$\underset{\sim}{\Omega}(V_1,V_2) = d\varpi(hV_1,hV_2). \qquad V_1,V_2 \in T_uO(N).$$

It has the invariance property, for $a \in O(H)$ and $V_1,V_2 \in T_uO(N)$:

$$\underset{\sim}{\Omega}(T_uR_a(V_1),T_uR_a(V_2)) = \mathrm{ad}(a^{-1})\underset{\sim}{\Omega}(V_1,V_2).$$

Also if Y_1 and Y_2 are horizontal vector fields on $O(N)$, i.e. sections of $HTO(N)$ then, since

$$d\varpi(Y_1,Y_2) = \tfrac{1}{2}\{Y_1(\varpi(Y_2)) - Y_2(\varpi(Y_1)) - \varpi([Y_1,Y_2])\}$$

we have

$$\underset{\sim}{\Omega}(Y_1,Y_2) = -\tfrac{1}{2}\,\varpi([Y_1,Y_2]) \tag{1}$$

where $[Y_1,Y_2]$ is the Lie bracket of the vector fields.

From this and Frobenius' theorem it follows that $HTO(N)$ is integrable (i.e. through each point u_o of $O(N)$ there is submanifold $H(u_o)$ with tangent space $T_uH(u_o) = HT_uO(N)$ for each $u \in H(u_o)$) if and only if the curvature form vanishes identically.

(D) There is also the *canonical 1-form* θ on $O(N)$. This is a 1-form with values in H

$$\theta:TO(N) \to H$$

defined by

$$\theta(V) = u^{-1}T\pi(V) \qquad V \in T_uO(N).$$

The *torsion form* $\underset{\sim}{\Theta}$ is the H-valued 2-form on $O(N)$ defined by

$$\underset{\sim}{\Theta}(V_1,V_2) = d\theta(hV_1,hV_2) \qquad V_1,V_2 \in T_uO(N)$$

it has the invariance property

$$\underset{\sim}{\Theta}(TR_a(V_1),TR_a(V_2)) = a^{-1}\underset{\sim}{\Theta}(V_1,V_2).$$

(E) For a vector $v \in T_xN$ we can choose $u \in \pi^{-1}(x)$ and then a unique *horizontal lift* $\tilde{v} \in HT_uO(N)$ such that $T\pi(\tilde{v}) = v$. The *curvature* and *torsion tensors* of the connection (using Kobayashi and Nomizu's conventions) are respectively

$$R: TN \oplus TN \to \mathbb{L}(TN;TN)$$

and

$$T: TN \oplus TN \to TN$$

defined by

$$R(v_1,v_2)(v_3) = 2u\underset{\sim}{\Omega}(\tilde{v}_1,\tilde{v}_2)u^{-1}(v_3)$$

and

$$T(v_1,v_2) = 2u\underset{\sim}{\Theta}(\tilde{v}_1,\tilde{v}_2)$$

for $\tilde{v}_1$ and $\tilde{v}_2$ horizontal lifts in $HT_uO(N)$ of v_1, $v_2 \in T_xN$. The invariance properties of $\underset{\sim}{\Omega}$ and $\underset{\sim}{\theta}$ ensure that the result is independent of the choice of u in $\pi^{-1}(x)$.

For the Levi-Civita connection (our main interest) $T \equiv 0$, by definition.

It is clear from the definition that $\langle R(v_1,v_2)v_3,v_4\rangle_x$ is skew-symmetric in v_1,v_2 and in v_3,v_4. See for example (Milnor 1963), for proofs of the other identities (valid when $T \equiv 0$):

$$\langle R(v_1,v_2)v_3,v_4\rangle_x = \langle R(v_3,v_4)v_1,v_2\rangle_x$$

and

$$R(v_1,v_2)v_3 + R(v_3,v_1)v_2 + R(v_2,v_3)v_1 = 0.$$

§2 HORIZONTAL LIFTS, PARALLEL TRANSLATION, COVARIANT DERIVATIVES AND GEODESICS

For each $u \in \pi^{-1}(x)$ a piecewise C^1 curve $\sigma:[a,b) \to N$ with $\sigma(a) = x$ has a unique *horizontal lift* $\tilde{\sigma}:[a,b) \to O(N)$ with $\tilde{\sigma}(a) = u$. This is a piecewise C^1 curve satisfying $\pi\tilde{\sigma} = \sigma$ which is *horizontal* i.e. when defined

$$\dot{\tilde{\sigma}}(t) \in HT_{\tilde{\sigma}(t)}O(N).$$

Equivalently it satisfies $\tilde{\omega}(\dot{\tilde{\sigma}}(t)) = 0$. If σ is C^r for $1 < r < p-1$ so is $\tilde{\sigma}$.

For $v \in T_xN$ and $\sigma, \tilde{\sigma}$ as before, the *parallel translate* of v along σ to $\sigma(t)$ is $//_t^\sigma(v)$ where

$$//_t^\sigma(v) = \tilde{\sigma}(t)\, u^{-1}(v).$$

Again the invariance property of $\tilde{\omega}$ ensures that this does not depend on the choice of u in $\pi^{-1}(x)$.

When σ is C^r and $v:[a,b] \to TN$ is a C^r curve with $v(t) \in T_{\sigma(t)}N$ for each t, i.e. v is a *vector field along* σ the *covariant derivative* of v along σ is the vector field $\frac{Dv}{\partial t}$ along σ given by

$$\frac{Dv}{\partial t} = //_t^\sigma \frac{d}{dt} [(//_t^\sigma)^{-1}\, v(t)]$$

$$= \tilde{\sigma}(t) \frac{d}{dt} (\tilde{\sigma}(t)^{-1} v(t)).$$

Clearly for fields v and w along σ we have

$$\frac{d}{dt} \langle v(t), w(t) \rangle_{\sigma(t)} = \langle \frac{Dv}{\partial t}, w(t) \rangle_{\sigma(t)} + \langle v(t), \frac{Dw}{\partial t} \rangle_{\sigma(t)} \qquad (2)$$

Moreover v is parallel along σ iff $\frac{Dv}{\partial t} \equiv 0$.

By definition σ is a *geodesic* if $\frac{D\dot{\sigma}(t)}{\partial t} \equiv 0$. Since

$$\theta(\dot{\tilde{\sigma}}(t)) = \tilde{\sigma}(t)^{-1}\, (T\pi(\dot{\tilde{\sigma}}(t)))$$

$$= \tilde{\sigma}(t)^{-1}\, (\dot{\sigma}(t))$$

we see that σ is a geodesic if and only if any horizontal lift $\tilde{\sigma}$ satisfies

$$\frac{d}{dt}\, \theta(\dot{\tilde{\sigma}}(t)) = 0 \qquad t \in [a,b).$$

(F) A vector field V on N determines a map

$$\bar{V}: O(N) \to H$$

by

$$\bar{V}(u) = u^{-1}\, V(\pi(u)).$$

Similarly a section A of $\mathbb{L}(TN;TN)$ determines, and is determined by, a map

$$\bar{A}:O(N) \to \mathbb{L}(H;H)$$

which satisfies

$$\bar{A}(u,a) = a^{-1}\bar{A}(u)a \qquad a \in O(H)$$

viz.

$$\bar{A}(u) = u^{-1} A(\pi(u))u \qquad u \in O(H).$$

The *covariant derivative* ∇V of V is the section of $\mathbb{L}(TN;TN)$ defined by

$$\nabla V(x)(v) = u\, d\bar{V}(\tilde{v})$$

where $\tilde{v} \in HT_uO(N)$ is a horizontal lift of $v \in T_xN$. Alternatively we can consider ∇V as a vector bundle map $\nabla V:TN \to TN$. Observe that for a curve σ in N we have

$$\frac{DV}{\partial t} = \nabla V(\sigma(t))(\dot{\sigma}(t)) \tag{3}$$

The covariant derivative ∇A, e.g. $\nabla^2 V = \nabla\nabla V$, can be defined similarly as a section of

$$\mathbb{L}(TN:\mathbb{L}(TN;TN)) \cong \mathbb{L}(TN,TN;TN).$$

<u>Proposition</u> <u>For the Levi-Civita connection, if</u> $v_1, v_2 \in T_xN$

$$\nabla^2 V(v_1,v_2) - \nabla^2 V(v_2,v_1) = R(v_1,v_2)V(x)$$

<u>Proof</u> Take $u_0 \in \pi^{-1}(x)$. For i = 1,2 let V_i be the horizontal vector field such that $V_i(u)$ is the horizontal lift of $u \circ u_0^{-1} v_i$ for all u in O(N). Observe that

$$\begin{aligned} 0 = \underset{\sim}{\theta}(V_1,V_2) &= d\theta(V_1,V_2) \\ &= \tfrac{1}{2}\{V_1\theta(V_2)) - V_2(\theta(V_1)) - \theta([V_1,V_2])\} \\ &= -\tfrac{1}{2}\theta([V_1,V_2]). \end{aligned}$$

Thus $[V_1,V_2]$ is vertical and so $[V_1,V_2](u_0) = A^*(u_0)$ for

$$A = \tilde{\omega}([V_1,V_2](u_0)) = -2\,\underset{\sim}{\Omega}(V_1(u_0),V_2(u_0))$$

by (1). Now

$$\nabla^2 V(v_1, v_2) = u_0 d(\overline{\nabla V})(V_1(u_0))(u_0^{-1} v_2)$$

$$= u_0 V_1\left(\overline{\nabla V}(-)(u_0^{-1} v_2)\right)$$

$$= u_0 V_1(V_2(\bar{V}))(u_0).$$

Thus
$$\nabla^2 V(v_1, v_2) - \nabla^2 V(v_2, v_1) = u_0 [V_1, V_2](\bar{V})(u_0)$$

$$= u_0 A^* (\bar{V})(u_0)$$

$$= u_0 \frac{d}{dt} \bar{V} (u_0 \cdot \exp tA)\Big|_{t=0}$$

$$= u_0 \frac{d}{dt} \exp(-tA) u_0^{-1} V(x)\Big|_{t=0}$$

$$= 2 u_0 \underset{\sim}{\Omega}(V_1(u_0), V_2(u_0)) u_0^{-1} V(x)$$

as required. //

Similarly, or as a consequence, if $S:[a,b) \times [c,d) \to N$ and $V(s,t) \in T_{S(s,t)}N$ for $(s,t) \in [a,b) \times [c,d)$ then if V is C^2, for the Levi-Civita connection

$$\frac{D}{\partial s}\frac{DV}{\partial t} - \frac{D}{\partial t}\frac{DV}{\partial s} = R\left(\frac{\partial S}{\partial s}, \frac{\partial S}{\partial t}\right) V(s,t) \tag{4}$$

A similar approach yields the formula

$$[V,W](x) = \nabla W(V(x)) - \nabla V(W(x)) - T(V(x), W(x)) \tag{5}$$

where the torsion T will be zero in the Levi-Civita case.

Covariant derivatives of other tensors are defined in the same sort of way. For example, with abuse of notation let G temporarily denote the metric tensor $G(v_1, v_2) = \langle v_1, v_2 \rangle_x$ for $v_1, v_2 \in T_x N$. Then G determines

$$\bar{G}: O(N) \to \text{Pos}(H)$$

by

$$\langle \bar{G}(u)\alpha, \beta \rangle = \langle u(\alpha), u(\beta) \rangle_x \qquad u \in \pi^{-1}(x), \qquad \alpha, \beta \in H.$$

Then, by definition,

$$\nabla G: TN \to \mathbb{L}(TN, TN; R)$$

is given by

$$\nabla G(v_1)(v_2,v_3) = \langle d\bar{G}(\tilde{v}_1)u^{-1}v_2,u^{-1}v_3\rangle .$$

However since u is an isometry

$$\langle \bar{G}(u)(\alpha),\beta\rangle = \langle\alpha,\beta\rangle$$

i.e. $\bar{G}(u) \equiv 1$ and so $\nabla G \equiv 0$. This is sometimes used as the defining property of a Riemannian connection.

Also for a vector field v along a curve σ in N, if ξ is a 1-form on N we have

$$\frac{d}{dt}(\xi(v_t)) = \nabla\xi(\dot{\sigma}_t)(v_t) + \xi(\frac{Dv}{\partial t}) . \qquad (6)$$

(See also (9) and (11) below.)

All of these constructions stem from the fact that the pull back bundle $\pi^*(TN) \to O(N)$ has a natural trivialization, as therefore do the pull backs of all weakly associated bundles to TN.

§3 LOCAL DESCRIPTIONS, CHRISTOFFEL SYMBOLS

(A) The orthonormal frame bundle is a submanifold of the full frame bundle $GL(N) \to N$ where $GL(N)_x$ consists of the continuous linear isomorphisms $u:H \to T_xN$. Replacing O(H) by GL(H) and o(H) by $\mathbb{L}(H;H) = g\ell(H)$ we can uniquely extend ϖ to an $\mathbb{L}(H;H)$-valued 1-form ϖ_0 on GL(N) which retains the corresponding properties to axioms (i) and (ii) for ϖ.

Our chart (U,ϕ) for N determines a section $s(\phi)$ of GL(N) over U:

$$s(\phi)(x) = (T_x\phi)^{-1} \qquad x \in U.$$

This determines a 1-form on U with values in $\mathbb{L}(H;H)$ namely

$$s(\phi)^*(\varpi_0) = \varpi_0 o Ts(\phi).$$

In turn this gives a map

$$\Gamma:\phi(U) \to \mathbb{L}(H;\mathbb{L}(H;H))$$

by

$$\Gamma(x)(v) = \varpi_0 \circ T_x s(\phi) \circ (T_x\phi)^{-1}(v).$$

For $H = \mathbb{R}^n$ and $e_1,\ldots,e_n$ the standard basis of $\mathbb{R}^n$ we can set

$$\Gamma^i_{jk}(x) = \langle \Gamma(x)(e_j,e_k),e_i \rangle$$

Then Γ^i_{jk} are the classical *Christoffel symbols*.

(B) Using the chart (U,ϕ) we can work in local coordinates to assume here that U is open in H and the frame bundle is $U \times GL(H) \to U$ with $s(\phi)(x) = (x,1)$ for $x \in U$. Then if $(v,0) \in H \times \mathbb{L}(H;H) = T_{(x,1)}(U\times GL(H))$ we have

$$\varpi_0(v,0) = \Gamma(x)v$$

while the extensions of axioms (i) and (ii) to the connection form ϖ_0 show in turn that

$$\varpi_0(v,A) = \Gamma(x)v + A \qquad (v,A) \in T_{(x,1)}(U \times GL(H))$$

and

$$\begin{aligned}\varpi_0(v,B) &= \mathrm{ad}(a^{-1})\varpi_0(v,Ba^{-1})\\ &= \mathrm{ad}(a^{-1})\Gamma(x)v + a^{-1}B \qquad (v,B) \in T_{(x,a)}(U \times GL(H))\end{aligned}$$

Thus a lift

$$\tilde{\sigma}:[a,b) \to U \times GL(H)$$
$$\tilde{\sigma}(t) = (\sigma(t),a(t))$$

is horizontal if and only if

$$\dot{a}(t)a(t)^{-1} = -\Gamma(\sigma(t))(\dot{\sigma}(t)). \tag{7}$$

If $v:[a,b) \to H$ is a vector field along σ and $\tilde{\sigma}$ is the horizontal lift of σ then

$$\frac{Dv}{\partial t} = a(t)\,\frac{d}{dt}\,(a(t)^{-1}v(t))$$

and so by (7)

$$\frac{Dv}{\partial t} = \frac{dv}{dt} + \Gamma(\sigma(t))(\dot{\sigma}(t))(v(t)) \tag{8}$$

For $H = \mathbb{R}^n$ if $\sigma(t) = (\sigma^1(t),\ldots,\sigma^n(t))$ and $v(t) = (v^1(t),\ldots,v^n(t))$

$$\frac{Dv^i}{\partial t} = \frac{dv^i}{dt} + \sum_{j,k} \Gamma^i_{jk}(\sigma(t))\dot{\sigma}^j(t)v^k(t).$$

We are now tied in (apart from the sign of R!) with the very clear classical treatment in Chapter II of (Milnor 1963).

Similarly if ξ is a 1-form on U, so we can consider ξ as a map $\xi:U \to \mathbb{L}(H;R)$, we have

$$\nabla\xi(x)v = D\xi(x)(v) - \xi(x)\circ\Gamma(x)v \tag{9}$$

for $x \in U$ and $v \in H = T_xU$.

Finally, in our local chart, from (5), (8) and (3) we see

$$T(v_1,v_2) = \Gamma(x)(v_1,v_2) - \Gamma(x)(v_2,v_1) \qquad v_1,v_2 \in T_xU = H \tag{10}$$

In particular if $H = \mathbb{R}^n$, for the Levi-Civita connection we have $\Gamma^i_{jk} = \Gamma^i_{kj}$.

§4 LAPLACE-BELTRAMI OPERATOR

We take $H = \mathbb{R}^n$. A C^r map, $r \geq 1$, $f:N \to \mathbb{R}$ has a *gradient* grad f, or ∇f, which is a vector field on N defined by

$$\langle\nabla f,v\rangle_x = df(v) \qquad v \in T_xN.$$

Also a C^r vector field V on N has a *divergence* div $V:N \to R$ defined by

$$\text{div } V(x) = \text{trace } \nabla V(x)$$

$$= \sum_i \langle\nabla V(x)v_i,v_i\rangle_x$$

where $v_1,\ldots,v_n$ is an orthonormal basis for T_xN, $\langle\ ,\ \rangle_x$. The divergence can also be defined in terms of the rate of change of the Riemannian volume element measure d(vol) under the flow of V. (Locally d(vol) is given by $\sqrt{|\det G(x)|}dx$.) From this follows the divergence theorem

$$\int_N \text{div } V \, d(\text{vol}) = 0$$

under fairly general conditions (in particular if N is compact). Orientations are irrelevant. Since $(\text{div } fV)(x) = \langle\nabla f(x),V(x)\rangle_x + f(x)\text{div } V(x)$ this shows that the operators div and $-\nabla$ are formal adjoints.

The *Laplace-Beltrami* operator Δ is defined on C^2 maps $f:N \to R$ by

$$\Delta f = \text{div grad } f:N \to R$$

or

$$\Delta f = \text{trace } \nabla\nabla f.$$

From above we see that Δ is formally self-adjoint and negative semi-definite.

To compute Δ in local coordinates we use the chart (U,ϕ) to identify U with an open set of $\mathbb{R}^n$ as before. In (6) take σ with $\sigma(a) = x$ and $\dot{\sigma}(a) = v_1$ and $v(t) = //_\sigma^t v_2$ to see, using (3) and then (2), that

$$\nabla df(v_1)(v_2) = \frac{d}{dt} df(v(t))\Big|_{t=a}$$

$$= \langle \nabla(\nabla f)(x)v_1, v_2\rangle$$

Therefore by (9), taking $v_1 = v_2 = G(x)^{-\frac{1}{2}}e_i$ and summing over i:

$$\Delta f(x) = g^{ij}(x) \frac{\partial^2 f}{\partial x^i \partial x^j} - g^{ij}(x)\, \Gamma^k_{ij}(x) \frac{\partial f}{\partial x^k} \tag{12}$$

using the summation convention, where $(g^{ij}(x))_{i,j}$ is the matrix of $G(x)^{-1}$. Another useful formula is

$$\Delta f(x) = g(x)^{-\frac{1}{2}} \frac{\partial}{\partial x^i} \{g(x)^{\frac{1}{2}} g^{ij}(x) \frac{\partial f}{\partial x^j}\} \tag{12'}$$

for $g(x) = \det G(x)$. This can be seen from the measure theoretic definition of the divergence.

§5. EXPONENTIAL MAP, NORMAL COORDINATES, MANIFOLDS WITH NON-POSITIVE CURVATURE

(A) Let N have a C^{p-1} connection. For simplicity we will assume that the connection is *complete* i.e. for each $x \in N$ and $v \in T_xN$ the unique geodesic γ with $\gamma(0) = x$ and $\dot{\gamma}(0) = v$ can be defined for all time. Then we write $\gamma(1) = \exp_x v$ to obtain a C^{p-1} map

$$\exp_x : T_xN \to N$$

B

for each x in N. It turns out that $\gamma(t) = \exp_x(tv)$, for $t \in \mathbb{R}$.

This *exponential map* should not be confused with the exponential map of a Lie group. The latter comes from the Lie group structure and not a priori from a connection. However it *is* the exponential map, from the identity, of the trivial left invariant connection (i.e. the one such that all left invariant vector fields have vanishing covariant derivatives). This connection will be a Riemannian connection for any left invariant metric, but in general it will have torsion and so not be a Levi-Civita connection. Nevertheless in some special cases the two exponential maps do agree: see the discussion in §10D of Chapter IX.

For a Levi-Civita connection completeness is equivalence to metric completeness with respect to the Riemannian distance function $d(x,y)$ which is the infimum of the lengths $L(\sigma)$ of all piecewise C^1 curves $\sigma:[a,b] \to N$ with $\sigma(a) = x$ and $\sigma(b) = y$ for

$$L(\sigma) = \int_b^a |\dot{\sigma}(t)|dt.$$

When $\dim N < \infty$ it is also equivalent to the statement that every (metrically) bounded subset has compact closure, and it implies that any two points can be joined by a geodesic segment with $L(\sigma) = d(x,y)$, for N connected.

(B) To differentiate $\exp_x$ set $\gamma_s(t) = \exp_x t(v+sh)$ for v,h in T_xN and set

$$V_s(t) = \frac{\partial}{\partial s}\gamma_s(t)$$

so that

$$V_0(1) = T_v(\exp_x)(h) \in T_{\exp_x v}N.$$

Then

$$\frac{D}{\partial t}V_s(t) = \frac{D}{\partial s} /\!/_t^{\gamma_s}(v+sh)$$

and applying equation (4) of §2 we see that $V(t) \equiv V_0(t)$ is a *Jacobi field along* $\gamma \equiv \gamma_0$, i.e. satisfies

$$\left.\begin{aligned} &\frac{D^2}{\partial t^2}V(t) - R(V(t),\dot{\gamma}(t))V(t) = 0 \\ \text{with}\quad &V(0) = 0,\ \frac{DV}{\partial t}(0) = h \end{aligned}\right\} \qquad (13)$$

In particular this shows that $T_0(\exp_x)$ is the identity map of T_xN. The inverse function theorem then shows that there is an open neighbourhood N_x of

0 in T_xN such that $\exp_x$ restricts to a C^{p-1} diffeomorphism of N_x onto an open neighbourhood U_x of x in N. This gives a C^{p-1} chart $(U_x, (\exp_x|N_x)^{-1})$ for N, with model T_xN. The corresponding coordinates are called *normal coordinates*. In them the geodesics emanating from x are represented as straight lines through 0. Usually N_x is chosen star-shaped from 0. Other derivatives of the exponential map are computed in (Eliasson 1967).

(C) From now on assume that N is a complete, connected finite dimensional Riemannian manifold with its Levi-Civita connection. A point y of N is *conjugate* to x along a geodesic γ with $\gamma(0) = x$ and $\gamma(b) = y$ if there is a non-zero Jacobi field V along γ with $V(0) = 0$ and $V(b) = 0$. By (12) this is so if and only if $D/_{\partial t}V(0)$ is in the null-space of $T_v \exp_x$ for $v = \dot{\gamma}(0)$. The *sectional curvature* determined by two orthogonal unit vectors A, B in T_yN is $\langle R(A,B)B,A\rangle_y$, and N has *non-positive curvature* if all its sectional curvatures are non-positive. If so

$$\langle R(A,B)B,A\rangle_y \leq 0$$

for all A,B in T_yN, for all y in N, and a simple argument from (12) shows that each point x has no conjugate points. This leads to:

Cartan-Hadamard Theorem. If the complete Riemannian manifold N has non-positive curvature then for each $x \in N$

$$\exp_x : T_xN \to N$$

is a covering map. In particular it is a diffeomorphism if also N is simply connected. //

For the details see (Milnor 1963), but remember his different sign convention for R(A,B)C! Note that in particular this says that for such N, if $x,y \in N$ there is a unique geodesic in every homotopy class of paths from x to y. This geodesic minimizes the lengths of all the paths in its homotopy class: an easy way to see this is to give T_xN the Riemannian metric which makes $\exp_x$ an *isometry* (i.e. makes $T_v \exp_x$ inner product preserving for each $v \in T_xN$) and so a *Riemannian covering* (i.e. a covering which is an isometry), then geodesics from x lift to straight lines in T_xN, $\exp_x$ preserves length of curves, and paths in N from x to y lift to paths in T_xN from 0 to points in $\exp_x^{-1}(y)$, each point in the inverse image corresponding to precisely one homotopy class.

§6. THE DIFFERENTIAL GEOMETRY OF SUBMANIFOLDS

B

Let M be a C^{r+1} submanifold of a Riemannian manifold P, $r \geq 2$. Give M the Riemannian structure induced from P, i.e. the inner product on T_xM is that induced on it by its inclusion into T_xP. Let ∇ and ∇' denote covariant differentiation in M and P respectively, using the Levi-Civita connections. The proofs of the following relationships between ∇ and ∇' can be found in Chapter VII of (Kobayashi & Nomizu 1963) Volume II.

For x in M let $T_xM^{\perp}$ denote the orthogonal complement of T_xM in T_xP. For each x in M there is a symmetric bilinear map

$$\alpha_x : T_xM \times T_xM \to T_xM^{\perp}$$

the *second fundamental form of* M *at* x, such that for every C^1 vector field V on M *Gauss's formula* holds:

$$\nabla' V(v) = \nabla V(v) + \alpha_x(V(x), v) \tag{14}$$

all $v \in T_xM$.

For each $x \in T_xM$ there is also the bilinear map

$$A_x : T_xM \times T_xM^{\perp} \to T_xM$$

such that for all $\xi \in T_xM^{\perp}$ and $v, w \in T_xM$ we have

$$\langle A_x(u,\xi), v\rangle_x = \langle \alpha_x(u,v), \xi\rangle_x .$$

Then *Weingarten's formula* says (in particular) that

$$\nabla'\xi(v) = - A_x(v, \xi(x)) + \text{a normal component} \tag{15}$$

for all C^1 fields ξ defined on M with $\xi(x) \in T_xM^{\perp}$ for all x, and for all v in T_xM.

The *mean curvature normal at* x is $\frac{1}{n}$ trace α_x in $T_xM^{\perp}$ where n = dim M, and M is a *minimal submanifold* of N if the mean curvature normal vanishes at each x in M.

APPENDIX C SOME MEASURE THEORETIC TECHNICALITIES

§1 CONVERGENCE IN MEASURE IMPLIES CONVERGENCE IN LAW

We will prove the following standard result: see §6B of Chapter VIII, for the terminology.

Theorem For every complete separable metric space M and finite measure space (Ω,F,μ) the map

$$L^0(\Omega,F,\mu;M) \to M(M)$$

$$f \mapsto f(\mu)$$

is continuous into the narrow topology.

Proof Let $\alpha_0:\Omega \to M$ be measurable and $f:M \to \mathbb{R}$ bounded and continuous. Let m be an upper bound for $|f(x)|$, $x \in M$. Since all finite measures on M are tight, for a given $\varepsilon > 0$ there exists a compact subset K of M with

$$\alpha_0(\mu)(K) > 1 - \frac{1}{5m}\varepsilon.$$

Also by Lemma 7B of Chapter VII there exists $\delta > 0$ such that if $x' \in M$ and and $x \in K$ with $d(x',x) < \delta$ then $|f(x) - f(x')| < \frac{1}{5}\varepsilon$.

Suppose $\alpha \in \mathcal{L}^0(\Omega,F,\mu;M)$ with

$$\mu\{\omega:d(\alpha(\omega),\alpha_0(\omega)) \geqslant \delta\} < \frac{1}{5m}\varepsilon.$$

Then

$$|\int_\Omega f\ d\alpha_0(\mu) - \int_\Omega f\ d\alpha(\mu)| = |\int_\Omega f\circ\alpha_0\ d\mu - \int_\Omega f\circ\alpha\ d\mu|$$

$$\leqslant I_1 + I_2 + I_3$$

where

$$I_j = \int_{\Omega_j} |f\circ\alpha_0 - f\circ\alpha|d\mu \qquad j = 1,2,3$$

for

$$\Omega_1 = \{\omega:\alpha_0(\omega) \notin K\}$$
$$\Omega_2 = \{\omega:d(\alpha_0(\omega),\alpha(\omega)) \geq \delta\}$$
$$\Omega_3 = \{\omega:\alpha_0(\omega) \in K \ \&\ d(\alpha_0(\omega),\alpha(\omega)) < \delta\}.$$

However clearly $I_1 \leq \frac{\varepsilon}{5m} \cdot 2m$ and $I_2 \leq \frac{\varepsilon}{5m}\, 2m$ while

$$I_3 \leq \int_{\Omega_3} \frac{\varepsilon}{5}\, d\mu \leq \varepsilon/5. \quad //$$

§2 EXISTENCE OF SAMPLE CONTINUOUS VERSIONS

(A) We shall use the *Borel-Cantelli lemmas* (at least the trivial one) and in any case they should be included for the sake of completeness:

Borel-Cantelli Lemmas Let $\{A_n\}_{n=1}^{\infty}$ be a sequence in F. Then

$$\sum_{n=1}^{\infty} \mu(A_n) < \infty \Rightarrow \mu(\overline{\lim_{n\to\infty}} A_n) = 0$$

where $\overline{\lim_{n\to\infty}} A_n = \bigcap_{N=1}^{\infty} \bigcup_{n=N}^{\infty} A_n$ is the event that A_n occurs infinitely often. In fact if $\{A_n\}_{n=1}^{\infty}$ is an increasing sequence of σ-sub-algebras of F with $A_n \in A_n$ for each n then, almost surely,

$$\overline{\lim_{n\to\infty}} A_n = \{\omega \in \Omega: \sum_{n=1}^{\infty} \mathbb{P}(A_{n+1}|A_n) = \infty\}.$$

Proof For the first part observe that, for all $N > 0$,

$$\mu(\overline{\lim_{n\to\infty}} A_n) \leq \mu(\bigcup_{n=N}^{\infty} A_n) \leq \sum_{n=N}^{\infty} \mu(A_n).$$

A proof of the second part (which we will not use) can be found in (Breiman 1968). //

(B) Rather than do as advertised in the text and prove the isonomy invariance of the existence of sample continuous versions we shall give the proof of the n-dimensional generalization of Kolmogorov's theorem following that given in (Meyer 1981c). We did not use the isonomy invariance and so leave it as an exercise (some readers might like to see in what generality they can prove it?).

<u>Theorem 2B</u> <u>Let</u> (M,d) <u>be a complete metric space.</u> <u>Suppose for</u>

$$x:[0,1]^p \to \mathcal{L}^0(\Omega,F;M)$$

<u>there exists</u> $\alpha,\beta,\gamma > 0$ <u>such that for all</u> $\delta > 0$ <u>and</u> s <u>and</u> t <u>in</u> $[0,1]^p$

$$\mu\{\omega \in \Omega: d(x_s(\omega),x_t(\omega)) > \delta\} \leq \beta\delta^{-\alpha}|s-t|^{p+\gamma}.$$

<u>Then</u> x <u>has a sample continuous version.</u>

<u>Proof</u> We can take the norm on $\mathbb{R}^p$ to be $|(s_1,\ldots,s_p) - (t_1,\ldots,t_p)| = \max_i |s_i-t_i|$. Set $T = [0,1]^p$ and for m = 1,2,... let Δ_m be the set of points in T of the form $(k_1 2^{-m},\ldots,k_p 2^{-m})$ where the k_i are integers in $[0,2^m]$. Set $\Delta = \bigcup_m \Delta_m$.

Take $\varepsilon = \frac{1}{2}\gamma\alpha^{-1}$ and define A_m by

$$A_m = \{\omega \in \Omega: \exists\ s,t \in \Delta_m \text{ with } |s-t| = 2^{-m} \ \&\ d(x_s(\omega),x_t(\omega)) \geq 2^{-\varepsilon m}\}.$$

By hypothesis, if s and t are in Δ_m with $|s-t| = 2^{-m}$ we have

$$\mu\{\omega: d(x_s(\omega),x_t(\omega)) > 2^{-\varepsilon m}\} \leq \beta\, 2^{\alpha\varepsilon m}\, 2^{-(p+\gamma)m}.$$

Also the number of points, $\#\Delta_m$, in Δ_m is 2^{mp} while there exists c such that for each $s \in \Delta_m$

$$\#\{t \in \Delta_m: |s-t| = 2^{-m}\} \leq c\beta^{-1}.$$

Thus

$$\mu(A_m) \leq c2^{\alpha\varepsilon m}\, 2^{-\gamma m} = c\ 2^{-\frac{1}{2}\gamma m}$$

and

$$\sum_m \mu(A_m) < \infty\ .$$

By the Borel-Cantelli lemma we can conclude that for all ω in a set Ω_0 of full measure there exists $m_0(\omega)$ such that for s and t in Δ_m

$$m \geq m_0(\omega)\ \&\ |s-t| = 2^{-m} \Rightarrow d(x_s(\omega),x_t(\omega)) \leq 2^{-\varepsilon m}.$$

Now fix ω in Ω_0 and suppose s and t are in Δ with $|s-t| \leq 2^{-(m+1)}$ where $m \geq m_0(\omega)$. Expand their coordinates as the finite sums

$$s_i = k_i\, 2^{-m} + \sum_{j>m} s_i^j\, 2^{-j}, \qquad t_i = \ell_i 2^{-m} + \sum_{j>m} t_i^j\, 2^{-j}$$

where $0 \leqslant k_i \leqslant 2^{-m}$, $0 \leqslant \ell_i \leqslant 2^{-m}$, and the s_i^j, t_i^j are 0 or 1. Let k and ℓ be the points with coordinates $k_i 2^{-m}$ and $\ell_i 2\text{-}m$. Then $|k-\ell|$ is 2^{-m} or 0 so

$$d(x_k(\omega), x_\ell(\omega)) \leqslant 2^{-\varepsilon m}.$$

Also the distance between the points with coordinates

$$k_i 2^{-m} + \sum_{j=m+1}^{r} s_i^j 2^{-j} \text{ and } \ell_i 2^{-m} + \sum_{j=m+1}^{r-1} s_i^j 2^{-j} \text{ is at most } 2^{-r}.$$

Therefore

$$d(x_s(\omega), x_k(\omega)) \leqslant \sum_{r=m+1}^{\infty} 2^{-\varepsilon r}$$

and similarly for $d(x_t(\omega), x_\ell(\omega))$. Thus if $m \geqslant m_0(\omega)$

$$s,t \in \Delta \;\&\; |s-t| < 2^{-(m+1)} \Rightarrow d(x_s(\omega), x_t(\omega)) \leqslant 2^{-\varepsilon m}(1 + 2^{-\varepsilon}).$$

This shows that the restriction of $x.(\omega)$ to Δ is uniformly continuous on the dense subset Δ of T and so has a continuous extension

$$\tilde{x}_{\cdot}(\omega): T \to M.$$

We now have a sample continuous version $\tilde{x}$, defined as above for ω in Ω_0, and taken as some constant if $\omega \notin \Omega_0$. It is a version since it agrees with x on $\Delta \times \Omega_0$ and our basic hypothesis on x implies immediately that x is continuous in measure. //

Corollary 2B Let (M,d) be a complete metric space. Suppose that for N = 1,2,...

$$x: \mathbb{R}^p \to \mathcal{L}^0(\Omega, \mathcal{F}; M)$$

when restricted to $[-N,N]^p$ satisfies the conditions of Theorem 2B, with the cube $[-N,N]^p$ replacing $[0,1]^p$. Then x has a sample continuous version.

Proof For each N the theorem implies that $x|[-N,N]^p$ has a sample continuous version x^N say, since clearly $[0,1]^p$ can be replaced by $[-N,N]^p$. However by separability of $[-N,N]^p$ we have $x_{\cdot}^{N+1}(\omega)\,[-N,N] = x_{\cdot}^N(\omega)$ almost surely. Therefore all the x^N agree where defined, outside of one set of measure zero and so can be combined to give the required version. //

§3 EXISTENCE OF MEASURABLE VERSIONS

(A) Let (X,z) be a stochastic dynamical system on our manifold M with $F(u):[0,\xi^u)\times \Omega \to M$ a maximal solution from $u \in M$. Let $\mathcal{F}_t$ be the σ-algebra generated by $\{z_s - z_0 : 0 \le s \le t\}$ together with all sets of measure zero in $\mathcal{F}$. In the notation of §3B of Chapter IX we shall say that a map

$$F:M \times \Omega \to C^{\Delta}([0,\infty);M^+)$$

is *measurable* if it is Borel (M) $*$ $\mathcal{F}$ measurable, and *strongly adapted to* $\{\mathcal{F}_t\}$ if for each $t \ge 0$ the map

$$\rho_t \circ F:M \times \Omega \to M^+$$

is Borel (M) $*$ $\mathcal{F}_t$ measurable, where ρ_t is the evaluation map at time t (*progressively measurable* might be a more suitable term). After some preliminary discussion we shall prove:

Theorem 3A There exists a measurable version

$$\tilde{F}:M \times \Omega \to C^{\Delta}([0,\infty);M^+)$$

of the solution flow which is strongly adapted to $\{\mathcal{F}_t\}$.

The proof is surprisingly tricky, especially in the case M infinite dimensional with (X,z) not complete. For finite dimensional M we could use the existence of continuous flows. One approach would be to use the general results of (Cohn 1972). Another method is indicated in (Ikeda & Watanabe 1981). However we shall make more use of our p.l. approximations, and of (Stricker & Yor 1978) where the finite dimensional complete case is considered.

(B) We start with a standard lemma.

Lemma 3B Let $g_m:P \to Q$, $m = 1,2,\ldots,$ be measurable maps from a measurable space $\{P,A\}$ to a complete metric space (Q,d). Set

$$\mathcal{D} = \{x \in P: \lim_{m\to\infty} g_m(x) \text{ exists}\}.$$

Then $\mathcal{D}$ is in A and $\lim_{m\to\infty} g_m: \mathcal{D} \to Q$ is measurable.

Proof To see that $\mathcal{D}$ is in A observe that

$$\mathcal{D} = \{x \in P: \overline{\lim_m} \sup_{p,q \geq m} d(g_p(x), g_q(x)) = 0\}.$$

Now set $g = \lim_{m\to\infty} g_m$ on $\mathcal{D}$ and let χ be the characteristic function of an open subset U of Q. Then

$$g^{-1}(U) = \{x \in \mathcal{D}: \overline{\lim_m} \chi(g_m(x)) = 1\}$$

and the lemma is proved. //

C. Proof of Theorem 3A We shall be unnecessarily precise about the construction of $\tilde{F}$. For m = 0,1,2,... let $\mathbb{D}_m$ be the set of reals of the form $k2^{-m}$ for $k \in \{0,1,2,\ldots\}$, and set $\mathbb{D} = \cup\{\mathbb{D}_m : m = 0,1,\ldots\}$. In the notation of Theorem 10 of Chapter VII, let Π_m be the partition of $[0,\infty)$ with set of partition points $\mathbb{D}_m$ and write $z^m = z_{\pi_m}$. Let F^m be the flow of (X,z^m). Then, for each m,

$$F^m : M \times \Omega \to C^{\Delta}([0,\infty); M^+)$$

is measurable and $\rho_t \circ F^m$ is Borel (M) $* F_t$ measurable provided $t \in \mathbb{D}_m$, since we are just solving ordinary differential equations depending measurably on the parameter ω in Ω.

Let $F: M \to \mathcal{L}^0(\Omega, F_\cdot; C^{\Delta}([0,\infty), M^+)$

be a version of the solution flow of (X,z), with ξ^u the explosion time from $u \in M$ and $\Omega_t^u = \{\omega \in \Omega : t < \xi^u(\omega)\}$. By Theorem 10 of Chapter VII, for any $t \geq 0$ we can follow (Stricker & Yor 1978) and inductively define

$$n_k : M \to \{1,2,\ldots\} \qquad k = 0,1,2,\ldots$$

by $\quad n_0(u) = 1$

and

$$n_k(u) = \inf\{m > n_{k-1}(u): \sup_{p,q\geq m} \mu\{\omega \in \Omega_t^u : \sup_{0\leq s\leq t} d(F_s^p(u,\omega), F_s^q(u,\omega)) > 2^{-k}\} \leq 2^{-k}\}$$

where d is a complete metric on M, extended so that $d(\Delta,u)$, $d(\Delta,\Delta)$, and $d(u,\Delta)$ are taken infinite for all u in M. Then n_k is Borel measurable for each k. Also if $g_s^k(u,\omega) = F_s^{n_k(u)}(u,\omega)$, then for all $k \geq 1$

$$\mu\{\omega \in \Omega_t^u: \sup_{0\le s\le t} d(g_s^{k+1}(u,\omega), g_s^k(u,\omega)) > 2^{-k}\} \le 2^{-k}.$$

Therefore, by the Borel-Cantelli Lemma, §2A, $\{g_s^k(u,\omega)\}_k$ converges uniformly in $s \in [0,t]$, with probability one on Ω_t^u.

If we take an enumeration of $\mathbb{D}$ we can apply the diagonal argument to obtain an increasing sequence $\{m_k\}_{k=0}^{\infty}$ of measurable maps $m_k: M \to \{1,2,\ldots\}$ such that if $G_s^k(u,\omega) = F_s^{m_k(u)}(u,\omega)$ then, with probability one, $\{G_s^k(u,\omega)\}_k$ converges uniformly on $[0,t]$ on Ω_t^u *for all* $t \in \mathbb{D}$. Define $\Omega(u,t)$ and $\mathcal{D}_t$ by

$$\Omega(u,t) = \{\omega \in \Omega: \lim_{k\to\infty} G_s^k(u,\omega) \text{ exists uniformly on } [0,t]\}$$

$$\mathcal{D}_t = \{(u,\omega) \in M \times \Omega: \omega \in \Omega(u,t)\}.$$

If $t \in \mathbb{D}$ set

$$\tilde{F}(u,\omega)(t) = \lim_{k\to\infty} G_t^k(u,\omega) \qquad (u,\omega) \in \mathcal{D}_t$$

and

$$\tilde{F}(u,\omega)(t) = \Delta \qquad \text{otherwise.}$$

For general $t \ge 0$ set $t_m = \max\{s \in \mathbb{D}_m : s \le t\}$, so that $t_m \uparrow t$ as $m \to \infty$; and then set

$$\tilde{F}(u,\omega)(t) = \lim_{m\to\infty} \tilde{F}(u,\omega)(t_m)$$

if the limit exists in M, and set it equal to Δ otherwise.

Clearly we have

$$\tilde{F}: M \times \Omega \to C^{\Delta}([0,\infty); M^+).$$

Also if $t \in \mathbb{D}$, the lemma, with $Q = C([0,t];M)$, implies that $\rho_t \circ \tilde{F}(u,\omega)$ is Borel $(M) * F_t$ measurable. We can then take $Q = M$ and $P = \bigcap_m \mathcal{D}_{t_m}$ to see that this also holds for general t. Thus $\tilde{F}$ is strongly adapted.

Now fix $u \in M$. To show that $\tilde{F}(u,.) = F(u)$ almost surely it suffices, by continuity of the paths, to show a.s. equality at all t in $\mathbb{D}$. Therefore fix t in $\mathbb{D}$. If $\omega \in \Omega_t^u$ we have probability one that $(u,\omega) \in \mathcal{D}_t$ and therefore that $\tilde{F}(u,\omega)(t) = F(u)(\omega)(t)$ by Theorem 10 of Chapter VII. The proof will be completed by showing that if $\omega \in \Omega(u,t)$ then $\omega \in \Omega_t^u$ with probability one.

To do this take $\varepsilon > 0$. By Lemma 7B of Chapter 7 there is a compact K in M and an $\Omega_0 \subset \Omega(u,t)$ with $\mu(\Omega_0) < \varepsilon/2$ such that $F_s(u,\omega) \in K$ for $0 \le s \le t$ if $\omega \in \Omega(u,t)$ with $\omega \notin \Omega_0$. Take an open neighbourhood U of K which admits a uniform cover in M and let τ be the first exit time of F(u) from U.

By Theorem 6 of Chapter VII, $t \wedge \tau < \xi^u$ a.s. For $r \in \mathbb{D}$ set

$$Z_r = \{\omega \in \Omega: t \wedge \tau(\omega) < r < \xi^u(\omega)\}.$$

From what we already know $\tilde{F}(u,\omega)(\tau(\omega) \wedge t)$ is equal to $F(u)(\omega)(\tau(\omega)\wedge t)$ a.s. on Z_r for each r, and therefore almost surely on Ω. However $\tilde{F}(u,\omega)(\tau(\omega)\wedge t)$ lies in K for ω in $\Omega(u,t)$, $\omega \notin \Omega_0$. Therefore, a.s.,

$$\omega \in \Omega(u,t) \ \&\ \omega \notin \Omega_0 \Rightarrow t < \tau(\omega) \le \xi^u(\omega).$$

Since $\varepsilon > 0$ was arbitrary this yields what we wanted: with probability one

$$\omega \in \Omega(u,t) \Rightarrow t < \xi^u(\omega). \quad //$$

Remark For F as constructed in the above proof define $\xi: M \times \Omega \to [0,\infty]$ by $\xi(u,\omega) = \sup\{t: \tilde{F}(u,\omega)(t) \in M\}$. In our construction $\tilde{F}(u,\omega)(\xi(u,\omega))$ may lie in M. This is not usually allowed to happen in the 'space of explosive paths'. Observe that for $t \ge 0$ if $\tilde{\mathcal{D}}_t = \{(u,\omega): t \le \xi(u,\omega)$ then

$$\tilde{\mathcal{D}}_t = \bigcap_m \mathcal{D}_{t_m} \in \text{Borel}(M) * F_t$$

c.f. the definition of a *wide sense stopping time* in (Dellacherie & Meyer 1978). Therefore $\{(u,\omega): t < \xi(u,\omega)\}$ is in $\bigcap_{s>t} (\text{Borel}(M) * F_s)$ and if we were to redefine $\tilde{F}$ so that $\tilde{F}(u,\omega)(\xi(u,\omega)) = \Delta$ we could lose strong adaptness to $\{F_t\}_t$.

However if we had defined $\tilde{F}$ by approximating from *above* by elements of $\mathbb{D}$ we could have got over this difficulty at the expense of being only strongly adapted to $\{F_{t+}: t \ge 0\}$ where

$$F_{t+} = \bigcap_{s>t} F_s.$$

This would not generally be considered much of a sacrifice: see (Dynkin 1965; Dellacherie & Meyer 1978).

REFERENCES

Arnold, L. (1974) Stochastic Differential Equations: Theory and Applications. New York, London: Interscience.

Airault, H. (1976) Subordination de processus dans le fibré tangent et formes harmoniques. C.R. Acad. Sc. Paris, Sér. A, 282 (14 juin 1976), 1311-1314.

Asorey, M. & Mitter, P.K. (1981) Regularized, continuum Yang-Mills process and Feynman-Kac functional integral. Commun. Math. Phys., 80, 43-58.

Azencott, R. (1974) Behaviour of diffusion semigroups at infinity. Bull. Soc. Math. France, 102, 193-240

Azencott, R. (1980) Grandes deviations et applications. In Ecole d'Eté de Probabilités de Saint-Flour VIII-1978, ed. P.L. Hennequin, pp. 2-176. Lecture Notes in Maths 774. Berlin, Heidelberg, New York: Springer-Verlag.

Azencott, R. et al. (1981) Géodésiques et diffusions en temps petit. Séminaire de probabilités, Universite de Paris VII. Astérique 84-85. Société mathématique de france.

Bauer, H. (1972) Probability Theory and Elements of Measure Theory. New York: Holt, Rinehardt and Winston.

Baxendale, P. (1976) Measures and Markov processes on function spaces. In Journées Géom. dimension infinie [1975-Lyon] ed. N. Desolneux-Moulis. Bull. Soc. math. France, Mémoire 46, 131-141.

Baxendale, P. (1980) Wiener processes on manifolds of maps. Proc. Royal Soc. Edinburgh, 87A, 127-152.

Baxendale, P. (1981) Stochastic flows and Malliavin calculus. In Proc. 20th I.E.E.E. Conference on Decision and Control.

Belopol'skaja, Ja. I. (1975) Diffusion processes in vector bundles. Teor. Verojatnost. i. Mat. Statist. 12. English Translation: Theor. Probability and Math. Statist. 12 (1976), 1-10.

Berger, M., Gauduchon, P. & Mazet, E. (1971) Le Spectre d'une Variété Riemannienne. Lecture Notes in Math. 194. Berlin, Heidelberg, New York: Springer-Verlag.

Bessaga, C. (1966) Every infinite-dimensional Hilbert space is diffeomorphic with its unit sphere. Bull. Acad. Polon. Sci., Sér. sci. math. astr. et phys., 14, 27-31.

Bichteler, K. (1981) Stochastic integration and L^p theory of semi-martingales. Ann. Probab., 9, no. 1, 49-89.

Bismut, J.M. (1980) Flots stochastiques et formula de Itô-Stratonovich généralisee, C.R. Acad. Sci. Paris t. 290, sér A, 483-486.

Bismut, J-M. (1981a) A generalized formula of Itô and some other properties of stochastic flows. Z. Wahrscheinlichkeitstheorie verw. geb, 55, 331-350.

Bismut, J-M. (1981b) Mécanique Aleatoire. Lecture Notes in Maths. 866. Berlin, Heidelberg, New York: Springer-Verlag.

Blagoveščenskii, Ju. N. & Freidlin, M.I. (1961) Some properties of diffusion processes depending on a parameter. DAN 138, Soviet Math., 2, 633-636.

Blanchard, Ph. & Sirugue, M. (1981) Treatment of some singular potentials by change of variables in Wiener Integrals. J. Math. Phys. 22, no. 7, 1372-1376.

Bourguignon, J.P. & Brezis, H. (1974) Remarks on the Euler equation. J. Funct. Anal., 16, 341-363.

Breiman, L. (1968) Probability. Massachusetts, California, London, Ontario: Addison-Wesley.

Caldwell, I.C. (1976) A relation between a lift of connections and a lift of stochastic dynamical systems. M.Sc. dissertation. Mathematics Institute, University of Warwick, Coventry CV4 7AL.

Carmona, R. & Simon, B. (1981) Pointwise bounds on eigenfunctions and wave packets in N-body quantum systems. V. Lower bounds and path integrals. Commun. Math. Phys., 80, 59-98.

Caverhill, A.P. (1982) A pair of stochastic dynamical systems which have the same infinitesimal generator, but of which one is strongly complete, and the other is not. Preprint: Mathematics Institute, University of Warwick, Coventry CV4 7AL, England.

Carverhill, A.P. & Elworthy, K.D. (1982) Flows of stochastic dynamical systems: the functional analytic approach. Preprint: Mathematics Institute, University of Warwick, Coventry CV4 7AL, England.

Chatterji, S. (1968) Martingale convergence and the Radon-Nikodym theorem in Banach spaces. Math. Scand., 22, 21-41.

Cheeger, J. & Ebin, D. (1975) Comparison Theorems in Riemannian Geometry. Amsterdam: North Holland.

Choquet-Bruhat, Y., DeWitt-Morette, C., & Dillard-Bleick, M. (1977) Analysis, Manifolds and Physics. North-Holland.

Clark, J.M.C. (1973) An introduction to stochastic differential equations on manifolds. In Geometric Methods in Systems Theory, ed. D.Q. Mayne & R.W. Brockett, Dordrecht, Boston, London: Reidel.

Cohn, D.L. (1972) Measurable choice of limit points and the existence of separable and measurable processes. Z. Wahrscheinlichkeitstheorie verw. Geb., 22, 161-165.

Colin de Verdière Y. (1973) Spectre du Laplacien et longueurs des géodésiques périodiques I. Compositio Math., 27, 83-106.

Daletskii, Yu. L. & Shnaiderman, Ya. I. (1969) Diffusions and quasi-invariant measures on infinite dimensional Lie groups. Funktsional' nyi Analiz i Ego Prilozheniya, 3, 88-90. English translation: Funct. Anal. Appl., 3, 156-8.

Darling, R.W.R. (1982) Martingales on manifolds and geometric Itô calculus. Ph.D. Thesis: Mathematics Institute, University of Warwick, Coventry CV4 7AL, England

Debiard, A., Gaveau, B. & Mazet, E. (1976) Théorèmes de comparaison en géometrie riemannienne. Publ. RIMS. Kyoto Univ., 12, 391-425.

Dellacherie, C. & Meyer, P.-A. (1978) Probabilities and Potential. Mathematics Studies, 29. Amsterdam, New York, Oxford: North-Holland.

Dellacherie, C. (1980) Un survoi de la théorie de l'intégrale stochastique. Stoch. Proc. Appl., 10, 115-144.

DeWitt, B.S. (1965) Dynamical Theory of Groups and Fields. New York, London, Paris: Gordon and Breach.

DeWitt-Morette, C., Elworthy, K.D., Nelson, B.L., & Sammelman, G.S. (1980) A stochastic scheme for constructing solutions of the Schrödinger equation. Ann. Inst. Henri Poincaré, Section A, 32, no. 4, 327-341.

DeWitt-Morette, C. & Elworthy, K.D. (1981) editors. New Stochastic Methods in Physics. Physics Reports, 77, no. 3, 121-382.

Dodziuk, J. (1980) Maximum principle for parabolic inequalities and the heat flow on open manifolds. Preprint: Math. Institute, University of Oxford & Math. Dept., Queens College (CUNY), Flushing, N.Y. 11367.

Doss, H. (1980) Quelques formules asymptotiques pour les petites perturbations de systemes dynamiques. Ann. Inst. Henri Poincaré, 16, no. 1, 17-28.

Dowell, R.M. (1980) Differentiable approximations to Brownian motion on manifolds. Ph.D. Thesis: Mathematics Institute, University of Warwick, Coventry CV4 7AL, England.

Dowker, J.S. (1975) Covariant Schrödinger Equations. In Functional Integration and its Applications, ed. A.M. Arthurs, pp. 34-52. Oxford: Oxford University Press.

Dugundji, J. (1966) Topology. Boston: Allyn and Bacon.

Dunford, N. & Schwartz, J.T. (1957) Linear Operators, Part 1. New York, London, Sydney: Interscience.

Dynkin, E.B. (1965) Markov Processes, Vols I & II. Die Grundlehren der Math. Wissenschaften, 121, Berlin, Göttingen, Heidelberg: Springer-Verlag.

Dynkin, E.B. (1968) Diffusion of tensors. Dokl. Akad. Nank SSSR, 179, 1264-1267. English translation: Soviet Math. Dokl. 9, (1968), no. 2, 532-535.

Ebin, D.G. & Marsden, J. (1970) Groups of diffeomorphisms and the motion of an incompressible fluid. Ann. of Math., 92, No. 1, 102-163.

Edwards, S.F. & Gulyaev, Y.V. (1964) Path integrals in polar coordinates. Proc. Roy. Soc. A,279, no. 1377, 229-235.

Eells, J. & Elworthy, K.D. (1971) Wiener integration on certain manifolds. In Problems in Non-Linear Analysis, ed. G.Prodi. pp. 67-94. Centro Internazionale Matematico Estivo, IV Ciclo. Rome: Edizioni Cremonese.

Eells & Malliavin, P. (1973) Diffusion processes in Riemannian bundles. Unpublished.

Eells & Elworthy, K.D. (1976) Stochastic dynamical systems. In Control Theory and Topics in Functional Analysis, Vol. III, pp. 179-185. Vienna: International Atomic Energy Agency.

Eliasson, H. (1967) Geometry of manifolds of maps. J. Diff. Geom. 1, 169-194.

Elworthy, K.D. (1974) Gaussian measures on Banach spaces and manifolds. In Global Analysis and its Applications, Vol. II, pp. 151-166. Vienna: International Atomic Energy Agency.

Elworthy, K.D. (1975) Measures on infinite dimensional manifolds. In Functional Integration and its Applications, ed. A.M. Arthurs, pp. 60-68. Oxford: Oxford University Press.

Elworthy, K.D. (1978) Stochastic dynamical systems and their flows. In Stochastic Analysis, ed. A. Friedman & M. Pinsky, 79-95. London, New York: Academic Press.

Elworthy, K.D. (1981) Stochastic Methods and differential geometry. In Séminaire Bourbaki vol. 1980/81. Lect. Notes in Math. 901, pp. 95-110.

Elworthy, K.D. & Truman, A. (1981) Classical mechanics, the diffusion (heat) equation and the Schrödinger equation on a Riemannian manifold. J. Math. Phys. 22, no. 10, 2144-2166.

Elworthy, K.D. (1982) Stochastic flows and the C_0 diffusion property. To appear in 'Stochastics'.

Elworthy, K.D. & Truman, A. (1982) The diffusion equation and classical mechanics: an elementary formula. To appear in: Proc. Int. Workshop 'Stochastic Processes in Quantum Theory and Statistical Physics' CIRM - Marseilles, June 1981. Lecture Notes in Physics. Springer-Verlag.

Ferebee, J.B. (1972) Parallel translation and diffusions on the tangent bundle, Ph.D. dissertation, Princeton University.

Feynman, R.P. & Hibbs, A.R. (1965) Quantum Mechanics and Path Integrals. New York: McGraw-Hill.

Fisk, D.L. (1963) Quasi-martingales and stochastic integrals. Tech. Rep. 1, Dept. Math., Michigan State Univ.

Friedman, A. (1975) Stochastic Differential Equations and Applications. In 2 volumes. London and New York: Academic Press.

Gangolli, R. (1964) On the construction of certain diffusions on a differentiable manifold, Z. Wahr. verw. Geb., 2, 406-419.

Gaveau, B. & Trauber, P. (1981) Une construction de la quantification euclidienne du Champ de Yang-Mills régularisé. J. Funct. Anal. 42, 356-367.

Gaveau, B. & Vauthier, J. (1981) Intégrales oscillantes stochastiques: l'équation de Pauli. J. Funct. Anal. 44, 388-400.

Gikhman, I.I. & Skorohod, A.V. (1972) Stochastic Differential Equations. Berlin, Heidelberg, New York: Springer-Verlag.

Goldberg, S.I. (1962) Curvature and Homology. New York, London: Academic Press.

Gorman, C.D. (1960) Brownian motion of rotation, Trans. Amer. Math. Soc., 94, 103-117.

Greene, R.E. & Wu, H. (1979) Function Theory on Manifolds which Possess a Pole. Lecture Notes in Maths., 699. Berlin, Heidelberg, New York: Springer-Verlag.

Hille, E. & Phillips, R.S. (1957) Functional Analysis and Semigroups. Providence, Rhode Island: American Math. Soc. Colloquium Publications.

Hirsch, M.W. (1976) Differential Topology. Graduate texts in Mathematics no. 33. Berlin, Heidelberg, New York: Springer-Verlag.

Ibero, M. (1976) Intégrales stochastiques multiplicatives et construction de diffusions sur un groupe de Lie. Bull. Sc. math., 2e série, 100, 175-191.

Ichihara, K. (1980) Curvature, geodesics and the Brownian motion on a Riemannian manifold II. Preprint: Dept. of Applied Sciences, Faculty of Engineering, Kyushu University, Fukuoka, Japan.

Ikeda, N., Nakao, S. & Yamato, Y. (1977) A class of approximations of Brownian motion. Publ. RIMS, Kyoto Univ., 13, 285-300.

Ikeda, N. & Manabe, S. (1978) Stochastic integral of differential forms and its applications. In Stochastic Analysis, ed. A. Friedman & M. Pinsky. pp. 175-185. London, New York: Academic Press.

Ikeda, N. & Watanabe, S. (1981) Stochastic Differential Equations and Diffusion Processes. Tokyo: Kodansha. Amsterdam, New York, Oxford: North-Holland.

Itô, K. (1950) On stochastic differential equations on a differentiable manifold 1. Nagoya Math. J., 1, 35-47.

Itô, K. (1963) The Brownian motion and tensor fields on a Riemannian manifold. Proc. Internat. Congr. Math. (Stockholm, 1962), pp. 536-539. Djursholm: Inst. Mittag-Leffler.

Itô, K & McKean, H.P. (1965) Diffusion processes and their sample paths. Die Grundlehren der Mathematischen Wissenschaften, Band 125. New York: Academic Press. Berlin, Heidelberg, New York: Springer-Verlag.

Itô, K. (1975) Stochastic parallel displacement. In Probabilistic methods in differential equations, ed. M.A. Pinsky, pp. 1-7. Lecture Notes in Maths, 451. Berlin, Heidelberg, New York: Springer-Verlag.

Jameson, G.J.O., (1974) Topology and Normed Spaces. London: Chapman & Hall.

Jørgensen, E. (1978) Construction of B.M. & O.U. process... Z.f.W., 44, 71-87.

Kendall, W.S. (1982) Brownian motion and a generalised little Picard's theorem. To appear in Trans. Amer. Math. Soc.

Kingman, J.F.C. & Taylor, S.J. (1966) Introduction to Measure and Probability. Cambridge: Cambridge University Press.

Knight, F.B. (1981) Essentials of Brownian Motion and Diffusion. Mathematical Surveys No. 18. Providence, Rhode Island: American Math. Society.

Kobayashi, S. & Nomizu, K. (1963) Foundations of Differential Geometry. Volume 1. New York, London: John Wiley, Interscience.

Kobayashi, S. (1967) On conjugate and cut loci. In Studies in Global Geometry and Analysis, ed. S.S. Chern, pp. 96-122. MAA Studies in Math., 4, Mathematical Soc. of Amer. with Prentice-Hall

Kohn, R.V. (1975) Integration over stochastic simplices. M.Sc. dissertation, Mathematics Institute, University of Warwick, Coventry CV4 7AL

Kuiper, N.H. (1971) Variétés Hilbertiennes: Aspects Géométriques. Séminaire de Mathematiques Supérieures-Été 1969. Montréal: Les Presses de l'université de Montréal.

Kunita, H. (1980) On the decomposition of solutions of stochastic differential equations. In Stochastic Integrals, ed. D. Williams, pp. 213-255. Lecture Notes in Maths. 851. Berlin, Heidelberg, New York: Springer-Verlag (1981!)

Kunita, H. (1981) On backward stochastic differential equations. To appear in 'Stochastics'.

Kunita, H. (1982) Stochastic differential equation and stochastic flows of homeomorphisms. Preprint: Dept. of Applied Science, Faculty of Engineering, Kyushu University, Fukuoka 812, Japan.

Lang, S. (1962) Introduction to Differentiable Manifolds. New York, London: Interscience.

Malliavin, P. (1974) Formule de la moyenne pour les formes harmoniques. J. Funct. Anal., 17, 274-291.

Malliavin, P. (1976) Stochastic calculus of variation and hypoelliptic operators. Report no. 13, Institut Mittag-Leffler.

Malliavin, P. (1977a). Un principe de transfert et son application au Calcul des Variations, C.R. Acad. Sc. Paris, 284, série A, 187-189.

Malliavin, P. (1977b) Champ de Jacobi stochastiques. C.R. Acad. Sc. Paris, 285, série A, 789-792.

Malliavin, P. (1978a) Géométrie Differentielle Stochastique. Séminaire de Mathématiques Supérieures. Université de Montréal.

Malliavin, P. (1978b) Stochastic Calculus of variation and hypoelliptic operators. In Proc. Intern. Symp. S.D.E., Kyoto 1976, ed. K. Itô. pp. 195-263. New York, Chichester, Brisbane, Toronto: Wiley-Interscience.

McKean, H.P. Jr., (1960) The Bessel motion and a singular integral equation. Mem. Coll. Sci. Univ. of Kyoto Ser. A. 33, Math., no. 2, 317-322.

McKean, H.P., Jr. (1960a) Brownian Motions on the 3-dimensional rotation group. Mem. Coll. Sci. Kyôto Univ., 33, 25-38.

McKean, H.P. Jr., (1969) Stochastic Integrals. New York: Academic Press.

McShane, E.J. (1974) Stochastic Calculus and Stochastic Models. New York, London: Academic Press.

Metivier, M. & Pellaumail, J. (1980) Stochastic Integration. London and New York: Academic Press.

Meyer, P.A. (1976) Un Cours sur les Intégrales Stochastiques. In Séminaire de Probabilités X. Proceedings 1974/75, ed. P.A. Meyer. Lecture Notes in Mathematics, no. 511. Berlin, Heidelberg, New York: Springer-Verlag.

Meyer, P.A. (1981) A differential geometric formalism for the Itô calculus. In Stochastic Integrals, ed. D. Williams, pp. 256-270. Lecture Notes in Maths. 851. Berlin, Heidelberg, New York: Springer-Verlag.

Meyer, P.A. (1981a) Géométrie stochastique sans larmes. In Séminaire de Probabilités XV 1979/80, ed. J. Azéma & M. Yor, 44-102. Lecture Notes in Maths. 850. Berlin, Heidelberg, New York: Springer-Verlag.

Meyer, P.A. (1981b) Geometrie differentielle stochastique (bis). To appear in: Séminaire de Probabilités XVI 1980/81. Lecture Notes in Maths. Berlin, Heidelberg, New York: Springer-Verlag.

Meyer, P.A. (1981c) Flot d'une equation differentielle stochastique. In Séminaire de Probabilités XV, 1979/80, eds. J. Azéma & M. Yor, 103-117. Lecture Notes in Maths 850. Berlin, Heidelberg, New York: Springer-Verlag.

Michel, D. (1979) Formule de Stokes stochastique.Bull. Sci. Math. (2), 103, no 2, 193-240.

Milnor, J. (1963) Morse Theory. Annals of Math. Studies, 51, Princeton: Princeton University Press.

Milnor, J. (1976) Curvatures of left invariant metrics on Lie groups. Advances in Math., 21, 293-329.

Mizrahi, M.M. (1979) Correspondence Rules and Path Integrals. In Feynman Path Integrals, ed. S. Albeverio et al., Lecture Notes in Physics, no. 106, pp. 234-253. Berlin, Heidelberg, New York: Springer-Verlag.

Mohammed, S.E.A. (1981) Markov solutions of stochastic functional differential equations. Proc. 2nd Int. Conf. on Differential Delay Systems and related topics, Poland, (to appear). School of Mathematical Sciences, University of Khartoum, Khartoum, Sudan.

Nelson, E. (1958) An existence theorem for second order parabolic equations. Trans. Amer. Math. Soc., 88, 414-429.

Nelson, E. (1967) Dynamical Theories of Brownian Motion. Mathematical Notes. Princeton: Princeton University Press.

Omori, H. (1974) Infinite Dimensional Lie Transformation Groups. Lecture Notes in Maths. 427. Berlin, Heidelberg, New York: Springer-Verlag.

O'Neill, B. (1966) The fundamental equations of a submersion. Michigan Math. J., 13, 459-469.

Palais, R.S. (1968) Foundations of Global Non-linear Analysis. New York: Benjamin.

Parthasarathy, K.R. (1967) Probability Measures on Metric Spaces. London and New York: Academic Press.

Pinsky, M. (1978) Stochastic Riemannian geometry. In Probabilistic Analysis and Related Topics, 1, ed. A.T. Bharucha Reid. London, New York: Academic Press.

Pinsky, M.A. (1981) Homogenization and stochastic parallel displacement. In Stochastic Integrals, ed. D. Williams, pp. 271-284. Lecture Notes in Maths. 851. Berlin, Heidelberg, New York: Springer-Verlag.

Pitman, J. & Yor, M. (1981) Bessel processes and infinitely divisible laws. In Stochastic Integrals, ed. D. Williams, pp. 285-370. Lecture Notes in Maths. 851. Berlin, Heidelberg, New York: Springer-Verlag.

Protter, Ph.E. (1979) A comparison of stochastic integrals. Ann. Prob., 7, no. 2, 276-289.

Rogers, L.C.G. (1981) Stochastic Integrals: basic theory. In Stochastic Integrals, ed. D. Williams, pp. 56-71. Lecture Notes in Maths. 851. Berlin, Heidelberg, New York: Springer-Verlag.

Rogerson, S.J. (1981) Stochastic differential equations with discontinuous sample paths and differential geometry. Control Theory Centre Report No. 101, University of Warwick, Coventry, England. Submitted to 'Stochastics'.

Schulman, L.S. (1975) Caustics and multivaluedness: two results of adding path amplitudes. In Functional Integration and its Applications, ed. A.M. Arthurs, pp. 144-156. Oxford: Oxford University Press.

Schwartz, J.T. (1969) Nonlinear Functional Analysis. New York, London, Paris: Gordon and Breach.

Schwartz, L. (1973) Radon Measures on Arbitrary Topological Spaces and Cylindrical Measures. Tata Institute Studies in Mathematics 6. Bombay: Oxford University Press.

Simon, B. (1979) Functional Integration and Quantum Physics. London, New York: Academic Press.

Steenrod, N. (1951) The Topology of Fibre Bundles. Princeton: Princeton University Press.

Streater, R.F. (1981) Euclidean quantum mechanics and stochastic integrals. In Stochastic Integrals, ed. D. Williams, pp. 56-71. Lecture Notes in Maths. 851. Berlin, Heidelberg, New York: Springer-Verlag.

Stricker, C. & Yor, M. (1978) Calcul stochastique dépendant d'um paramètre. Z. Wahrscheinlichkeitstheorie verw. Geb., 45, 109-133.

Stroock, D.W. & Varadhan, S.R.S. (1979) Multidimensional Diffusion Processes. Berlin, Heidelberg, New York: Springer-Verlag.

Sunada, T. (1982) Trace formula and heat equation asymptotics for a non-positively curved manifold. To appear in American J. Math.

Totoki, H. (1961) A method of construction of measures on function spaces and its applications to stochastic processes. Mem. Fac. Sci. Kyushu Univ. Ser A, 15, 178-190.

Ustunel, A.S. (1981) Some applications of stochastic integration in infinite dimension. Preprint: 2, Bd. Auguste Blanqui Paris 75013, France.

Vainberg, M.M. (1964) Variational methods in the study of nonlinear operators. San Francisco: Holden Day.

Vauthier, J. (1979) Théorèmes d'annulation et de finitude d'espaces de 1-formes harmoniques sur une variété de Riemann ouverte. Bull. Sc. Math., 103, 129-177.

Wallach, N. (1972) Minimal immersions of symmetric spaces into spheres. In Symmetric Spaces, eds. W.M. Boothby & G.L. Weiss, pp. 1-40. New York: Marcel Dekker.

Watanabe, S. & Yamada, T. (1971) On the uniqueness of solutions of stochastic differential equations. J. Math. Kyoto Univ., 11, 156-167.

Williams, D. (1980a) Review of "Multidimensional Diffusion Processes" by D.W. Stroock and S.R.S. Varadhan. Bull. A.M.S. 2, no. 3, 496-503.

Wu, H. (1979) An elementary method in the study of non-negative curvature. Acta Math., 142, 57-78.

Yau, S.-T. (1975) Harmonic functions on complete Riemannian manifolds. Comm. Pure Appl. Math., 28, 201-228.

Yau, S.-T. (1976) Some function-theoretic properties of complete Riemannian manifolds and their applications to geometry. Indiana Univ. Math. J., 25, No 7, 659-670.

Yau, S.-T. (1978) On the heat kernel of a complete Riemannian manifold. J. Math. pures et appl., 57, 191-201.

Yosida, K. (1968) Functional Analysis. (Second Edition). Grundlehren der math. Wissenschaften, 123. Berlin, Heidelberg, New York: Springer-Verlag.

INDEX

NOTATION AND ABBREVIATIONS

$O(n)$, $O(H)$, 274

$O(N)$, 285

$p_t(u,B)$, 209

P_t, P_s^r, 209, 234, 249

$p(s,u;t,-)$, 233, 238

$\mathbf{P}_q(G;H)$, 89

$\mathbf{P}(D|B)$, 236

R^*, 40

R, 289

Reg([a,b]), 97

r.ℓ., 113

sup {α:α ∈ A}, 42

S^{n-1}, 274

$S(\Pi)$, 23

S.D.S., 111

$S(T)$, $\underset{\sim}{S}(T)$, 88

$S(t,m)e$, 145; $S^i(t,m)$, 158

$\underline{t}$, 11

tr = trace, 62

T_xN, TN, 275; T^*N, 278

T_xf, Tf, 278

VTO(N), 286

X_e, 145

X_Λ, x_t^Λ, 113

$\underline{z}_t$, 11

β, 27

$\gamma_{\underline{t}}$, 15

δ, 22

δX, 139; δF, 140; δP_t, 216

Δ, 131; Δf, 296

$\underset{\sim}{\Delta_j}t$, $\underset{\sim}{\Delta_j}z$, 23

$\underset{\sim}{\delta^2}X$, 203, 204; $\underset{\sim}{\delta^3}X$, 205, 207

θ, $\underset{\sim}{\Theta}$, 288; θ_{y_0}, 265

μ = basic prob. measure

$\mu_{\underline{t}}$, 14

μ(Π), 23

ξ^u, 135, 193; ξ^K, 194

Π, 23

$\rho(t)$, $\rho_{\underline{t}}$, 14, 17

σ{h} = σ-alg. gen. by h, 7

τ(U), 41; τ_t^Λ, 113

ϕ_*X, 113; $\phi_*(A)$, 279

χ_A = characteristic f'n of A

$(\Omega^R, F^R, \gamma^R)$, 17

Ω_t, $\Omega_t(\xi)$, 40; Ω_t^u, 135; Ω_t^Λ, 113

$\underset{\sim}{\omega}$, 287

$\underset{\sim}{\Omega}$, 288

$\| \ \| = \| \ \|_{\mathcal{L}^2}$, 3, 24

$\| \ \|_{\mathcal{L}^p} = \| \ \|_{L^p}$, 3, 24

$| \ |_2$, 87

RAYMOND H. FOGLER LIBRARY